CLOSURE

We are lost, both as individuals and as a culture. For over 2,000 years we have believed in the possibility of a single true account of the world. Now this age, the age of truth, is coming to a close. As a result there is much unease. In the new relative, post-modern era, there is no unique history, no agreed morality, and no uncontested knowledge. In their place a mass of alternative and sometimes incompatible theories, from 'chaos' and 'string' theory to 'fuzzy logic' and 'consilience', proposing a theory of everything. *Closure* is a response to this crisis: a means to understand our experience and our circumstances in an age without truth. It is a radically new story about the nature of ourselves and of the world.

Instead of seeing the world as a thing, a universe, whose truths we might uncover through for example the procedures of science, *Closure* proposes that we regard the world as open and it is we who close it through our stories. The resulting framework offers solutions to the central questions of contemporary philosophy: the character of language and meaning, of the individual and consciousness, of truth and reality. As a theory of knowledge *Closure* has dramatic consequences for our under-standing of the sciences, changing what we think science does and how it is able to do it. It also accounts for why we need and desire both art and religion. It reshapes our understanding of ourselves and the organisation of society, our goals and our capacity to achieve them. But above all it makes sense of where and who we are.

A superb new account of how order is created out of disorder, *Closure* is an exhilarating work of conceptual geography.

Hilary Lawson is a philosopher, journalist and documentary film-maker. He is the author of *Reflexivity: the Post-Modern Predicament* and *Dismantling Truth: Reality in the Post-Modern World*.

CLOSURE

A story of everything

Hilary Lawson

London and New York

First published 2001
by Routledge
11 New Fetter Lane, London EC4P 4EE

Simultaneously published in the USA and Canada
by Routledge
29 West 35th Street, New York, NY 10001

Routledge is an imprint of the Taylor & Francis Group

Typeset in Joanna by Taylor & Francis Books Ltd
Printed and bound in Great Britain by TJ International Ltd,
Padstow, Cornwall

British Library Cataloguing in Publication Data
A catalogue record for this book is available from the British Library

Library of Congress Cataloging in Publication Data
Lawson, Hilary.
Closure : a story of everything / Hilary Lawson.
Includes bibliographical references and index.
1. Philosophy. I. Title.
B72 .L378 2001
110–dc21   00-054369

ISBN 0–415–01138–8 (hbk)
ISBN 0–415–13650–4 (pbk)

CONTENTS

CONTENTS

PREFACE

I want to tell you a story. It is not a fictional story, but then nor is it a factual story. Rather it is a story to hold still that which cannot be held at all. It is a story about the nature of the world and ourselves; a story about what it is to be human. It is the story of closure.

The story of closure is a philosophical story in the narrow sense that it addresses questions posed by philosophers such as the nature of language and meaning, of the individual and of identity, but it is also philosophical in the broader sense that it provides an overall account of our circumstances. It offers a framework that can be used to make sense of where and who we are. There was a time when the stories of religion were the primary source for such an overall perspective, now more typically we look to the stories of science. In their place, the story of closure provides a new framework, a new geography, by which to understand ourselves and our world.

This account of closure is a response to the chaos and confusion that surrounds us. For we are lost. Lost in a world that has no map, not because it has been mislaid or forgotten, but because we can no longer imagine how such a map could be constructed. In our post-modern relativistic age we find ourselves adrift in a sea of stories that cannot be fathomed nor anchor found. For we find ourselves in a world without certainties; without a fixed framework of belief; without truth; without decidable meaning. We have no unique history, but a multitude of competing histories. We have no right or moral action but a series of explanations for behaviour. We have no body of knowledge, but a range of alternative cultural descriptions. It is not simply that our thoughts and beliefs are seen to be relative to experience, culture, history, and language, but that without access to facts that are not vitiated by the perspective of the observer we have had to abandon the very possibility of neutrality or objectivity in their traditional sense.[1] Without the possibility of neutrality or objectivity we have in turn lost the capacity to give a description of things, people or events which is not at once at risk of being overturned or abandoned in favour of an alternative perspective. Without the possibility of being able to give such an account of our circumstances we have thereby become unable to give an account of what we mean by what we say, for we have no fixed point from which to identify any particular meaning.

Faced with this chaos of ideas, the account given of closure does not propose that we return to the false certainties of the past. Instead, it offers a framework that accepts the limitations of the stories that we tell about the world and ourselves, but at the same time offers us a map when we thought no map was possible. In order to find this map we have to embark on a journey away from the familiar categories of our current thinking. It is a journey that is required because from our current patterns of thought there are no solutions to be found. It is not possible to rearrange or reorder our concepts to escape the current confusion because these concepts have embedded within them the source of the malaise. Instead, we have to find a different way of holding the world altogether.

Instead of seeing the world as a thing, a universe, whose truths we might uncover through for example the procedures of science, *Closure* proposes that we regard the world as open and it is we who close it through our stories.

One way to understand this story of closure is to see it as a description of a process that underlies experience, the behaviour of individuals, and the operation of society. This process, the process of closure, is the means by which we are able to identify things from the flux of the world and thereby create a reality which we can understand and manipulate. I will argue that it is this process of closure that makes consciousness and language possible, drives human endeavour, and determines the way we intervene in the world. Seen in this light, the story of closure offers a theory about the operation of the human organism both individually and collectively. It does so not by reducing the mind to a mere mechanism, but by finding in the body that which is not mechanism. It is a theory which as a result casts light on the pattern of human development and the way individuals interact. Furthermore, it accounts for the character of both thought and desire, and as such has the potential to have practical application, not least perhaps in aiding our attempts to build an intelligent machine.

At the same time the story of closure, in addition to being a theory about the biological system that is the human being, is also a theory about the nature of stories. The account given of closure is in this light a description of language and a description of the way stories are created. It is an account of what these stories can achieve and what they cannot achieve;. what they enable us to understand about the world and how they enable us to intervene in it. It is an account that does not rely on our having a special access to the truth, to how things are, to explain the success and the failure of our theories be they scientific or otherwise. For the world is not taken to be a thing which might in principle be fully and accurately described. In this respect, the story of closure could be regarded as offering an account of language that does not rely on the notions of representation, correspondence, or reference to tie words to the world.

Seen as a theory of stories, the story of closure uncovers the underlying process driving the structure of knowledge determining both its limitation and its potential. It shows how it is possible for our theories to enable successful intervention in the world and draws attention to the constraints on that success. It is an account that has widespread consequences for our understanding of science, changing what we think science does and how it is able to do it. It also has implications for those spheres of activity that are traditionally placed outside of knowledge, such as art and religion. For it finds in the practical and down to earth that which is esoteric; and in the esoteric that which is at once accessible. It could as a result be said that it brings to a close the opposition of the factual and the romantic, of the practical and the mystical, of science and art, and in doing so accounts also for why we both need and desire art and religion.

These two aspects of the story of closure are embedded in each other. On the one hand, closure as a description of the operation of the human machine — both individually and collectively — is at the same time a description of the means by which we are capable of generating stories that enable us to understand and intervene in a world that is not already divided into things and is instead open. While on the other hand, closure as a theory about the character of language and stories is also an account of how it is possible that we should be able to provide a theory to describe the operation of the human machine and human society, even though this theory is itself but a story. These two ways of understanding the story of closure are therefore not so much two different aspects of the theory but two faces of the single notion of closure; a notion that is gradually uncovered as the story unfolds.

There is a final, and largely unseen, aspect of the story of closure. For the story of closure is also a theory that seeks to account for its own possibility. This self-referential constraint is a hidden motor driving and directing the story of closure. The account of closure in describing the operation of the human machine and human society is itself the product of such a machine and such a society. Similarly, as a theory of stories, it is itself an example of that theory for it is itself another story. The story that is told, the account given of the operation of the human machine and the theory of stories, is therefore at the same time a description of how it is that the story can be told at all. Another way of understanding the story of closure is therefore to see it as the story of how the story of closure is itself possible. So it is that the story of closure is a bootstrap theory: it uses itself to account for itself.

It will be apparent that the story of closure has an unfashionably broad sweep. Instead of seeking to escape the conceptual abyss that faces us by a reordering of familiar terms, it proposes a journey into an unfamiliar landscape. Lacking known landmarks, it will require some effort and some sympathy on the part of readers. I would contend however that any theory that seeks to overcome the present crisis in understanding, and the paradoxes in which it is enmeshed, will need to discard not only our current account of the relationship between ourselves and the world, between language and the world – if any such account could be said to exist – but to discard the very notions of language and the world themselves. At least in the sense that they are commonly understood. Any such theory will as a consequence need to offer a new account of what it is for us to describe the world, and therefore a new story of what it is to be human.

The story of closure offers therefore a central principle by which to understand human experience and language, both at the level of the individual and of society. Although grand in design, the story of closure is however modest in its claims. For it is a theory that sees theories as stories by which to hold the world. It therefore makes no pretence to provide a definitive or final account. Not least because from the perspective of the theory no such definitive solution is possible. There will in the future be other solutions and other philosophies; but, for the time being, it does seem to me that the framework of closure offers the only viable response to the chaos of thought and meaning that currently faces us.

PROLOGUE

There are many summaries, many paraphrases, that might be given of
Closure. It could be said that *Closure* is a theory about how we make sense of
the world, in a world that is open and not closed. Or that it is a theory
about the operation of the human machine, that identifies a single process
enabling both experience and thought. Ot that it outlines a theory of
language that does not rely on the notion that language refers to things in
the world. Yet although these descriptions are appropriate they are also
misleading. For the story of closure is one that requires us to abandon
terms such as 'reality', 'language' and the 'world' in favour of a new terminology. These new terms allow us to escape the deep-seated paradoxes of
the present in which we are currently enmeshed. In so doing *Closure*
uncovers a new landscape that enables us to explain afresh our circumstances and where we are.

At the outset therefore attempts to summarise the story of closure
inevitably operate with the familiar categories of our current thinking and
as such are inaccurate. For they seek to describe a new geography with
distinctions that apply to our present location. Couched in the landscape of
home, the recognisable is offered in place of the unseen, with the result
that those aspects of the foreign that are conveyed are also those in some
measure already familiar. Yet it is that which makes the foreign unfamiliar
which is of significance and which remains undisclosed.

This prologue does not seek therefore to offer an introduction that
would be a summary of the story of closure – to attempt to do so would
suggest that instead of the journey that is proposed an easy shortcut was
available. Instead it aims to demonstrate why such a journey is required.
Why our current thinking is in such disarray, and why if a solution is to be
found our current terminology needs to be abandoned and a new vocabulary adopted. Having identified why we cannot remain in our current
location, it then goes on to propose a starting-point for the journey ahead
and to indicate how it might be possible to proceed.

Some readers may feel they do not need to be convinced that the journey is necessary. Others may readily accept that a new framework is required. In such cases little may be lost in turning directly to Part I, for it is there that the story of closure begins in earnest. For those who are more sceptical, who do not recognise our current circumstance as one riven by paradox and confusion, or who are not convinced that drastic manoeuvres are required, the remainder of the prologue is divided into two sections. The first sets out to describe our current circumstances and demonstrates why this location is unsustainable; the second proposes where we might begin our search for an alternative.

THE HISTORY OF A MISTAKE

There is little reason to embark upon an extended and potentially difficult journey to a distant and currently unknown land unless our present location is thought to be at least undesirable in important respects. I will argue that the framework of contemporary thought is not only undesirable but is enmeshed in a predicament so insistent and destructive that it is not sustainable at all.

The cutting edge of this predicament has been apparent in the writings of philosophers, but initial signs of its destructive force can be found throughout our culture. It is found in our acceptance of the perspectival and relative character of our knowledge and beliefs, and at the same time our refusal to accept the consequences of this recognition. In, for example, our desire to uphold moral behaviour despite our acceptance that others adhere to different moral codes; in our desire to believe that science might uncover the ultimate laws of the universe and yet our suspicion that science is not itself value free; in our recognition that there are as many histories as there are points of view, yet our conviction that certain events cannot be denied as having taken place.

In the face of the contemporary predicament many have argued that we should retreat into some supposedly safe haven in the past. Into a less complicated world, a world without perspectives, a world that allows for some elementary observations, some simple neutral facts, into a world that enables objectivity. The case will be made however that such a retreat is not an option. The reason I will put forward is that the origins of the contemporary predicament can be traced to the outset of Western culture. For it can be seen to be embedded in the project to provide an accurate description of reality. Despite the remarkable successes of this project – science and technology being perhaps the most telling example – I will argue that from its inception the project carried an inherent flaw. A flaw that will inevitably bring about its failure. A flaw that has its mathematical counterpart in Gödel's theorem, and its scientific counterpart in the Copenhagen Interpretation of quantum theory. It is a flaw which stems from our misunderstanding the nature of the world and has now in the form of the contemporary predicament come to threaten our whole system of thought.

If we cannot stay where we are, nor can we retreat to a safe haven in the past, we must seek a way forward. Before proceeding however, as a conclusion to this section, it will be necessary to engage in a brief excursion to examine claims that a mathematical or logical solution can be employed to evade the flaw in the great project of Western culture and thereby remove the paradoxes of the contemporary predicament. It will be shown that these supposed solutions are not solutions at all but mere logical sleights of hand. As a result we have no alternative but to seek an entirely different location altogether.

The contemporary predicament

> The circle of self-reference in which contemporary thought has been increasingly enmeshed, typified by rhetorical self-denials and the use of inverted commas, is not sustainable

The end has been a long time coming, but now it is here it is all of a rush. Truth, in the sense of the possibility of a correct description of an independent reality, has had a good innings, but its time is over. It is not however the abandonment of truth in itself which is of concern, but the threat to meaning with which it is accompanied. It is as if we have fallen into an Alice in Wonderland rabbit-hole that has no beginning and no end. We have become lost, not as an adult is lost in a city that is not known well but which can nevertheless be negotiated, but lost as a child in a world which we not only do not know, but in which we cannot imagine how we might be able to find our way to somewhere that was known. Such is the contemporary predicament. A circumstance in which we have become unable to express what we seemingly wish to say, with the result that it is no longer apparent what could be said at all.

Those who already find themselves caught in this predicament will at once be aware of its vertiginous and unsettling character and the desirability of an alternative. There will however be many who do not recognise this description, either of their own views or more generally the state of our culture, and with this in mind some further explanation is required.

There have been many influences that have led to the contemporary predicament but perhaps the primary one has been the increasing recognition of the importance of context. For if once it was believed that claims could be made that were unequivocally and uncontentiously true it now appears that we wish to express their particular perspectival character – a perspective that is limited by the historical, social, cultural, and above all linguistic context. As a consequence, facts, whose truth is supposedly independent of context, and which provide us with pleasantly reassuring nuggets of certainty, have been in retreat. The retreat from facts can be seen to have been under way for a long time but it is in the last century that the pace has quickened. In the interests of brevity an attempt will be made to offer a very summary account of this broadly based phenomenon.

Amongst philosophers, it was for example still possible for G.E. Moore at the beginning of the twentieth century to propose the existence of moral facts but, in the analytic or English-speaking tradition, it was not long before the notion of moral facts began to look anachronistic. In what can be regarded as an attempt to maintain the sanctity of facts there were those who sought to identify a strict distinction between facts and values: a distinction which left matters of morality, aesthetics, and religion beyond the reaches of truth or falsity. For a while, this distinction, promoted by the logical positivists and encouraged by the writings of the early Wittgenstein, allowed its supporters to argue that through a combination of observation and logical deduction, along with the precise defining of our terms, a body of knowledge could be constructed based on a secure foundation of agreed facts. Such a stance can in retrospect be seen as a temporary respite in an irreversible tide. The next layer of facts to come under attack were cultural and historical facts. These were gradually undermined, no doubt in part by the cultural fallout from Frazer's *Golden Bough* along with stirrings of anthropological relativism with tales of Trobriand Islanders and Hopi Indians.[1] Over the next few decades the advance of relativism became more apparent and with the arrival of Kuhn's account of scientific paradigms[2] the case can be made that the way was open not only for the theories of science to look uncertain but also for the facts and observations on which they rested to be placed in jeopardy.[3] The archetype of a fact, found in the strict and supposedly precise observations of science, was itself to come under scrutiny and came to be seen by some not as an accurate description of an independent reality but as itself the product of a particular model and a particular conceptual framework.[4] Since then it would appear that the relativist momentum has been unstoppable. It is now not uncommon for it to be argued that there are no facts that can be identified independently of culture and society, of perspective and theory, and increasingly there are those who find in the retreat from the certainties of the past, an opportunity to proclaim the value of alternative traditions and cultures, and a means to denounce what are seen to be the tired and outdated canons of the West.

A case could be made that in the wider European philosophical tradition the importance of context and the resultant erosion of truth began rather earlier. In the mid nineteenth century the historicism of Hegel and Marx already relativised truth to a particular time and a particular society, although both sought a means to ensure that their own philosophy was deemed to have escaped the perspectival character applied to others; and more than a century ago, by explicitly abandoning an attachment to truth in a realist sense,[5] Nietzsche was perhaps the first to adopt the contemporary outlook. In doing so, he also carried through the self-referential consequences of such a perspective. At the time this aspect of his work was largely ignored but in the context of post-structuralism and post-modernism it has come to be centre stage.

These philosophical developments have mirrored, and it could be argued have perhaps to some extent led, a broader cultural awareness of the erosion of truth, in the sense of the possibility of knowledge of an independent reality. As a result there have been those who have inveighed against the growing tide of relativism claiming that it threatens to undermine all that is valuable in our culture,[6] arguing that if we deny the possibility of a viewpoint that is independent of culture, society, and individual preference, we will find ourselves at the whim of prejudice. So the argument runs: we stand at the end of a great tradition, which has provided us with a tolerant, liberal environment that has husbanded the valuable and discarded the worthless. It has done so on the basis of an adherence to empirical, rational thought and endeavour. If it is accepted that there is only perspective, all of this is at risk. For there can be no agreed method for advance, nor any notion of what progress would comprise, and as a consequence we will be at the mercy of those who can shout loudest and longest in the pursuit of their own ends and their own values.

Such a response, however, has the plaintive ring of an establishment under threat. If doubts about relativism were restricted to an assertion of the importance of what is currently regarded as the mainstream tradition they could perhaps to a large degree be ignored. A further argument has been proposed: namely that the problem with the erosion of truth is not so much that it threatens the accepted tenets of the past, but that it offers no stopping place, no point at which a line can be drawn. In its initial phase the relativising of truth can be used to challenge the dominant authority of an established belief, but in due course it undermines the basis of its own challenge as well. If all is perspective, why should any one perspective prevail, including the perspective that 'all is perspective'? How as an individual, or as a society, can we choose between one perspective and another if the ground on which the choice is made is itself only available from a particular view? While this argument is perhaps more persuasive than the mere assertion of the value of the Western tradition it still relies on the notion that a point of view needs to be defended on grounds that appeal to the notion of an independent reality that can be approached through rational thought. Such an argument need not be accepted by those who wish to endorse the abandonment of truth.

I wish to argue however that there is a more telling argument in response to the erosion of truth. It is an argument that points to the underlying nature of the contemporary predicament. For the problem with the erosion of truth is not that we are unable to find an ultimate ground for our claims, disconcerting though that may seem to some, but that the erosion of truth leads to the undermining of meaning, with the consequence that the meaning of what we seemingly wish to express itself becomes unclear. This undermining of meaning can be seen to follow from the identification of the importance of the context of language and the problem of self-reference that follows in its wake.

A preliminary indication of the nature of the problem can be found in general claims about the nature of truth that typify the contemporary perspective. Such claims may be expressed in a variety of forms such as: 'there is no truth'; or 'there are no ultimate truths'; or 'truth is dependent on context'. In each case the claim is at once paradoxical. As with the ancient liar paradox,[7] the assertion 'there is no truth' if applied to itself denies its own truth, and thus destroys the meaning that we at first attach to it. All of these claims have the characteristic that the self-reference of the assertion undermines its meaning, for what it asserts denies itself. If there is no truth, we cannot know that there is no truth for that after all would then be true. Similar arguments apply to any claim that denies its authority by drawing attention to its general perspectival character. Examples of such claims would include the statements: 'Everything we express is limited by language'; or 'we cannot step outside of language'; or 'we find ourselves within a particular conceptual scheme'; or 'we cannot escape the ideology of our time or our class'; or 'this is only my view or perspective'. As a result it has been argued that the very notion of a view or perspective of the world, or a conceptual scheme, or an ideology, is itself paradoxical and meaningless if the view, perspective, conceptual scheme, or ideology is understood in such a way that it is not possible to stand outside of it.[8]

An all-embracing relativism can be seen therefore to be incoherent for through its claims it denies its capacity to make those claims. Nor does the paradox simply invade a few general relativistic claims that could be discarded. It is because the paradox applies to claims that characterise the outlook as a whole and are thus a summary of the overall stance, that the impact of the paradox applies to all views held by someone adopting a relativist position. For any individual claim, however limited in character, such as 'snow is white' for example, is from a relativist perspective not capable of asserting a truth about the world independent of context. Instead it is to be understood as if with the parenthesis 'from my point of view'. 'Snow is white is true from my point of view' is however also not capable of asserting a truth and so requires a further parenthesis. There can therefore be no end to the additions and thus no means of determining the meaning of this or any other claim by reference to an independent reality.

Those who adopt a relativist stance get by because they either ignore such paradoxes, or implicitly limit the relativism so that there is an arena from which at least the relativist perspective itself can be stated. A weak relativism is adopted which denies truth in a particular context but retains the notion of truth to give the claim itself meaning. The case I wish to make however is that the underlying conceptual shift which has brought about the gradual abandonment of what were once taken for facts will not allow this as a stopping place. If the attack on truth were limited to a social, or cultural, relativism it could perhaps be contained. It is unsustainable because the erosion of truth is intimately linked to the contextualisation of language and meaning.

The recognition of alternative perspectives to our own as the result of a difference in historical, cultural, or social factors is in itself not a matter that needs to be of concern. It is at once apparent that others have different views to ourselves and the identification of this phenomenon on a social scale is simply an extension of a self-evident circumstance. What turns the identification of alternative perspectives from being innocuous to being a threat to our understanding in general is the abandonment of the assumption that the terms in which these perspectives are expressed are themselves transparent. So long as language is thought to enable a simple description of the world which can be judged to be correct or incorrect, the identification of alternative outlooks merely has the consequence that some views are seen to be closer to the truth than others, or to have identified aspects of the world that others have overlooked. If language refers to things, or the relation between things, a profusion of perspectives merely requires a careful identification of their alternative claims and a determination of those that are accurate and those that are not.

Much of the work of twentieth-century philosophy, particularly in the

English-speaking world, has been concerned to attempt to provide just such an account of language. It is the failure of this project, the failure to provide a credible realist account of the relationship between language and the world which has made the identification of different perspectives significant. For if language cannot be understood to refer in some way to a realm that is independent of language, an alternative perspective no longer simply provides a different version of the world to our own, but is itself the vehicle of an alternative world. This in turn has the consequence that the perspectives are no longer comparable since it is not possible to judge each against an independent reality; and without an independent reality against which to compare the perspectives it is no longer possible to determine the accuracy or validity of the claims put forward independently of the social and linguistic context in which the claims are made. It is therefore the failure of realism and the adoption of non-realist accounts of language that gives the relativist argument its force, and which at the same time has led to the contemporary predicament.

Some will argue that it is premature to claim that relativism and non-realism have become the dominant perspective of our culture. Later I shall in the most general terms indicate why in principle the realist project cannot succeed, but for the moment I will largely take as read the presumption that the project of uncovering what really exists, implicit within the empiricist or materialist strands of analytic philosophy, has been seen to fail. Since arguments to this effect have been powerfully expressed elsewhere it seems pointless to elaborate another version of them.[9] At its most general the case can be summed up by saying that there has been no satisfactory account of the means by which language hooks onto the world nor is there any realistic hope of such an account emerging.[10] Rather than itemising the failures of realism therefore, it is the consequences of non-realism on which I shall focus.

Relativist or post-structuralist positions are frequently adopted either without an awareness of their self-referential problems, or on the assumption that these are minor concerns that can for the present be put to one side. It is because the problems of self-reference are easily overlooked and are only brought to the fore by a determined pursuit of the consequences of the beliefs in question, that it is only in those with a rigorous turn of mind that the full impact of this reflexivity is made evident. It is thus precisely in the writings of those who seek to express a non-realist perspective with some care: the philosophers Nietzsche, Wittgenstein, Heidegger, and contemporary figures such as the French post-structuralist, Jacques Derrida, and the American philosopher, Richard Rorty, that it is also easiest to identify the paradoxical character of the contemporary predicament.

The contemporary predicament is initiated therefore by a desire to maintain a relativist or non-realist stance. It consists in the fact that although we have been led to take up this position we can find no means by which such a stance can coherently be expressed. We are relativists. We are non-realists. Yet we can find no means of saying so that is not at once self-denying and paradoxical. As a result we are forced into a series of moves to avoid the failure of self-reference, currently typified by rhetorical self-denials and the use of inverted commas as if to deny, and not yet deny fully, what is at once stated. A case can be made that it is for similar reasons that Wittgenstein, in his later writing, sought to avoid making any general claims about the nature of language and its relation to world, and why Derrida has avoided settling on any single description of the operation of language. Of course there have been followers of these philosophers who have sought to remove the seemingly unnecessary complexity of the texts in question and provide instead a theory that could with relative ease be applied. Wittgenstein or Derrida cannot however be reduced to the belief that we are exploring our language game, or through deconstruction uncovering the undecidability of meaning, precisely because such beliefs cannot be expressed without undermining themselves. If we find ourselves trapped in a language game, there can be no means of stating such a circumstance for the statement will need to step outside the language game for it to have the meaning intended. If meaning is undecidable, this also cannot be stated, since such a statement indicates that meaning is decidable after all. Nor can deconstruction as a method or technique have the purpose of uncovering this circumstance since it is not clear what would thereby be uncovered. Wittgenstein's avoidance of any general philosophical claims, and Derrida's continual reinvention of his own vocabulary can be seen therefore as the means by which each has sought to come to terms with the self-referential consequences of non-realism.

The problem with these and other available responses that have been offered in an attempt to express a non-realist position, if for the moment we allow the notion that such an unsayable stance might have the character of a position, is that however much the texts in question manoeuvre to avoid the aporia of self-referential paradox, the paradox remains as vigorous and insistent as ever. It will be argued therefore that no amount of avoidance or deferment solves the puzzle, but rather serves to make the puzzle ever more perplexing.

In order to make apparent the insistent and destructive character of the problem of self-reference two general strategies into which the various responses fall will be distinguished. The first of these, which I shall refer to as the structural strategy, is for the non-realist text to seek to show through the structure of the text itself what it is unable to say directly. This may consist in the text making claims about the nature of language that are successively abandoned, thus suggesting that in the claim and the abandonment of the claim the reader is able to catch sight of the underlying character of language, or rather to catch sight of the impossibility of expressing in language the underlying character of language. Derrida's writings would be an example of such an approach. Alternatively it may consist in the single-minded avoidance of claims that are self-referentially paradoxical thereby presenting a text that appears to make no general assertions at all. Yet the text functions by encouraging the reader to catch on to a strategy which implicitly involves the non-realist outlook that cannot be expressed. Wittgenstein's *Philosophical Investigations*, for example, might be described in this manner. A further version of this strategy is for the text to express evidently opposing views. By doing so the text implicitly denies that its claims are to be taken as statements that set out to describe the world and instead suggests an alternative relationship between the text and its meaning. Some have interpreted Nietzsche in this light.

There is a problem with the descriptions that have been given of these structural strategies. For the nature of these strategies is such that if they are successful the descriptions given are not merely simplistic but actually undermining of the strategies themselves. If it was possible to say what these texts were seeking to do or express, it would not be necessary for their authors to have engaged in the manoeuvres described. For example if one could state that the character of language was inexpressible and that in order to indicate this the text would engage in a series of descriptions each of which would fail, it would not be necessary for the text to undergo such a strategy. For the claim that the nature of language is inexpressible is precisely one of those general claims about language which the theories themselves deem to be inexpressible. Similarly if one could declare that the text was avoiding general philosophical claims about the nature of language and the world because such claims cannot be expressed, it would not be necessary to write the text in the form of a therapy to overcome such concerns. The characterisation that has been given therefore of the structural strategy and its various forms is at once, from the perspective of the texts concerned, a misleading description of their intent. The underlying intent of the texts cannot be expressed for that is the very reason an alternative and structural strategy has been employed.

However I want to argue that the inability to provide a characterisation of the structural strategy which is not misleading is in itself an indication that the strategy is problematic. For at its most straightforward, if the text is able to show something through its structure it will be possible to describe what is thereby shown, and if it is not possible then nothing has been shown. The texts imply that it is possible to convey something that cannot be said as if we could have some non-linguistic understanding of the nature of language. If however this were the case it would not be possible to say so, nor would it be possible for such a thought to be entertained, for such knowledge would itself be an example of the type of understanding that must lie outside of that which can be stated in language. In order to understand these texts the reader has implicitly to formulate the stance that the text avoids expressing. If this does not take place, the reader is left with merely copying the surface manoeuvring without an understanding of its purpose. Perhaps there have been followers of Derrida and Wittgenstein, for example, who have adopted the vocabulary and textual manoeuvres without appreciating the motivation behind the approach, but if this is the case it could hardly be said that the text has succeeded, or that the philosophical stance has been conveyed.

The avoidance of the presentation of a theory, be it in the form of contradictory assertions, in the manner of Nietzsche, the successive employment of alternative descriptions each of which undermines itself, as with Derrida, or the simple absence of any general philosophical claims at all, as with Wittgenstein, would appear therefore either to be in bad faith — a smokescreen for an underlying theory which is implicit but unsaid — or we have no apparent means of determining what to do with the text and what meaning to assign it. We can offer a whole series of descriptions of these texts, but no one of them can be maintained nor can we endorse all of them in conjunction. We can only provide content to these texts if we illicitly allow ourselves an overview of what the texts are seeking to achieve. Perhaps this overview is that 'we are lost in the web of language', or 'we are trapped in our own language game', or 'we are unravelling the tradition from within' or 'we are at play'. But if we are truly lost we cannot know this to be the case for to know that we are lost is precisely to have escaped from the web of language and ascertained where we 'really' are; if we are unravelling the tradition we cannot have identified this procedure for to have done so would be to take part in the tradition; and if we are at play the play must itself be playful in which case we cannot claim to be at play. We can read these texts and believe that we have identified what they wish to show, but as soon as we have made such an identification it cannot be held. Either we illegitimately imagine that these texts are expressing some view, or it is unclear how we are to determine any meaning or purpose to the text at all.

Each version of the structural strategy attempts to avoid the presentation of an overall theory, with the intended consequence that the texts concerned do not make claims that are at once self-referentially inconsistent. The absence of such claims does not however mean that the problem is solved. For in order to understand the text, to understand where it is coming from and what it is seeking to achieve, the claims that are not stated in the text must be assumed on the part of the reader. It is no good to pretend that the reader has simply to catch on, as if the non-realist position is akin to riding a bicycle, for it is unclear what the reader is expected to catch on to. Having climbed a ladder to a non-realist position the ladder cannot be thrown away leaving an unproblematic text.[11] For the ladder is the means of determining what the text is seeking to express. No amount of deferring, denial, rhetorical play, or the simple avoiding of general claims, can be sufficient therefore to halt the reader from an attempt to find a meaning, or meanings, in the texts through which to comprehend them. Once however an implicit meaning is provided it is at once undermined through its own self-reference. We understand these texts therefore by not understanding them.[12] We allow ourselves to hold some part of the texts to provide an overview, or to presume an implicit overview, but if we are true to the rigours of the texts themselves there is no part of the text that can be held and no overview that can be implied. As a consequence there is also no means of knowing how they can be understood or used or communicated. The structural strategy appears to have a response to self-reference but it remains deeply mired in the reflexive web.

The other strategy that can be regarded as having been employed in response to the problem of self-reference has been for the text to offer itself in a non-assertoric mode. As with the previous structural strategy whereby the text seeks to demonstrate through its structure what it cannot express directly, this non-assertoric strategy can take a variety of forms. Since the text is incoherent if it is taken to state a non-realist position, for the self-referential reasons that have been outlined, the non-assertoric text explicitly abandons the attempt to state something in favour of an alternative mode of expression. One form of this strategy adopted by the later Heidegger[13] and more recently by Richard Rorty[14] is to propose that the text is poetic. In being poetic the text both seeks to demonstrate its non-realism and avoids the circularity of stating a position which is at once not a position.

The problem with the non-assertoric strategy is similar in form to the problems that beset the structural strategy. If the text is genuinely poetic it cannot be understood to be expressing a point of view, or be translated into a method for acting or intervening in the world, for if such an understanding or translation were possible the poetic stance could be abandoned in favour of simply stating such a position. Yet those who adopt the non-assertoric strategy are presumably seeking to influence our understanding in some way, in which case a view is being expressed.

If for example we approach Rorty's text in a traditional manner, namely that it is trying to tell us something about the world, we can determine the main points of the argument and the seeming intention and meaning of the text. Rorty cannot however be wishing to tell us something about the world since his 'theory' precisely advocates the abandoning of such a task. Instead we must regard the text as poetically expressing the poeticisation it encourages. Yet if the text is treated as being engaged in poetic expression it then becomes unclear what we are to do with it, or how we are to provide the text with any particular content.

In practice, unlike a poem, Rorty's text has all the appearance of trying to convince us of something. There are certainly a large number of views that are expressed in a manner which is not self-evidently poetic. When, for example, Rorty seeks to defend his abandonment of the correspondence theory of truth he says: 'Truth cannot be out there – cannot exist independently of the human mind – because sentences cannot so exist, or be out there. The world is out there, but descriptions of the world are not.'[15] All of which appears to be a very definite description of our metaphysical circumstances, if for a moment we overlook the destructive self-reference in which the claim is at once embedded. If however we accept Rorty's advocacy of a poeticisation that is itself poetic, we are given no guidance as to how we can provide the text with content. If we are to understand that Rorty is not actually asserting that 'truth cannot be out there', but is instead engaged in poetry, what constraints are there that might limit what we could draw from the remark?

The non-assertoric strategy, as with the structural strategy, appears to suggest that there is a way language relates to the world that cannot be said but is to be understood in some other way, as if to hint at a realism that is not expressible in language. It is seemingly caught however, between two unsatisfactory outcomes. The hints and suggestions would appear to have the consequence either that despite denials there remains an underlying theory, a thesis which means that realism has not after all been abandoned; or, there is no such theory in which case it is unclear how the text can have meaning and thus any particular consequence. If Rorty's text is more inclined to raise the suspicion that an underlying theory remains, Heidegger's later texts being more explicitly poetic in character are more open to the criticism that the text lacks content. For if there is no underlying theory, if the text is itself poetic, how is content to be provided? How is the text to avoid the charge that it engages in empty mysticism? It is not sufficient simply to abandon assertoric meaning in favour of poetic gestures, or any other characterisation of the text such as use, or redescription, for in order to give content to such a characterisation the text will either provide a theory in which case it will remain caught in the cycle of paradoxical self-reference or it will fail to provide a means by which the text can have an identifiable meaning or purpose.

As I understand it, Rorty's reply is that we find ourselves at a particular juncture, with a particular vocabulary and its set of literal metaphors, and as such we do not need an explanation to understand what he is saying. Such a reply however has already provided the explanation, has already given us our metaphysics, within which we can interpret Rorty's perspective. As with those he describes as being engaged in ironist theory, Rorty wishes to provide us with a perspective which denies the possibility of authority. Recognising the reflexive problems of such a proposal his solution is to opt out of philosophical or scientific language in favour of literature or poetry. The problem with such an approach is that if such a solution was a solution he could not tell us about it.

These two strategies, the structural and the non-assertoric, which have been employed in response to the problems of self-reference that beset relativism and non-realism, are not incompatible. Elements of each strategy can be found in a number of the philosophers to whom reference has been made. Indeed, it could be argued that the structural strategy is a particular example of a form of non-assertoric expression, and to this extent the two strategies are really one. What I have attempted to briefly demonstrate however is that the destructive cycle of self-reference that these strategies set out to avoid is insistent and pervasive, and that despite the sophistication of the philosophers who have employed these strategies the predicament remains.

The critique that has been offered is not intended as evidence that these strategies and philosophies are simply mistaken. It is because the philosophers in question have worked through the contemporary predicament as thoroughly as they have that the character of the predicament has been brought to the surface. The predicament, which was initiated by the recognition of the importance of context, has become apparent through the attempts to abandon a realist notion of truth. It is because the denial of a realist truth is now so ubiquitous that the predicament can be said to characterise current thinking.[16] So it is that we find ourselves in a hall of mirrors where nothing is as it seems. Where not only is there no bedrock, no ground to our views, but the absence of such a ground is itself unsayable, with the consequence that we face a radical collapse of meaning.

The contemporary predicament, for the reasons that have been briefly outlined, is not a sustainable location. Nor is it clear how any of the currently available theories might be developed to deal with the problem.

The Great Project and its failure

> The great enterprise of knowledge itself, the dream of the Enlightenment and one of the driving forces of Western culture, is flawed from the outset and carries within it the seeds of its own self-destruction.

If the contemporary predicament is not sustainable, and stems from the abandonment of truth in a realist sense, the first and most likely response is to seek to retain the notion of an independent reality that can be accurately described. The attempt to provide a realist account of the relationship between language and the world and implicitly therefore the possibility of uncovering what might really exist may show no signs of success[17] but there would appear to be good reason for seeking to retain at least the possibility of such a theory. I shall argue however, that realism even as a hypothetical goal, of whatever form or however limited, is not an option, for the destructive self-reference that has been identified in non-realism, and which typifies the contemporary predicament, has its roots in the project of uncovering a true picture of an independent reality.

The assumption that knowledge of the world is possible and that new knowledge can be acquired is so ingrained in our culture that one can easily overlook the grandness of the metaphysical story on which it relies. Indeed it has not been uncommon for realists to claim that no metaphysical claims are involved and that the stance is merely an expression of common sense. While realism reflects the widespread belief that language gives us the capacity to describe things as they are, such a belief implicitly incorporates the grand notion that we are capable of uncovering the essential character of the world. I shall describe this notion, with perhaps a hint of caricature, as the 'Great Project' of Western culture. The Great Project in its unmodified form has consisted in the belief that it is possible to make steps, however small and painstaking, towards a true, final, and complete account of the world. Although philosophers have contributed to the Great Project, it is science and its achievements that have been its primary propagandist. Pursuit of the Great Project in the form of science has largely been assumed to consist in the attempt to make careful and gradual progress, through observation and deduction, towards an account which however limited and circumscribed would be a small advance towards a true account of the world.

For many of the philosophers who can be regarded as adherents to the Great Project, the task of the philosopher is to be engaged in an important but second-order activity. The underlabourer metaphor may have a somewhat falsely modest ring to it, but certainly from this outlook the philosopher is the structural engineer rather than the architect of the edifice, engaged in an attempt to prove the foundations and ensure the continuing solidity of the building as new layers are added. As a consequence of these deliberations some have wished to modify the more grandiose aims of the project. A modest modification proposes that while a complete and true account of the world may remain the goal of our endeavours we need not be committed to the view that such a goal is attainable. We may make ever finer alterations to our description of the world in the attempt to approach the goal but we should not hold to the belief that the goal can be reached. Our views about the world can thus be shown to be mistaken but we cannot know that they will always hold true.[18] A more extensive modification involves the limiting of knowledge in some respects thereby abandoning the notion of a complete and true account, even as an ideal, in favour of the possibility of a limited arena of knowledge. These modifications do not however constrain the main thrust of the Great Project, since it is still possible to engage in the successive extension of our understanding.

The possibility of the Great Project is a philosophical dream, but as it has been indicated it is not a dream exclusive to, or even primarily held by, philosophers. Uncovering the true nature of reality has been perhaps the driving motivation behind much scientific work and the characteristic that has been used to distinguish science from technology. In the concluding sentence of his highly successful attempt to popularise contemporary scientific theory Stephen Hawking even comes close to suggesting that a significant portion of the Great Project is nearing completion when he proposes that science is on the verge of providing a complete theory of the physical world, and that when it does so we will have uncovered 'the mind of God'.[19]

Although only infrequently explicit, the Great Project, it can be argued, has until the relatively recent appearance of non-realism, been the dream of the European tradition since the Enlightenment.[20] The case can be made that it has sustained an attachment to scientific progress, and thereby enabled an order of economic change previously unseen. Furthermore, it has been the source of an assumed social and historical superiority. For in addition to military and material success, the assumption of knowledge can be seen to have sustained the belief in 'civilisation'. In the light of the Great Project, Western culture has been able to regard itself not merely as being more economically successful than previous or alternative cultures but more advanced, having begun the slow acquisition of those modestly eternal truths known as facts and the placing of them within a theoretical framework. In addition, the possibility of social and historical progression, although not a necessary consequence of a belief in the possibility of the Great Project, can be seen to rely on the possibility of a framework of knowledge. For if we have the criteria by which to assess culture against some fixed points and the means to observe it, it then becomes possible for social progress to be discerned.

The case will however be made that there is an inherent flaw in the Great Project. It will be argued that the goal of the Great Project is unachievable because the goal is itself inconsistent. As with the contemporary predicament, the inconsistency in the Great Project stems from a paradox of self-reference. It is a flaw that I shall maintain cannot be obviated and which can be briefly expressed: a complete and true account of the universe is not possible because if it is complete it will be self-referential, and if it is self-referential it cannot also be true.[21]

In order to provide a complete account of the world, such an account must along with an account of everything in the world account for itself. Although therefore the theory that is the goal of the Great Project may distinguish between the observer, the theory, and the universe, any such initial distinction must be subsumed at a more general theoretical level into a single layer in which the observer, the theory, and the universe are each embedded. Without such a manoeuvre the account of the universe, provided by the theory, cannot be complete for the theory is not itself part of the universe which it has described. In order to be complete, the theory, the observer and the universe need at some level to be part of a single whole. Each of the options available have been extensively explored. Materialism involves embedding the observer and the theory in the universe; idealism the embedding of the theory and the universe in the observer; and the so-called linguistic turn in philosophy has frequently appeared to involve the observer and the universe being embedded in the theory. One can argue over whether various perspectives fall into one category or another. Whether, for example, empiricism is a form of idealism or materialism. The principle is however unavoidable: if the notion of the possibility of a complete account of the world is to be retained, dualism or a tripartite division is not sustainable, instead some type of monism must be adopted.[22]

The general principle can be illustrated in the materialist context of science. In order to provide a complete account of the physical world it will be necessary to give an account of how the theories of science themselves, as part of the physical world, are also the outcome of the laws which the theories express. A complete theory will thus need to be self-referential, so that in addition to providing the laws of the universe, those laws will need to be capable of providing an account of how human observers, as a certain combination of physical constituents on a planet in one part of the universe, will necessarily formulate at a certain point in time, through physical activity in their brain, the true theory of the universe. Such a theory would then be self-referential for the observer and the theory would not be distinguished from the remainder of the universe and would therefore be governed by the same laws.

The problem is that although a monist theory, of this sort, might in principle be capable of providing an account of its own existence, it is no longer apparent how this theory can be recognised as true. If, adopting the materialist form of the paradox, a scientific theory emerged which was itself the product of its own laws in what sense could it be distinguished from any other product of those laws? It would make no more sense to say of this theory that it was true than it would to say of any other product of the universe that it was false. Products of a monist universe cannot be true or false, they merely exist. The brain state of intelligent beings that reflects the true theory of the universe needs to be distinguishable from any other combination of physical states by virtue of its being true, but in a purely material universe we cannot give an account of the relation between any one physical state and the universe as whole, two necessarily different states, which would identify it as being true, or for that matter false. In summary form therefore the materialist paradox is that there can be no means of identifying any one physical state as being a view of the universe as a whole, whether true or false. Correspondingly, the idealist form of this paradox is that if the world is subjective it is not going to be possible to recognise that subjectivity.[23] While in its linguistic form it is that if the world is language it is no longer going to be possible to express this within language.[24]

Another way of expressing the materialist form of self-referential paradox is in the problem of the observer. The Great Project in its general form requires a theory that allows for no observer, since the observer must be part of the system that is being observed for otherwise the project will fail to be a complete account of the world. At the same time however a complete and true account of the world suggests an Olympian or God's Eye View of the universe, a view that would enable an independent observation of how things are. The contradiction in this notion has led some to suggest that it is not so much a view from on high but a view from nowhere.[25] Such a description covers up the inconsistency. It might be more accurately described as the No View View, for in this description is immediately shown the circularity: the Great Project both demands the observer and must reject the observer. In order to provide an account which is independent of historical and cultural relativism and of the subjectivity of the observer, the Great Project in its materialist form requires an Archimedean point from which to describe the universe. Such a perspective will not however be explained by the theory since the function of the Archimedean point is to distance the observer from the world irrevocably so that an unblemished view of the universe is made possible. If the perspective itself becomes a product of the system it describes it will no longer be a perspective on the universe but a part of it, and if it is part of it how is it to be distinguished as true?[26] A materialist version of the Great Project requires both to assert a distinction between the observer and the system to provide an Archimedean point, and to deny such a distinction in order to allow the Great Project to explain itself as a product of the universe.

The logical inconsistency in the notion of uncovering a true and complete account of the world has direct practical parallels in both scientific and philosophical theories. Quantum mechanics is a central theory in the current body of science. As such it can be regarded as being part of the attempt to provide a true description of the world. In its generally accepted form, namely Bohr's version of the Copenhagen Interpretation, its predictions however rely on a distinction between the observer and the system and thus exemplify self-referential paradox. For as long as quantum mechanics incorporates a distinction between the observer and the system, it is necessarily incompatible with the notion of a description of physical reality which is independent of observers.[27] That a central theory of science incorporates self-referential paradox does not entail the abandonment of metaphysical realism or the Grand Project for there is always the alternative of abandoning this version of quantum mechanics: a strategy which has at least had respectable historical support in the form of Schrödinger and Einstein. It is a problem for contemporary materialists that so central a scientific theory should currently be regarded as 'true', but in order to retain the Great Project one only has to propose variations to the theory, such as hidden variables or possible world interpretations, or that some future theory will prove the current version to be incorrect. The logical inconsistency of the Great Project has however the consequence that any attempt to provide a true story of the universe will necessarily incorporate theories which, like quantum mechanics, are at odds with the goal of the project itself.

To find a philosophical theory that illustrates the inconsistency in the notion of the Great Project we need only identify theories that have sought to provide a description of the relation between language and the world and that assume or propose the possibility of a true account of the world. The one which, due to its own internal rigour, perhaps most clearly illustrates the paradox in question, is Wittgenstein's early work the *Tractatus Logico-Philosophicus*, the conclusion of which provides an explicit form of the self-referential paradox with which we have been concerned. The problem with the famous final sentence, 'What we cannot speak about we must pass over in silence', is that the theory outlined in the *Tractatus* itself is one of those things that according to the theory cannot be said. In a more general manner it is an example of the circularity which has already been identified, namely that the attempt to provide in language a total symbolic system that describes the relation between language and the world must fail because language and its relationship to the world is not part of the world and therefore cannot be described by it.

The reflexive problems that have been identified in quantum mechanics and the *Tractatus* are not simply examples of theories that have gone wrong, but can be seen to be products of the underlying paradox that is incorporated into the Great Project from the outset. The paradoxes may not have appeared in this particular way and in these particular theories, but the argument that has been put forward suggests that they will necessarily appear in some part of the theory that would in principle constitute a completion of the Great Project for the reasons that have been outlined.

It would appear therefore that a return to realism as a means of escape from the contemporary predicament is not a possible strategy; but before accepting this outcome it is necessary, at least briefly, to identify why attempts to limit the Great Project do not provide a solution to the problem of self-reference. For if the attempt to provide a complete and true account of the world is not possible it might at first sight appear that the Great Project could be limited in some manner so that reflexive paradox can be avoided. If for example we could abandon the requirement to provide a complete account of the world could we not accept a partial description of an independent reality? Could not the problems of self-reference be avoided by making it no longer necessary for the theory to account for itself?

A summary explanation that can be offered to account for the failure of attempts to limit the Great Project is that either the limitation itself falls within its own limits in which case it succumbs to the paradoxes of self-reference, or it does not fall within its own limits in which case the Great Project has not been limited after all. An example of this further form of the self-referential paradox can be seen to have been held by the logical positivists. In an attempt to limit the arena within which knowledge was possible, and thereby allow for the possibility of a body of true meaningful statements, it was necessary to provide a definition of meaningful statements. The problem can be seen to be immediate, for any statement that sought to describe the criteria for meaningfulness was itself incapable of fulfilling such criteria. Suggested definitions went through a number of variants of the form: 'all meaningful statements must be empirically verifiable' but none managed to avoid reflexive paradox. The problem being that the statements themselves were not empirically verifiable. The statement of limitation itself therefore fell outside of its own limits and was not according to its own criterion a meaningful claim.

Writing a century before language became the dominant philosophical concern, the great German philosopher, Kant, is another example of a philosopher who sought to limit the Great Project. He did so by seeking to identify limits to our understanding. According to Kant we are only capable of knowledge of the world as it appears to us, not the world as it ultimately is. The task is however doomed to failure[28] for as Wittgenstein succinctly summarised 'in order to be able to draw a limit to thought, we should have to find both sides of the limit thinkable (i.e. we should have to be able to think what cannot be thought)'.[29]

Other attempts to limit the extent of the Great Project by making the goal of a complete and true account an unattainable ideal to which nevertheless progress can be made fall to similar self-referential paradox. If the goal of the Great Project is an ideal how can we know this to be the case, for such knowledge would be an example of a truth the possibility of which is seemingly denied? Either we take the claim that 'the Great Project is an unattainable ideal' as a truth, in which case this element of knowledge is not ideal and has been attained, or we do not take it as truth in which case it falls to reflexive paradox. If the claim is true and this element of knowledge is not ideal, in order to avoid immediate paradox we will require an account of the type of knowledge that is possible and not unattainable. This in turn will require a definition to limit the Great Project in the manner we have already considered and will therefore also fall to the paradoxes of self-reference.

The possibility of a true description of an independent reality, the assumption of realism, would appear therefore to be inconsistent. If we are seeking a solution to the crisis of meaning that follows the adoption of non-realism and the abandonment of truth, we are unlikely therefore to find a solution in a return to realism. Metaphysical realism is faced with the problem that the paradox of self-reference is embedded in the notion from the beginning. The great enterprise of knowledge itself, the dream of the Enlightenment and one of the driving forces of Western culture, is seemingly flawed from the outset and carries within it the seeds of its own self-destruction.

An illusory solution

> Russell and Tarski's solution to self-referential paradox succeeds only by arbitrarily outlawing the paradox and thus provides no solution at all.

Some have claimed to have a formal, logical, solution to the paradoxes of self-reference. Since if these were successful the problems associated with the contemporary predicament and the Great Project could be solved forthwith, it is important to briefly examine them before proceeding further. The argument I shall put forward aims to demonstrate that these theories offer no satisfactory solution to the problem, and that they only appear to do so by obscuring the fact that they have defined their terms in such a way that the paradox is not so much avoided as outlawed.

The problems of self-reference that we have identified are analogous to the ancient liar paradox. The ancient liar paradox stated that 'All Cretans are liars' but was itself uttered by a Cretan thus making its meaning undecidable. A modern equivalent of this ancient paradox would be 'This sentence is not true', and the more general claim that we have already encountered: 'there is no truth'. In each case the application of the claim to itself results in paradox. Similarly, the theory that is the goal of the Great Project cannot be expressed because when it is applied to itself it can no longer be held as true, and in the same way, the denial of literal meaning at the heart of the contemporary predicament cannot be expressed because when applied to itself it denies the denial of literal meaning.

Both Russell and Tarski have proposed solutions to this paradox, and offer the most important formal attempts to solve the problems of self-reference. Russell's Theory of Types, and Tarski's hierarchy of languages are both based on the principle of introducing a series of levels enabling the sentence to remain distinct from its meaning, or in Russell's case a set to be distinct from its contents, with the consequence that destructive self-reference can be avoided. If these proposals are effective they could be applied to the contemporary predicament and the Great Project enabling a denial of literal meaning that would not include the claim itself and similarly enabling the true theory that is the goal of the Great Project to avoid having to include itself as part of the theory.

Russell can be regarded as the first to formalise the paradox in modern times. In *Principia Mathematica* he sought to provide a symbolic language capable of describing mathematics and then wished to apply a system of logical analysis to language as a whole. His symbolic language that described mathematics was based around sets and the Russellian form of the self-referential paradox was generated by sets that were able to include themselves. Examples given by Russell of the paradox include the case of a barber who shaves everyone in a town who does not shave himself, or the catalogue of catalogues that do not name themselves. The paradox is made explicit when we consider whether the barber should shave himself or not, or whether the catalogue should name itself or not. In either case there is no possible solution. If the barber does not shave himself, he should do so for he shaves everyone in the town who does not shave himself; yet if he does shave himself he should not do so, since he only shaves those who do not shave themselves. A similar argument applies to the catalogue: if it includes itself it should not have done so, and if it does not include itself it should have done so. The theory which is the goal of the Great Project is thus also an example, for it is a theory which must include itself, but if it does so is no longer true. Russell's proposal is to generate a hierarchy of sets of different types. A set cannot include itself for it would be of a different type than its contents. Tarski adopts Russell's theory but applies it to language as a whole, proposing a hierarchy of languages, each higher layer of which refers to the layer below, thereby avoiding the possibility of language referring to itself.

I argued in *Reflexivity: The Post-Modern Predicament*[30] that the principle of a hierarchy of sets or languages fails to cope with circumstances in which the self-reference of the claim is essential to its expression. Russell wished to outlaw self-reference in order to avoid undecidability in his logic, but in the paradoxes we have considered the self-reference is itself a necessary part of the claim and cannot be simply outlawed. If claims such as 'there is no truth' are allowed only by denying their capacity to refer to themselves, such an outcome is not satisfactory for it would then appear that there is truth after all, only at a higher level in the hierarchy. In order to seek to avoid this outcome one would then be forced to extend the hierarchy of sets or levels indefinitely and still the problem remains. More recently, the American philosopher Hilary Putnam has also sought to contend, and with considerably greater detail and precision, that no satisfactory solution has been proposed to the ancient liar paradox and its modern logical counterpart of set theoretic and semantic paradoxes.[31] The problem as Putnam describes it is where the language used to describe the hierarchy is itself to be placed. The paradox is that one has to stand outside the hierarchy in order to formulate the statement that the hierarchy exists. This formulation of the problem stems however from the underlying problem outlined above that these supposed solutions do not really allow for self-reference at all.

The next few paragraphs are intended for those readers who would find a technical version of the argument that I have put forward more persuasive.

Adopting Tarski's hierarchy of languages one can formulate sentences that have the appearance of being self-referential. For example, a Tarskian version of 'This sentence is not true' would be:

(I) The sentence (I) is not true-in-L.

So Tarski's argument runs, this sentence is both a true sentence of the language meta-L, and false in the language L, because it refers to itself and is therefore, according to the rules of Tarski's logic and the hierarchy of languages, not properly formed. The hierarchy of languages apparently therefore enables self-referential sentences but avoids paradox.

More careful inspection however shows the manoeuvre to be engaged in a sleight of hand for the sentence as constructed only appears to be self-referential. It is a true sentence of the meta-language that makes an assertion of a sentence in L, but these are two different sentences − although they have superficially the same form. What makes them different is that the meaning of the predicate 'is not true' is different in each case. In the meta-language it applies the meta-language predicate 'true' to the object language, while in the object language it is not a predicate at all. As a consequence the sentence is not self-referential. Another way of expressing this point would be to consider the sentence in the meta-language. The sentence purports to be a true sentence in the meta-language, and applies the predicate 'is not true' to a sentence in L, not to a sentence in meta-L. Yet what is this sentence in L? It cannot be the same sentence for this is expressed in meta-L. The evasion becomes more apparent if we revise the example so that the sentence is more explicitly self-referential:

(I) The sentence (I) is not true-in-this-language.

Tarski's proposal that no language is allowed to contain its own truth-predicate is precisely designed to make this example impossible. The hierarchy of languages succeeds therefore only by providing an account of truth which makes genuine self-reference impossible. It can hardly be regarded therefore as a solution to the paradox of self-reference, since if all that was required to solve the paradox was to ban it, this could have been done at the outset.

The problem identified by Putnam regarding the description of the theory of the hierarchy of levels can be seen to be the reappearance of the problem of self-reference which is in practice outlawed within the theory itself. The supposed solutions offered by Russell and Tarski to the problems of self-reference that beset set theory and realist theories of meaning, and which also beset the theory that is the goal of the Great Project, thus provide no solution at all. For, the paradox is only avoided by arbitrarily making it impossible, in which case the problem simply reappears in the theory that has been employed to evade the issue.

A WAY FORWARD

Since we cannot stay where we are, and since a return to some form of realism is not a possible strategy, we must look elsew here if we are to find a means to escape the contemporary predicament. But where might we look, and how?

In an attempt to come to terms with the problems of self-reference non-realists have tended to concentrate on the mechanisms of language, and the inability of language to describe the world. Instead I shall propose as a first step the adoption of an alternative notion of the world. A notion in which the world is not held as a thing or a combination of things. For the moment such an outlook will be identified as the holding of the world as not-thing. At first sight this might appear to be an unlikely and unpromising proposal but in due course I will hope to demonstrate that it has value.

To begin with in the first part of this section I will primarily be concerned to show that the familiar and contrary notion that the world is a thing or a collection of things is much more questionable than we usually suppose. The aim of this argument will be to indicate why the converse, holding the world as not-thing, might therefore be a stance worth exploring. Some aspects of the arguments put forward in this part are rather technical in nature and some readers may prefer to skip these elements and proceed to the second part of the section which describes the task ahead given the starting point that has been proposed.

Having identified a starting point and a goal we will be in a position to set out on the project to find an alternative framework that might enable us to escape from the malaise into which we have fallen. It is to this task that the remainder of the book is then devoted.

The world as not-thing

> The world is not a list of things, nor is it itself a thing, no matter how complicated a list, no matter how complicated a thing.

From forest to tree, from beech to oak, from branch to leaf, we have things it would appear for every part and aspect of the world. Although we may not choose to do so, there is in principle, no corner left untouched, no crevice unfilled, no location in which some thing cannot be found.[32] In proposing that the world is held as not-thing, these distinctions are to be seen not as the outcome of distinctions in the world but as the outcome of language. It is not simply that we could have named these things differently, or have made different distinctions, but that the possibility of dividing the world into things at all is itself the outcome of a process of which language is a part. The world is not either divided into things or the result of the combination of things, nor is itself a thing.

I want, at this stage, to put forward two preliminary reasons for adopting this seemingly unlikely starting point. The first of these is that such a stance might provide an explanation of the prevalence of the paradoxes of self-reference; and the second, that since we can give no credible account of the nature of a thing it is unhelpful to consider the world and its contents as consisting of things. These reasons are not intended to function as a proof, but rather to indicate why such a starting point might be productive. A more powerful reason for adopting such a stance is to be found in the capacity of the theory of closure to account for our circumstances, but that reason will only become apparent as the theory itself unfolds.

The first reason for holding the world as not-thing is that the ubiquity of the paradoxes of self-reference can be seen not as some strange and inexplicable accident but the outcome of a mistake. If we hold the world as not-thing, it is to be expected that language cannot uncover the true nature of the world, for language provides a description of things. The method of rational empiricism that has underpinned Western culture, and on which science is based, seeks to ascertain whether its descriptions of the world – its theories – are accurate, by placing them up against reality – by testing them. Although powerful, the method ultimately fails, for any theory offers a description of things and their relations, and the world is always different since it is not a thing or combination of things. Furthermore, the process of rational empiricism in failing to provide a true description of the world has itself gradually brought to our notice the inability to describe the world independently of the language used to make the description. Rational empiricism thus starts with the assumption that there is a world which can be accurately described and uncovers our inability to provide such a description. Realism has thus uncovered its own mistake and we have increasingly adopted a non-realist stance. The step into non-realism however has been equally embedded in self-referential paradox due to its historical debt to realism. For although non-realism denies the possibility of realist truth we have currently no means of understanding in what such a claim could consist without an implicit reliance on a realism that the claim itself denies. A description of language as non-literal can only have force if we have an account of how such a claim can have content.

Holding the world as not-thing might appear therefore to enable an explanation for the failure of realism and the paradoxical character of the contemporary predicament, but it faces a similar self-referential problem. The empirical failure of realism cannot be explained on the grounds that the world is not a thing, if we are to understand by this that we have thereby described the true nature of the world. For to have done so would be at once reflexively paradoxical. As with similar non-realist claims, the assertion 'the world is not-thing' cannot itself be offered as a description of the world. It is for this reason that this outlook has been described as 'the holding of the world as not-thing'. Only when the theory of closure is elaborated will it become fully apparent what is intended by this formulation.

The second reason for holding the world as not-thing is at the same time a reason that obliquely provides a clue as to how the descriptions provided by language might themselves be interpreted. For while we take the notion of a thing for granted, the closer it is examined the more elusive it becomes. Although there are countless examples of things, as soon as an attempt is made to define a particular thing with precision we can neither empirically find a physical example, nor on further examination can we in principle envisage in what it could consist. The purpose in pursuing the nature of a thing will be to demonstrate this point: that the familiar notion of a thing which we take for granted is riven with problems and potential inconsistency. The seemingly simple proposal that the world is a thing, or consists of things, turns out to be highly questionable and certainly one that cannot be assumed will make a credible starting point for an account of ourselves and our relation to the world. As a result we should consider the unlikely alternative of proposing that we hold the world as not-thing. In addition, and perhaps more importantly, there will be an indication that the division of the world into things by language is itself in the limit not successful, and that as a consequence an account of language needs to provide for this outcome.

A first attempt to define the nature of a thing might propose that for something to be a thing it has to be one and it has to be the same: it has to be this singular thing and it cannot also be something else. Everyday material objects do not however satisfy these apparently elementary criteria. In the first instance everyday objects are not uniquely identifiable as some one thing: a particular example of a house, ostensively defined by pointing to it and saying 'that house', for example, is not uniquely a house. The house in question may also be a building, an environment, a structure, a habitat. If each of these 'things' was equivalent to the others we could retain the notion that objects were uniquely identifiable by regarding the various terms as synonyms. Such a trivial rejoinder is not possible. 'House' means something different from 'building' or 'structure', and when used to identify a particular object as in 'that house' it seeks to identify a different thing from 'that structure'. It certainly cannot be assumed that these different words have the same meaning, or refer to the same thing. Nor can examples be found of particular things that cannot be described as some other thing, for things are a function of the role they serve and the context in which they are found. However, if every particular thing is potentially an unlimited number of other things, the defining character of thinghood would appear to have been lost, since nothing could be said to be uniquely anything in particular.

Then again, not only can everyday material objects be described as alternative things, but it is equally the case that any material thing can be sub-divided into further things. While a cup is at the same time a container, it is also a handle and a bowl. However, if each material thing can be divided into further things, the thing is either something in addition to the combination of things of which it is constituted, or there must be elementary things of which all other things are constituted. Neither of these conclusions looks attractive. If the cup is something in addition to the handle and the bowl, what is this something that is other than its constituents? If, on the other hand it is proposed that everyday material objects are made of elementary things which are not themselves made up of anything else it is not possible to provide an example of such a material simple.[33] Nor as I shall later demonstrate is it possible to envisage a circumstance in which such a material simple could be identified.

It is widely believed that problems of this sort have been largely overcome by the development of modern logic. In this context one likely response would be to argue that the case presented has engaged in an elementary confusion between the thing and its identification, between reference and meaning. While any particular thing may be identified in

xl

innumerable ways, and can be described as consisting of innumerable other things, the thing in question is unique and could not be something else. The subject of a proposition need not uniquely label a thing with the consequence that there are as many things as subjects of propositions. The subject of a proposition is not a name for an individual thing but a set of criteria whose solution is found in the thing to which they refer. Thus 'that house' and 'that structure' can both refer to the same material thing, because each offers a different set of descriptions which in each case is satisfied by the physical entity in question. It can be argued therefore that the thing, the particular, is identified only as the solution to an abstract variable, x. Within such an account what appears as the name of a particular is instead understood as a function of x with a particular as its solution. The Quinean doctrine 'to be is to be the value of a variable'[34] can be understood in this light; a conclusion made possible by the original Fregean symbolism that introduced quantification and founded modern logic. As the result of such arguments the proliferation of things would appear to be contained. Instead of being committed to an array of different particulars each of which might be said to exist, we can say there exists an x such that x satisfies the conditions of being a house and a building and an environment and a structure and a habitat. Or there exists an x such that x is a cup, and a handle and bowl. Although in such an account we still use words which seemingly name things, these are not names of the object but are hidden descriptions. Thus names such as 'the house' and 'the building' can be considered as predicates of the abstract variable. 'That house' when analysed in this manner is understood to mean 'that thing which satisfies the conditions of being a house'. To propose that this thing might also satisfy a whole range of other conditions, does not therefore lead to a proliferation of things but merely a proliferation of descriptions of the one thing. We are thereby apparently able to maintain the notion that a thing is this thing and not something else.

There is however a cost to this response. The everyday objects with which we began have ceased to be things. In their place we have an ideal notion of a thing, the particular or object which is understood as that which satisfies the requisite criteria. Although everyday material objects imply a theoretical thing which lies behind the implied description provided, we are unable to give an account of the thing itself nor to indicate what makes it one and the same thing. Inevitably such a philosophical standpoint will face epistemological difficulties, since the logical simple recedes from view in much the same manner as a Platonic idea.[35] The epistemological difficulties associated with a retreat from everyday material objects to logical objects is not however the matter which shall concern us here. More salient to the question of the nature of a thing is the problem that the move from material objects to an underlying simple makes it no clearer what makes this notional logical, or material, simple a thing. It can be proposed that behind the material object called a house, building, and so forth, there is a thing which satisfies these descriptions, but in so doing we are left no clearer as to the nature of a thing nor what possible characteristic enables this thing to be identified as one thing and one thing only. Why is such an underlying simple incapable of being further subdivided? What is it that makes it unitary? The notion of the underlying simple merely supposes that it cannot be so subdivided, that it is just one. Yet not only do we not understand how this is in practice possible we are no clearer to understanding how this outcome is in principle imaginable.[36]

The proposal of a material simple that underlies everyday objects is reflected in the physical account of matter provided by science. The notion of smaller physical units that in combination make up familiar material objects is perhaps the commonplace understanding. A table is thus made of its relevant elements such as the top and legs, and these in turn are made of a material, such as wood, which in turn consists of a complex lattice of atoms. This hierarchy of things does not however get us any closer to understanding the nature of the thing. For it is always possible to ask of the thing in question, no matter how small, of what it is constituted. The atom is thus no closer to being one and the same, than the table with which we began. The same is equally true of the sub-atomic particles of physics. Whether the elementary particles are seen as quarks, leptons, strings, energy packets or force fields, it remains possible to ask of what this particle consists or to question the make up of the energy packet or force field.[37]

In proposing a material or logical simple we rely on an extrapolation from our everyday notion of material objects in order to determine what might be understood by such an account. It is thus supposed that the material or logical simple is itself and not anything else, in the same way that we usually suppose that one material object is not something else. As we have indicated however everyday material objects do not in fact have such a character. We cannot therefore understand material or logical simples by extrapolation from our experience of material objects. If elementary simples were to have the same characteristics as material objects they would in a similar fashion be capable of alternative descriptions. However in such a case, the elementary simple would no longer serve to solve the initial concern with the nature of a thing, for the problem re-emerges one layer further back.

The purpose of a logical or material simple is to provide a basis for our general understanding of a thing. Since this basis takes as its assumption the very issue that is in question, namely the nature of a thing, the matter is hardly brought to a conclusion. The elusiveness of that which we take for granted, the thing as material object, is not solved by providing an endpoint which assumes the character of a thing but fails to explicate what is involved in such a notion. Rather than assume as a consequence that we can replace our everyday things, material objects, with idealised elementary things, either as logical or material simples, we need to examine in what such an idealised notion might consist.

It has been argued that the criterion that a thing should be one thing and not something else does not apply in the case of everyday material objects but could it not apply to ideal simple objects? It is because simples are defined in such a manner that an example cannot be provided, that we can entertain the possibility that such simples have some undefined, and it begins to look undefinable, characteristic that makes them one and the same thing. The nature of material objects that makes it impossible to identify something that is both one and the same must however apply equally to material and logical simples.

If we consider the requirement of singularity, it has been argued that we are unable to give an example of a material thing that is one, because any material object can both be divided into smaller things and is itself part of something else. It can be seen that a similar conclusion applies in the context of supposed material or logical simples. In order for a material or logical simple to be uniquely one thing it could neither be divisible nor could it be part of some other simple thing. It must therefore be unrelated to all other simple things for otherwise it could be regarded as being made up of other things or itself as being part of something else. However if it is unrelated to any other simple thing it cannot be combined with any other simple or combination of simples to form a complex unit unless the complex is merely a combination of the simples. Material objects do not appear to satisfy this criterion. A chair, for example, is not fully described by being a combination of leptons, quarks and forces; to be a chair it also has to fulfil a certain function in human society. It is not clear how such a characteristic could be derived from the elementary particles alone. It can equally be argued that in order to be uniquely one, the thing must be indivisible, for otherwise it would be capable of being more than one. However, if something is to be indivisible it cannot have any spatial dimension, for otherwise a line could be drawn through it. It might be described as a geometrical point. However that which has no spatial dimension cannot be a material thing nor can a logical object or combination of such objects without spatial dimension be equivalent to a material thing.

Similar problems arise with the other criterion of being a thing, namely that it is the same as itself. Material objects are in large part made up of other things. A chair consists of the seat, the legs, the back, which in turn consist of the wood, the cloth, the material used for the stuffing, and so forth. If we are to imagine a material simple it must be itself through and through, otherwise it would be no longer simple and could be broken up into its constituent parts. Such a notion appears plausible because we can distinguish between objects that apparently consist of various parts, cars, houses, chairs, and those which seemingly do not, a log, a billiard ball, a shadow. By extension therefore it can be supposed that we can imagine a simple thing as approximating to a pure billiard ball. However, an actual billiard ball is not billiard ball throughout, indeed any part of the billiard ball taken on its own is quite specifically not a billiard ball, since it would not be conducive to playing billiards. The same argument would apply if instead of a billiard ball we were to consider a piece of billiard ball although it might be less apparent since the piece would be unlikely to have a characteristic that we commonly identify. It can be seen therefore that the requirement that the thing is the same thing throughout is to require identity between any parts of the thing. However, if there are parts they will not be identical. The criterion therefore that a thing is the same throughout is equivalent to the criterion that it cannot be divided into parts, and thus falls to the paradox identified with the criterion of oneness. The two criteria of a thing that it is one and the same are thus in this respect equivalent. The paradox involved in the notion of a thing can thus be summarised: a thing must be itself throughout; to be itself throughout it must be indivisible; to be indivisible it must have no spatial dimension; in which case it cannot be a material thing.[38]

It can be concluded therefore that in addition to being unable to produce an example of a thing, defined as that which is both one and the same, amongst material objects, we are also unable to provide an account of the characteristics such a thing would in principle have to have in order for it to be identified as a thing. As a result the notion of a logical or material simple is no less problematic than the notion of a material thing. It remains unclear therefore how the notion of a thing, as that which is one and the same, can be consistently applied.

The argument provided here does not seek to claim originality in drawing attention to the incoherence of the notion of a thing (although in the wake of the development of modern logic many have believed that the traditional problems of the existence of particulars, or things, has been solved)[39]. Classical philosophy extensively explored similar arguments; Wittgenstein in his later work drew attention to the impossibility of simples; and Heidegger devoted a book to the exploration of the notion of a thing and sought to demonstrate its elusive character.[40] However the inconsistency in the notion of a thing has largely either been ignored or thought to have merely mystical consequences. That our everyday notion of a thing, a notion which we employ all the time and which underlies our understanding of the physical world, should be so elusive is a puzzle therefore that many have wished simply to put to one side. After all it would seem we can get by perfectly well without having to solve it. Instead, for the present, I wish to use the arguments put forward here to suggest that it is at least worth exploring an alternative notion of the world; a notion of the world that is not based on things.[41] I wish to argue that the problematic nature of things should be taken as an indication of a mistake in our thinking and as a hint of how we might proceed. For in concluding that we can find no example of a thing nor can we conceive in what a thing would consist, it would seem sensible to abandon also the notion that the world is a thing or consists of things. Later, in the light of the theory which gradually unfolds, I will be in a position to give an account of why we find ourselves in this strange circumstance. For the theory that follows will make it apparent why we are unable to find an example of a thing. Furthermore it will explain why our inability to provide an example of a thing is no hindrance to our ability to use the notion of a thing to describe and explain what we take to be the world.

In proposing that we explore the idea that the world is not a thing nor a combination of things it is not being proposed that the world does not exist. As if, as in caricature versions of idealist philosophy, it is to be proposed that we create the world in our heads. To suppose that the world is not a thing is not to propose that there is no constraint upon us but merely to suppose that the world should neither be conceived as a single object that is one and the same, nor as a combination of independent objects each of which is considered as one and the same. In so doing there is no special critique being made of material or physical things – as if we are about to propose that the world is a thought or concept. For thoughts and concepts are equally things.

It is one matter however to propose that the world is not a thing or combination of things, it is another to explain what might be understood by holding the world as not-thing and furthermore to explain how from such a starting point we are seemingly able to understand the world and intervene in it effectively. Furthermore, 'holding the world as not-thing' cannot itself be taken as a statement about the true nature of the world. To make sense of this remark an account of how it might have meaning is required that does not rely on the identification of an independent reality. As an indication of how this might be achieved it should be noted that any inconsistency identified in the notion of a thing applies equally to language and its descriptions as it does to the world and its contents. If the world is not a thing or combination of things, neither is it possible for the descriptions of language, which divide the world into things, to be sustainable. Any account which seeks to explain how one could hold the world as not-thing, must also therefore give an account of how it is that any distinction proposed by language must itself fail.

The task

> If the theoretical, the metaphysical, the overview, cannot be
> excised, it must be grasped.

The case that has been made in this prologue began by arguing that our current thinking is beset with problems of self-reference so insistent, invasive, and destructive that alternatives must be sought. It went on to argue that the source of these problems stems from the outset of the empirical/rationalist project to provide an accurate description of the world, and that as a consequence a retreat into the seeming certainties of the past is unlikely to be a fruitful option. It has now been proposed that the flaw in the project to accurately describe the world, a project that could be argued to have typified Western culture since the Enlightenment, is to be found in the very notion of the world as a something which itself consists of things that might be described or whose relations might be described.

Although the general thrust of these arguments may be accepted, there will be those who are nevertheless suspicious of the notion of a new terminology and an account which has the hallmarks of a general theory, if not a metaphysics. Amongst contemporary philosophers, aside from those still pursuing a theory that attempts to describe accurately the relation between language and the world, Rorty and Derrida might be expected to propose criticisms along these lines. Rorty has taken up arguments similar to those put forward in *Reflexivity* and might therefore accept at least elements of the description offered of the contemporary predicament. It is possible however to identify a likely challenge to the story of closure which is shortly to be outlined, namely that it is an attempt at 'ironist theory' and thus a regressive move towards redivinising the world. An alternative and parallel critique employing Derridian language might be to charge that the vocabulary of closure reintroduces an unnecessary and undesirable logocentrism.

As a first and preliminary reply to such criticisms I would argue that we are all theorists, whether we admit to being so or not, and however ironic we choose to be. No amount of differing, displacement, poeticisation, metaphorical allusion, deconstruction, or simple avoidance is sufficient to squeeze from the text a residual theory, without at the same time squeezing from the text its capacity to be of use or value. If the theoretical, the metaphysical, the overview, cannot be excised, it must be grasped. Certainly the resultant theory cannot have the character of grand metaphysical accounts of old most easily caricatured with the words of the preface to Wittgenstein's *Tractatus*: 'the truth of the thoughts that are here communicated seems to me unassailable and definitive. I therefore believe myself to have found, on all essential points, the final solution of the problems.' A retreat into the past, into the project of turning philosophy into a science whose aim is to uncover the essential character of the world in a definitive if piecemeal fashion, is not a possible manoeuvre. Neither however is it a credible strategy to seek to avoid theory in general, since a theory however occluded and covered up will still be required if the account is to have meaning: even a denial of theory is a theoretical stance, an assertion that seeks to be pure surface is an assertion with depth, and a repudiation of metaphysics is a metaphysical claim.

The story of closure is driven by the need to provide an account that, in the worn phrases of our current vocabulary, provides a framework with which to explicate our relationship to the world, but which at the same time avoids the destructive self-reference prevalent in the currently available theories of our culture. It is because there is no apparent means by which to evade the destructive character of the contemporary predicament given our familiar conceptual framework, that an alternative vocabulary is proposed, a vocabulary that does not imply that the world is a thing, nor that the categories of language describe things. The story of closure is not therefore a reordering or redescribing of a known landscape which might thereby generate clarity, as if we are currently in a mist that with sufficient care could be dispelled – as if our current paradoxes could be solved by some minor theoretical adjustment – but is instead the provision of a new vocabulary along with an associated landscape with its own character and its own as yet unseen contours. A vocabulary which is, of course, designed to escape the weakness and retain the force of our current terminology and thereby offer a more effective and desirable means of navigating our way in what we currently term 'the world'.

Having now provided the grounds for a starting point, the remainder of this book sets out to offer an account of how from this unlikely beginning we are able to construct a world which appears to consist of things, and which enables us to understand and intervene effectively in what we take to be the world. Part I outlines the basic framework of such an account. Part II applies this framework to language, while Parts III, IV and V, apply the framework of closure to science and mathematics; art and religion; and to society. The purpose of these later Parts is to demonstrate the effectiveness of the framework and to extend and develop the story of closure. It is for this reason that the framework of closure seeks to offer an account of language and an explanation of the relationship between language and the world; an explanation for the effectiveness of science and mathematics and an indication of the potential the framework of closure has for aiding the development of new theories; a description of the functions and goals of art and religion and the relationship of these pursuits to those of supposedly factual disciplines such as science; and an account of society and the mechanisms of personal and institutional power.

Part I, which provides the outline framework of closure, is therefore central to the remainder of the book. On a cautionary note, readers may find its abstract character and that of the one following, which applies the framework to language, rather difficult for many new concepts are introduced. It is also possible that some of the distinctions made may at first seem arbitrary and the examples given, due to their brevity, may be interpreted as either misguided or inconclusive. If this is the case I would encourage readers to persevere to the later parts and chapters where the application of the framework to science, art, and society may clarify points that have been opaque and which at the same time will I believe provide convincing evidence for its adoption.

Part I

THE STRUCTURE OF CLOSURE

Introduction: the making of reality and ourselves

In the familiar everyday picture of the world, the world is divided into things: the sun and moon, the sea and sky, houses and people, tables and chairs. We are able to describe these things and the way they interact through language. We refine our account of the world by testing our views against reality. We throw out those descriptions that are not accurate, or modify them, so that our account of how things are is continuously improved upon. Something is understood to be true because it accurately reflects the way the world is, and is false because it does not do so.

Yet despite this everyday assumption, the Prologue concluded that the world cannot be a thing or consist of things. It cannot be so because when examined closely the notion turns out to be inconsistent. Moreover it was shown that this inconsistency leads to paradoxes and confusions which threaten to undermine our most central theories and beliefs. Yet if the world is not a thing or a combination of things it cannot be that we identify things in the world, nor that through language we describe these things or how they are related. As a result the state of things cannot account for the truth and falsity of our descriptions. If the world is not a thing or combination of things, then there is in the world no sun and moon, no sea, no sky, no houses or people, tables or chairs, no leaves, no bits of dust. In which case what are these things that in combination make up what we take to be the world? What is experience and language if not a reflection of reality?

Part I of the story of closure offers a preliminary answer to these questions by describing the underlying structure of closure. It is a structure which is then developed and applied throughout the remainder of the book. This first part is subdivided into three chapters. The first chapter introduces and defines the central terms and identifies the general characteristics of closure. The second chapter offers an account of the process of closure as it applies to human experience and language. The third chapter gives an initial description of the mechanism by which closure enables us to intervene in the world.

The case will be made that it is closure that gives us language, and thought, sensation, and experience. For the process of closure is the means by which our experience is constructed. It provides the content and form to reality through the realisation of individual closures. Moreover, it will be argued that not only is closure the basis of experience and language, it is the means by which we engage with others, the means by which we intervene in the world, the means by which we are able to make things happen.

1

AN OUTLINE FRAMEWORK

Openness and closure

It is through closure that openness is divided into things.

The words 'the world' suggest a place that awaits discovery, a place that can be charted and described, that can be catalogued and in part known. The world, however, for reasons that have already been made clear, is not a thing nor is it differentiated awaiting discovery. Yet it is not empty. Instead it is open, and in place of 'the world' I shall refer to 'openness'. I do so to avoid our slipping back into the familiar and mistaken notion of the world as a thing. Instead 'openness' indicates a site of possibility: a space that is not a thing or combination of things but is at the same time full. It might be called an undifferentiated flux, so long as the description does not encourage us to imagine that its character has after all been captured, thereby reducing it to a thing once again, even if on this occasion it is a moveable, changeable sort of thing.

At the outset, there remains a risk attached to the introduction of the term 'openness'. For as with all such terms our present habitual inclination is to provide a particular content, and thereby use the word as if it referred to a thing. To draw attention to this a line or cross could be put through the word in order to indicate that it was not to be held in this manner.[1] A strategy of erasure is however inappropriate, for such a strategy makes it look as though a word either refers to a thing or it does not refer to a thing, while as I will later argue it does neither. In due course the risk of misunderstanding will be avoided for it will be shown that 'openness' like all apparently referring terms has no discrete reference nor in principle could it have one.

Even though 'openness' does not refer to a thing the term can nevertheless be provided with content. In the context of the individual, openness can be conceived as the other of experience. Not as a collection of things

3

that are the external cause of inner experience but as the space within which experience takes place. In addition, openness can also be conceived as the other of language. Once again, not as that to which language apparently refers, but as the space within which the activity of language takes place. The issue of the relationship between language and the world has in many philosophical circles largely replaced previous concerns about the relations between the human subject and external reality. However, the account of openness is deliberately couched in both of these contexts. In either case there is a similar question to be faced: how is it that as individuals we perceive the world as consisting of things, if within openness there is no differentiation, and how is language capable of dividing openness into things or, perhaps it would be better to say, of fashioning within openness things, both material and abstract, collective and singular?

In its most general form the answer to these questions is to be found in the process of closure. It is through closure that openness is divided into things. Without closure we would be lost in a sea of openness: a sea without character and without form. For in openness there is no colour, no sound, no distinguishing mark, no difference, no thing. Yet openness is not nothing, it is infinitely dense with possibility, but it is not differentiated. It is closure that provides particularity and differentiation, and with it the pieces of reality, the material of the world. It is through closure that we are able to identify things, understand our circumstances, and intervene to a purpose. Language and perception are both the outcome of closure: the complex product of layers of closure that interact and combine. Yet closure does not describe or map openness, nor does it have either content or form in common with openness.

Through closure therefore there are things. Closure enables us to realise objects of every type and variety. Closure is responsible for our being able to describe the atoms of hydrogen and the molecules of water that make up the sea; for our being able to experience a sunrise over a field of corn; or hear the sound of a log fire and the warmth that it brings; it is closure that makes possible the kiss of a lover or the pain of injury; closure that allows the crossword puzzle and its solution; the words of language and the meanings they offer; Newton's theory of gravity and Shakespeare's sonnets; the state of peace and the activity of war; a society based on tyranny and a society based on democracy; the universe: its beginning and its end. Without closure there would be no thing.

Closure can be understood as the imposition of fixity on openness. The closing of that which is open. It is the conversion of flux into identity, the conversion of possibility into the particular. It is achieved by holding that which is diverse as one and the same. Such a process is not limited to human beings, or even animate beings, but it is at first perhaps easiest to understand the process of closure in the context of our own linguistic and perceptual closures where the closures involved are both new to the individual concerned and relatively unconnected to prior closures.

Suppose that we are looking at a random pattern of dots on a page. If asked what can be seen amongst the dots, we can imagine scanning the pattern looking for some combination of dots that allows the formation of an image of some sort. To begin with nothing may be seen other than the dots, but in due course let us suppose that an image of a face is identified. Having found the face the dots are no longer a random pattern. Instead we have the experience of seeing a face, of discerning perhaps the eyes and the nose, even an expression. The page of dots is now not what it was. The dots appear to be the same yet we see something which we did not previously see, which we can describe and identify and which was previously absent. This thing which we see is an example of a closure: the outcome of a process of closure.

Now it might be supposed that this face, this closure, was always there lost in the random pattern of dots waiting to be discovered. But what else could be found in the dots, in this imaginary version of drawing by numbers? If there are a hundred dots on a page and each dot can be linked to any of say ten dots surrounding it, there are, by many orders of magnitude, more combinations and shapes on this single page than there are atoms in the universe. There is therefore no practical limit to the number of possible images to be found. So what is happening here? It is not that the dots already contained these images and that we simply uncovered them ready and waiting for us, but that through closure we realised – we made real – particular shapes and images. It can be seen therefore that the page of dots is a page of unfathomable possibility, capable through closure of realising a world of almost infinite complexity.

In the context of this example, closure can be understood as a process which generates something from a space of possibility. The page of dots has unlimited potential but it has no concrete form until a process of closure has taken place and realised a particular thing. It is however only an analogy for the relationship between closure and openness, for the page of dots is not openness and is already the outcome of a complex process of closure. Nevertheless, if all human lifetimes had been spent examining the single page of dots still new things could be found. Each new shape could be added to those found previously thereby generating combinations of images whose inter-relation could itself become a basis for further closure offering stories made from the patterns. What was once a page of dots could thus become a multi-layered plethora of signs and images, pictures and stories, that could be extended and explored without limit.

So it is with the relationship between closure and openness in general; only more so. The page of dots in this example is itself the outcome of many prior layers of closure. It is through these prior layers of closure that it is possible to realise this part of openness as a page of dots. Over the next few chapters an attempt will be made to give a preliminary account of the detailed mechanism by which this takes place. It will as a result be seen that openness is already constrained both by the particular linguistic closure, 'the page of dots' itself, along with many layers of linguistic and non-linguistic closure which preceded it and which enabled this complex closure to take place.

As in the example of the dots and the myriad ways they can be combined, there is no practical limitation on the ways in which openness can be closed. All of the variety and detail of the world is provided through closure and in the realisation of things the unlimited character of openness is obscured, hidden behind a seemingly solid wall of known orderliness. Certain images that we have found in the page of dots that is openness have become central to us because they are linked to other images found elsewhere on the page and combine to tell an overall story. This story enables us to find our way around the dots and to refer seemingly precisely to each dot on the page. Having developed this complex web of closures the original unlimited possibilities held within the dots is gradually obscured, and our attachment to the images we have realised through closure grows so strong that we cannot conceive of alternatives. What we take to be reality is thus the complex web of closures we have come to use in order to make our way about in the world, and as we become accustomed to them and rely on them so the original possibilities held within openness fade from view.

We have the impression that there is no alternative than to see the world as we do, and to divide it into the familiar objects of everyday life; but instead it will be argued that the categories of language and the objects that make up reality are the result of closures and could have been otherwise, for in principle there can be no logical limit to the number of possible closures available. Although, as it will later be shown, seemingly unlikely closures can be realised, we are not in a position to adopt any closure we please, for we are constrained, on the one hand, by the historical legacy of previous closures held within the web of language, and on the other, by our physiology. The closures that make up language and perception are not therefore realised in isolation but in the context of a web of previous closures which serve to reinforce each other and the closure in question. The web of closures within which new closures are formed provides the framework or environment within which we operate. As a result of the familiarity of our closures and their self-reinforcing character the process of closure and the plasticity of openness is obscured. In combination these constraints are often sufficiently tight to give us the impression that there is no alternative to the closures adopted, and that these closures are demanded by the way the world is divided up, with the consequence that the particular closures we happen to have realised are often mistaken for a description of the world. In this sense we are imprisoned by our own closures and cut off from the diversity of openness. It is as if we are lost to the plot of one story we have constructed from things we have realised in the limitless page of dots that is openness and have yet to appreciate that there are whole libraries of alternative plots and characters. Occasionally through the apparent fixity of our particular closures we can glimpse the teaming mass of possibility of what might have been, and what might be, and thereby come to appreciate that openness is neither captured nor described by any particular closure or set or combination of closures.

Through the process of closure the character of openness is hidden but it is also through closure that reality is realised. Each closure provides something that we did not have previously, and at the same time obscures the openness from which that something was realised. Closure can be seen therefore to be in part a loss, an obscuring of openness, in the same way that catching sight of the face in the dots is also the loss of the other things that those collection of dots might have been. Yet without this loss we would have no-thing. While the obscuring of openness is an inevitable corollary of the provision of things, it is through the realisation of things that we are able to make some sense of where and who we are, and of how we might intervene in order to change our circumstances. Closure enables us to escape the flux of possibility through the provision of particularity; and while it obscures the potential of openness, without closure and the provision of things we would have no means of understanding, or intervening, to any particular effect.

The linguistic turn in philosophy has made it look as if an account of sensation and perception, in terms for example of physiology and brain processes, is part of the natural sciences and as such distinct from an account of language. I shall argue however that sensation and language are not different in kind but are both forms of closure. Nor are they the only forms of closure, for they are themselves only possible as the result of simpler, more elementary closures. Human experience is the result of many levels of closure within each individual, each of which has the capacity to interact with, and alter the character of, the others, and which in the case of high-level closures such as language are also able to interact with the closures of other individuals both present and past.

The account of closure outlined here is therefore at the same time a theory of language and a theory of perception. More generally it offers an account of organisms that intervene to a purpose in openness, of which humans are but one.

The mechanism of closure: material and texture

> Material is an enclosure that on the one hand takes place in openness, but which at the same time contains openness in the form of texture.

In due course, as the account of closure unfolds, an attempt will be made to describe with some precision the manner in which layers of closure interact in order to provide individual human experience and the social framework within which that experience takes place. Before it is possible to do so however it is necessary to explore further the nature of closure, its mechanism, and its characteristics.

Consider again the example of the page of dots. In the page of dots it was possible to find patterns or images. These patterns can be seen to be in addition to the initial perception of the page of dots. Nor are these patterns and images merely present awaiting discovery but are in some sense the product of the process of closure itself. It was argued that the initial page of dots contains an almost limitless number of possible patterns and images and is in this sense open. Closure in this instance consists in the process of realising these images or patterns. We can therefore consider the images and patterns as the outcome of the process of closure. This outcome of closure it will prove helpful to identify as 'material'.

Closure, of whatever form – and three basic forms will be identified: preliminary, sensory, and inter-sensory closure – consists in the provision of a particularity: a particularity which was not available prior to the closure. There is something present, or perhaps not present, as a result of the particularity. Thus when the dots are held as an image of a face, a new thing is created, and we perceive something that we did not perceive previously. The image of the face we find in the dots is neither the same as the dots themselves nor is it the same as an idealised image of a face, instead it is something new. All closure provides a particularity which is in addition to the context in which the closure was realised. Material is this 'thing in addition', and it is material that provides us with what we take to be reality.

It is not difficult to distinguish many different forms and types of material, but each is the product of closure. Sensations, the perception of physical objects, and meanings are all in this sense material. Thus the shapes and colours, the sounds, smells, and tastes, that provide the sensory elements of experience are material, realised through closure. So also are the individual physical things that we identify, and the world in which they are placed. Then again, in the context of the closures of language, any unit of meaning associated with a word or combination of words is material. In this context, if we are unable to make sense of language, either because we don't understand the individual words or because we are unable to understand their use in combination, it is because for the individual concerned closure has not taken place, and in the absence of closure no material has been realised. Similarly for perception, if we are unable to find a face in the dots when told to look for one, we can be said to lack this perceptual closure and have realised no material. When the face is 'found', the closure has taken place and material realised. As a consequence we see something that we did not see previously.

So what is involved in realising material? The provision of material can be seen to be the outcome of holding that which is different as the same in some respect. For example, in order to see a face in the dots we have to hold a set of dots together as one thing, namely a face. To do so we hold these different dots as the same in this respect. They are the same in virtue of all being part of a face. The material realised, the face, is the outcome of closure and the means by which these different things are held as one. The process of closure can be described therefore as the holding of that which is different as the same through the realisation of material. This principle can be seen to apply to all forms and types of closure, and all forms and types of material.

While all forms of closure consist in the provision of particularity in the form of material, the first or preliminary layer of closure realises material from openness, while subsequent layers of closure realise material from other forms of material themselves the outcome of prior closure. It is to these subsequent layers of closure that most of our attention will be devoted. In either case, closure consists in the holding of that which is different as the same thereby realising something in the form of material that was not previously present. In the case of preliminary closure the flux of openness is held as material, while in the case of later closure material realised from prior closure is organised into new patterns with the provision of new material. In saying that preliminary closure holds that which is different as the same it is not proposed that after all openness is already differentiated and preliminary closures holds these differences as one. Preliminary closure holds that which is different as the same in so far as it realises the same material on different occasions, and since openness is not differentiated these different occasions cannot stem from the same thing. Preliminary closure holds as the same therefore that which is different, but it is not different because openness is already differentiated but because openness cannot have the character implied by the preliminary closures. Later levels of closure hold different preliminary closures as the same in the form of new and higher-level closures. In this manner later levels of closure hold different things already realised by lower levels of closure as one new thing. Thus by holding that which is different as the same, closure realises material which is in addition to that which preceded it. When therefore we see a particular object, a person say or a house, we hold that which is different as one and the same thing. The person or the house can be regarded as consisting of countless different colours, shapes, textures and so forth, but through closure each of these are held as being part of one thing. Nor is the person or house simply a label for these different elements but is in addition to them. A computer could be given the visual data of the person taken by a camera but this would not have the consequence that the computer could identify that there was a person in the data. The holding of the different elements as one involves the provision of something new: the provision of material. The person or house that we see is therefore material provided by closure. In a similar manner when we suppose that a particular word has a specific meaning, we combine many different uses of the word into one thing. There are, to paraphrase Charles Peirce,[2] small chairs and large chairs, wooden chairs and metal chairs, there are armchairs and dining chairs, chairs that cross into benches, chairs that cross the boundary and become settees, dentists' chairs, thrones, theatre stalls, and seats of all sorts. The meaning of the word 'chair' is the closure by which we hold all of these different things as the same.[3]

It is because material is 'in addition' to that which preceded it, and not merely a manipulation of that which preceded it, that closure does not eradicate or exhaust openness, but instead provides a means of holding openness as something. Closure not only does not eradicate openness, but openness is held within closure. While preliminary closure provides particularity from openness and subsequent closure is based on material, these subsequent closures are nevertheless still realised in the context of openness because the array of prior material which provides the basis for the closure is not itself differentiated and is open. Openness therefore is not eradicated by material but is still present in a different form. The array of material provided by preliminary closure is open for it does not necessitate any particular further closures. The same principle applies to each further layer of closure. This can be seen in the example of the dots on the page. The perception of 'dots on a page' is the outcome of prior closures – both sensory and linguistic – but the array of dots although itself the product of closure is not itself differentiated and is open to further closures. Subsequent closure realises new material which is in turn capable of further closure. At each stage the material realised through closure can be said to contain openness because further alternative closures are possible. The face that is found in the dots can therefore itself be further differentiated in many different ways through the provision of new closures which divide the face into sets of new things. These might consist of things such as eyes or nose, or of expressions, or of shapes and angularities. Material, although it provides particularity, remains available to further closure; it remains open.

Although 'in addition' to openness, the realisation of material can often appear to exhaust openness. We divide the world into physical things – a form of material, such as trees, and cars, houses and people. This material, the outcome of many layers of closure, makes it look as if there is nothing other than the material. When we see a car, the perception of a car can appear to exhaust that thing, to be in this respect all there is. As if this material – which is in addition to the prior closures from which it was realised – fully describes openness. Yet the car is no different from the face in the dots. There are countless alternative closures, although embedded as it is within a framework of other closures these are not always easily accessible. We can suggest easily realised alternative closures by saying that the car is also a vehicle, a machine, an object of desire, a collection of materials, a shiny surface, a thing of value, a tradeable object, a product. It can be regarded as all of these things and a limitless number of others, some of which are accessible from our current framework of closure and a far greater number that are not accessible to us. These different closures are not merely different ways of describing the same thing, as if instead of calling the dots, 'dots', we had used a different word, but are different things offering different material. These different things are often not incompatible but allow different ways of understanding and thus different ways of intervening. It is often supposed that there is an underlying physical substrate on which all of these different things rely. It will be argued however that there is no precedence in physicality, for these descriptions be they in the form of objects, or atoms, are themselves closures.

The product of closure, material, does not describe openness for there is a remainder, a residue that is not exhausted by the material. A remainder or residue that is almost everything. For in holding that which is different as the same, or in holding one or more pieces of material – one or more things – as another, there remains that which is different. In the case of preliminary closure the remainder is inaccessible for it is in openness, but for subsequent closure the remainder is found in the prior material from which the closure was realised. The identity provided by subsequent closure thus contains within it that which is not identical. The remainder not exhausted by material is not anything in particular, it is open, but the manner in which it is open is influenced by the material. This remainder will be referred to as 'texture'. We can think of material therefore not so much as a little nugget of self-same stuff, but as an enclosure which on the one hand takes place in openness, but which at the same time contains openness. Texture is the outcome of prior material seen through the material realised by the closure in question. All closure, aside from preliminary closure, therefore realises a circle of material within which is found texture, which is open. In referring to a 'circle of material' it is not proposed that there is here a geographical or perceptual circle rather it is an attempt to indicate that openness is found within material as well as material being realised within openness. The openness found in the context of material remains open but its plasticity has in respect of the material been constrained, and it is as a result of this constraint that it can be said to have texture.

Each successive layer of closure realises material and texture, which in turn can become the basis for a further layer of material which will incorporate the prior material as texture. In this hierarchy of closure openness is seemingly gradually squeezed out, yet it can never be eradicated for it is held within the texture of the new material realised, and at each layer of closure alternative closures could have been provided which would realise alternative material and texture.

When looking at the page of dots, having realised the closure 'face' we generate material which we are able to see as a face. In this case we can imagine the material consisting of an outline shape. It does not however fully describe this part of the pattern of dots. Within this outline shape new shapes can be found, and parts of the outline will be capable of further closure. There remains that which is open within the material 'face' in the same way that there was openness prior to material being realised, only it is now found in the context of this realised material. If new material is provided in the form of the eyes, the mouth, the ears and so forth, so also is further texture: the angle of the mouth, the distinctness of the ears. Yet again within the new material texture is found enabling further closure. In each case the new material offers up new texture which offers the possibility of further closure, but in each case the texture provided is found in the context of its material. As additional material is realised so openness becomes increasingly textured. The entire web of closures that provides the environment within which we operate, and which shall be referred to as 'space', is therefore highly textured.

Aside from preliminary closure, all closure realises both material and texture, but it is only material that is capable of being known and it is only material that provides us with something, something that can be identified, manipulated, destroyed, built upon, and changed. Knowledge, at least in its conventional sense, is concerned therefore to eradicate residual texture. For the pure differentiation that would be the endpoint of closure would be perfectly known. Thus through material we can know of a country, its size, its population, its geography. The hypothesis of knowledge is that through the provision of ever more detailed closure it would be possible to list everything about that country. It might have to be a large catalogue – it would have to list the size, and shape and colour of every stone and leaf, it would have to describe every house and town, and every year of its history, it would have to tell of the lives of all living and all dead. In the context of material such a catalogue appears in principle to be possible even if in practice it is not attempted. Yet texture cannot be eradicated in this manner. An equally large catalogue could be produced for even the smallest and least distinctive object in the realm. And it would still not be known for there would remain ever further aspects of the object that could be uncovered. In this manner the natural world has the character of fractals in which every part can be further unpacked to uncover an equally complex reality. This consequence is not an unfortunate empirical accident that hampers our attempts to know the world fully, but in the context of the framework that has been outlined it can be seen rather as an inevitable consequence of the character of closure and openness, of material and texture. Although each additional closure realises new material, it must at the same time generate new texture, with the consequence that openness is always present and cannot be eradicated. The provision of material results in an increasingly textured openness, but its open character remains. For openness to have been overcome by closure, material would have to have exhausted openness but a world of material alone would be a world without content, without substance, a world of form akin to a world of mathematics.[4] Closure provides the discreteness, the things of the world, but without texture, without openness, these would have no content.[5]

The balance between openness and closure is a delicate one. In order for closure to realise material the flux of openness must be held at bay. The page of dots cannot be everything at once. If closure is to take place and material is to be realised there must be a modicum of stability. It is for this reason that closure seemingly exhausts openness. If closure was not assumed to exhaust openness in this respect it would not be possible to hold openness as this thing as opposed to something else. Of all the innumerable possibilities amongst the dots we perceive at any one time only one set of patterns or images.[6] In order to maintain particularity closure requires therefore that material excludes openness. At the same time the material realised through closure generates its own texture. This texture is itself open and is potentially a threat to the initial closure. The closure is maintained on the basis that if carried through further closure could eradicate the remaining texture. Thus in order to see a face in the dots, we must see it exclusively as a face and not as something else. The material realised by the closure 'face' however generates texture which is open. The openness held within the texture is compatible with the material on the basis that if further closure was applied it would be capable of realising further material that would not conflict with the initial closure and that in the limit all texture and thus openness could be eradicated. Closure is thus a self-enhancing activity. In order to be retained each closure requires further closures. With each new closure there is new material but that material carries within it further texture requiring further closure. This hierarchy of closure does not need to be completed for the initial closure to be sufficiently stable for material to be realised. If it did need to be completed no closure could ever take place, but for material to be realised the initial closure needs to proceed on the basis that subsequent closures could be completed if required thereby eradicating texture. Any individual closure is realised in a context which assumes that the completion of closure can in principle go through if it was followed up, while the structure of openness and closure ensures that such a completion of closure is not possible. It is for this reason that all closures are essentially fragile.

One of the immediate consequences of the relationship between openness and closure is that for every distinction we can find a further distinction, for every thing we can find other things of which it is constituted. In each case, for closure to occur material must be assumed to have exhausted the textured openness of the local space. Yet in each case although material is provided so also is texture. Examples of this relationship between openness and closure can of course be found throughout discourse but the assumption of complete closure and the provision of material to close gaps that appear, is perhaps most easily demonstrated in our answers to the question: what is the world made of? This simple child-like question brings to the fore the presuppositions of closure. The question assumes the answer will consist of something in the manner of a closure. So that the world is made of some thing or some combination of things. A current everyday response might be to reply in terms of atoms and molecules. Most closures are able to hide from the failure of closure because the failure is not an issue, what matters is the provision of material that can be used. On this occasion the failure of closure is more insistent for the initial question has only to be repeated for it to become apparent that we have only a thin circle of material. What are atoms made of? A few decades ago the answers would have come to an end with protons, neutrons and electrons. Now we can continue with a variety of types of quarks and leptons. There is no end to the regress. No matter how elementary the particle it would appear that we can always repeat the question.[7] Each time we ask the question, 'And what is that made of?', we are asking for more material to exhaust the texture that lies within the preceding closure. The answers fail because closure can never be complete since the world is open. Openness will always remain, and is inextricably bound up with closure. For while closure realises material it also generates texture. If we could come to an end, if closure could be completed, we would have lost texture; it would be as if we had lost the substance of the world. When some contemporary physicists, having lost the materiality of sub-atomic particles, talk of mathematics as providing the ultimate material of the world they have precisely sought to complete closure and in the process have lost everything; for as I shall argue in a later chapter mathematics is an idealised form of material that has no texture and thus has no content.

The requirements of closure have many consequences. If atoms did provide an endpoint, an answer to 'What is the world made of?', we would either have to imagine them as atom through and through, or as an empty sphere. Otherwise further closures would be required. There would be something else other than atoms of which they themselves were constituted. A case can be made that it is for this reason that when physicists believed that atoms did provide such an endpoint, and corresponding theories were taught in schools, the account given of the atom was in terms of one of these two options. Either the atoms were portrayed as solid indivisible balls or as empty spheres. Such a description can be regarded as the outcome of the requirements of closure. This attempt to maintain closure tried to exclude openness altogether; as if something could be just atom, as if the circle of material could exhaust texture. The exclusion of openness is however not sustainable. The consequent 'discovery' of more elementary particles forced the abandonment of the atom as an endpoint. The failure of the closure was only made evident by this empirical development, it was not its consequence. Its failure was inherent in the initial closure.[8] Even if such other elementary particles had not been discovered the notion of the atom as endpoint either as solid ball or empty sphere is not coherent as we shall demonstrate in later chapters.[9]

Atoms are of course only one answer that has been given to the question: 'what is the world made of?' There have been and there continue to be many others: such as thought, sensation, God. However, it can be seen that any closure offered in answer to this question must overcome the problem that it will in turn generate texture and thus fail as an answer. The question asks for a complete closure, but such a closure is not possible. One means of seeking to avoid the texture held within material is to pretend to have found building bricks which are small enough to have sucked out all residual texture. Such an account which was hardly convincing in the case of atoms is no more so for those with a positivist turn of mind who wish to go for sensation, or sense-data, or some such variant. One alternative is to retreat to a closure that instead of seeking to eradicate all texture, offers a closure that offers no differentiation at all. Such closures can be seen in notions such as God, Being, or Spirit. While this strategy avoids the inevitable problems that follow from seeking to eradicate texture, it suffers instead from having failed to provide material. For if everything is included, we have only openness, and so can provide no identity, or identifying character.[10]

From our familiar everyday categories that assume a referential relationship between language and the world it appears mysterious that so simple a question as 'What is the world made of?' should not only be difficult to answer but be unanswerable. In the context of closure however it is explicable. The question seeks an answer, an answer which must be the outcome of closure. Closures generate material, but the world does not consist of material. To answer the question we would have to escape from closure, and that we cannot do and still have an answer. To say that we cannot escape from closure makes it look as if we are trapped, but although we cannot escape closure, we are also immediately in touch with openness. Not an openness that is a mystical other, but an openness that flows from the character of closure. Closures are realised in the context of an environment that is open, and are themselves open. It can in this sense be said that closures take place in openness and hold openness within them – but this openness is not a thing which might be imagined as a transcendent other. Openness is shown in the impossibility of the completion of closure and the provision of texture. Openness in this sense is available to us, not as a thing but as a characteristic of our space, involved in closure and held within the circle of material. It is however only through closure that we have material and a world for us to inhabit; a world in which we can intervene.

The characteristics of closure

Since there are no bounds to closure, and there are no closures that do not in the limit fail, all closure requires a framework of stability.

We have already encountered in a provisional manner the three characteristics of closure. These are firstly, that it is unlimited; secondly, that all closure under examination fails; and thirdly, that stability is a requirement for closure. It will be shown that these characteristics are responsible for the character of language and experience and in turn for the character of knowledge and social organisation. They can at the same time be seen to be an immediate consequence of the initial relationship between openness and closure.

Closure is unlimited in the sense that there are no bounds to the number of possible closures that can be realised. Closure has nothing in common with openness, and therefore no closure is either necessary nor is any closure an adequate account of openness. Other closures are always possible. If openness was already pre-packaged, if it was already divided into things, then closure could be a form of identification, a labelling of that which was already present. In such a case closure would not be unlimited for there would only be as many closures as there were distinctions in the world. There is however no limit to the number of ways that the flux of openness can be divided up since openness is undifferentiated.

The unlimited character of closure means that there are always alternative closures, alternative ways of holding that which appears to be the same. We have seen this to be the case in the page of dots, but the same principle applies to all closure. Everything is also something else. A glass is also a container, a vessel. For a child it might be a hat, for another a weapon. We are tempted to assert the dominance of the closure 'glass' as if these other closures are just metaphors. We do so in order to maintain the stability of our space, and thus to avoid closure being overwhelmed by the flux of openness. These alternative closures are not however different in kind from the closure 'glass'. They may be less common, but when they are realised they are just as real. There is no means of distinguishing the supposedly real thing from the metaphorical thing. What we take to be the real thing is merely the most common way the thing is seen, or described.

A first objection to this argument might be to say that 'glass' is different in kind from all the other examples of closure – hat, weapon, container – because 'glass' identifies the object by its substance and not by its use. Such a counter argument would however seem to be at odds with at least some of the conventional uses of the word. In many instances substance would appear to be of secondary importance to the use of the object, for example it is not uncommon for glasses to be made of transparent plastic – used in bars and at public events for example. This sense of 'glass' – as a transparent container for drinking – is no different in kind from the other examples of closure.

However, it may further be argued that instead of a variety of alternative closures what we have here is simply a variety of descriptions of the same underlying physical thing. 'Glass', 'hat', 'weapon', are just different descriptions of the same physical object, and there are of course as many descriptions as there are uses of this object. Even if for a moment it is accepted that the physical object has priority over its uses, a position that this account of closure does not support, there are still different descriptions of this underlying physical substance. Is it a clear transparent material in a certain shape perhaps, or a lattice of silicon atoms? These are new alternative closures. They are descriptions of use perhaps to the physicist in the same way that 'the hat' may be of use to the child, but they are no more of an endpoint than the previous closures, and no more capable of being placed on a different level. For each of these physical descriptions has many alternatives. None of the closures can be identified as the real thing, for the real thing is always elusive, that something that lies behind each and every closure, each and every description. There is however no thing that lies behind closure only that which is open. Each closure can be regarded as offering one way of closing openness, one way of holding a reality. The limitation on our capacity to hold the 'it' which is the glass as something else comes only from the attachment we have to other current closures and the availability of other closures, both of which are a consequence of our network of current closure – our space. The glass could be a table for a box of matches, or a chair for a teddy-bear. It could be a mountain for an ant, or death for a spider. It could even be an example for a philosopher. It could also be something else entirely, in a language we don't know with categories that have no parallel with ours. It can be all of these things because the 'it' behind the closures is not a physical thing, or for that matter any other type of thing.

In one sense the example of a glass is not typical. For a glass is an artefact, an object designed for a purpose. It might be argued that because those who made the object intended it to be used as a glass, it is a glass rather than anything else. Such an argument would appear to have the uncomfortable consequence that the intentions of others plays a part in determining what is real, but more importantly there is nothing special about artefacts which enables an unlimited set of closures to be realised. A stream, for example, can be a barrier or a defence, an energy source, a liquid, a boundary or a mirror. Nor can it be argued that its description as a stream is somehow primary to the other ways it which it can be held. There is a common presumption that physical matter is somehow prior. As if we could identify the material thing and then separately and in addition its uses and purposes. The stream however can be seen as water, which can be seen as hydrogen and oxygen, which can be seen as collections of sub-atomic particles. Even if the sub-atomic particles are seen as packets of energy, and the energy as perturbations in four-dimensional space-time, it is still possible to ask what these perturbations consist of, what constitutes a packet of energy. There is no end to the possible physical descriptions. It is not possible therefore to identify a primary material thing which is somehow prior to all the other ways in which the 'thing' can be held. Furthermore even if these physical descriptions could be completed they would miss what else this thing can be. For the collection of hydrogen and oxygen atoms does not tell us that this thing is a stream. The stream does not tell us that it is a barrier. The barrier does not tell us that it is a defence. And none of these indicate that it is a mirror. The stream, the barrier, the defence, the mirror are all different things through which we are able to hold openness in different ways. No matter what closure we choose there can be no limit to the alternative ways of holding this 'thing'. We may perhaps be tempted to respond that 'the thing' is all of these things. Although they may not all be compatible, in some sense 'it' is all of these things, but it could also be an unlimited number of other things as yet unidentified and perhaps unidentifiable from the current available closures. 'It' may therefore be all of these things but we are no closer to saying what it is. Behind the closures there is no ready-made something, or previously delineated category, that awaits accurate description or proper labelling. Behind closure there is openness and as a result there is no limit to the closures available, nor is there a boundary to the plasticity of our world.

The second characteristic of closure that has been identified is that all closures fail, in the sense that closures are not capable of being equivalent to, or replicating, openness. We become aware of the failure of closure to reflect the character of openness in two ways. On the one hand closures fail because the material in question need not be this material, it could be something else and the alternative material may threaten and supersede the current closure. On the other hand closure fails because the texture generated by material offers new closure which in turn potentially undermines current closure. So long as a closure is not undermined by alternative or additional closures it need not be abandoned. However all closures are at risk of failure and will be seen to fail if the closure is pursued. Once again this outcome follows directly from the structure of openness and closure, namely that closure is not openness, nor does it describe an element of openness. As a result closure cannot be equivalent to openness.

One consequence of the inability to secure closure from failure, the inability to find a closure or closures that equate to openness, is that there is always more to the world, there are always further distinctions which could be made, further aspects that could be perceived. Each of the closures so far offered as an alternative to 'glass' has its own material with its own texture and its own branching network of potential further closure. As a glass we can look for its capacity, its aesthetic quality, its clarity; as a weapon, its ability to inflict damage, its thickness, its weight; as a lattice of atoms, the types of atoms and their bonding structure. Each closure offers its own branching chain of closures, its own reality. Each one is capable of being further explored with new closure found and new material realised.

If an attempt is made to secure closure, its failure will become apparent. As a glass we can place it alongside other things that through closure are held as the same. This particular glass thus takes its place alongside all other glasses, from cultures both present and past. While it is placed alongside these other objects it is at the same time different from them. For this thing will be different in some respect. If we seek to secure closure and thereby eradicate this difference, further distinctions will need to be introduced to remove those characteristics of the glass that are not the same. We can realise new closure in an attempt to identify this thing precisely. Thus we can identify new material such as its style or its type of manufacture, and as a result provide a set of closures which perhaps by a combination of location, age, manufacture, aims to identify this and every other glass without residue. Yet there is no end to this process of closure. For with each closure there is not only new material but new texture which enables further closure. The categorisation of glasses can always be provided with more detail − further distinguishing characteristics that separate this glass from others, yet still held within the new material will be texture which is not this thing and which contains yet further difference. The same pattern of nested closure applies to each of the unlimited alternatives to the closure 'glass'. The process cannot be completed because there is no end to closure, for no closure is capable of exhausting openness. (We could of course seek to identify the glass uniquely by naming it, but what would the name identify? The 'it' referred to by the name would prove to be as elusive as the thing that is the closure 'glass', for it would not be apparent what characteristics the name was seeking to identify, and as soon as these were identified the 'it' would prove to be otherwise.) No thing and no aspect of any thing is the same as any other, for the similarity is the consequence of closure and the world is not closed, it is open. Not only are there no bounds to the possibilities of closure due to its unlimited character but there is no end to closures that follow from any one of these possibilities for no closure can be made secure from failure.

The failure of closure, the inability to bring closure to an end in the search to describe openness, can be seen to result on the one hand in the branching network of closure which we take to be knowledge and in the other in the impermanence and changing nature of that knowledge. All closure may be abandoned in favour of an alternative, and in addition each closure is in part undermined by its own texture. Not only does closure fail to exhaust openness but if we seek to identify the material realised we can provide no content. It is as if the circle of material is squeezed out by its own texture. The closer we examine material the less we can find in what it consists. So it is that for each material thing, what makes it a thing becomes more puzzling the more we probe. Is a thing not simply a combination of its elements, in which case can we not do away with the thing and retain only the elements? Is not a glass, its use, its shape, its appearance? And if it is only these things what is added by the material realised in the closure 'glass'? The circle of material realised through closure has no thickness, it marks out texture but when examined closely itself has no content.

The failure of closure has the consequence that there can be no final resting place. No closure under examination will cease to generate further closure that may in turn undermine the closure in question. In so far as philosophers have seen themselves as attempting to complete closure, to provide a set of closures that might provide a final account of some aspect of the world, they have thus set themselves an impossible task. Moreover it is precisely in the attempt to complete closure that its failure is made apparent. Equally the attempt to find a core of certainty, a bedrock of knowledge on which our systems of belief could be safely based can be shown to have no more chance of success.[11] No closure can be safe, for there is no closure that can avoid failure. Scientists in addition to philosophers have been inclined to imagine that they are in the process of providing secure closures, despite the empirical and historical evidence to the contrary. The inevitable failure of closure means that there can be no scientific observation or scientific theory that could be other than provisional. So far from uncovering the mind of God[12] scientists can therefore be regarded as playing out the consequences of their own particular set of closures – some of which can be known in advance from the character of closure in general.

The failure of closure is also shown in the character of texture, a character which shall be described as 'active'. The failure of closure can be said therefore to result in the provision of activity. We might call it change, but change implies the movement from one thing to another, while the activity generated as the result of the failure of closure is not a transition. Closure realises material which is offered as self-same, it is this thing and not something else. In order to be self-same, material must be fixed, it must be still, it must remain the same thing. However since the world is open it cannot be held still. The failure to hold the world still is shown as activity which is exhibited in texture.

It has been argued that the process of closure through the provision of material seeks to hold openness as this thing; but it cannot be so held for openness is always other than this thing. This other is texture and in contrast to the fixity of material is active. We can observe this in the context of everyday physical objects, for none of them is capable of remaining the same. The activity of texture is sometimes described as change – the object is not as it was because it has become different; or as disintegration – the object is no longer the same because it no longer exists. The activity generated by the failure of closure is not however necessarily physical activity. The manner in which material seeks to hold the world still is not in a physical sense. For example the material realised by the closure 'rain' at once contains movement. The provision of fixity can be seen to come not from holding the water still, but holding the falling water as a something. The activity generated by the failure of closure is not therefore to be confused with the movement contained in this instance within the material but is the result of the impossibility of the rain remaining the same.

In this sense all material can be said to generate activity. In the context of the physical world this activity can be witnessed as change or disintegration. From a perspective in which the world is a thing to be described it appears accidental that nothing remains fixed. Some things obviously change, cars get old and battered, trees grow and extend their roots, rooftops glisten caught in the sunshine after rain, but even rock changes in time, a short time by comparison to the aeons that straddle the stage of the world to which our current closures seemingly refer. Not only does every-thing change but everything disintegrates. From the point of view of science that each thing should change and disintegrate seems to be an empirical coincidence. As if we could go round to check. Yet not only can we find no physical thing that does not change, we cannot imagine it either. To do so would be to imagine a closure that did not generate activity: in the context of the physical world an object that could be held still would have to be cut off from the rest of the world. It could not be on earth, for wherever it was it would interact with other material, if it was placed inside something the container, whether natural or artificial would over time move and disintegrate. A meteor in deep space seems closer. But the meteor could never be so isolated that no light reached its surface from other stars. In addition the meteor would move in relation to stars no matter how distant. To try to imagine material without activity is to imagine a complete closure. We attempt to achieve this complete closure by trying to cut off the material from the world, from openness, by sealing it in a container, by placing it in deep space. We cannot succeed however, for the closure must take place within openness, and by being necessarily incomplete the residue of openness will be exhibited as texture. We cannot imagine material without activity because the world cannot be closed, and the failure of any particular closure must appear in texture as activity.

The world realised through material has necessarily the appearance of stability. Look closer and this stability hides a seething flux. We have the impression of permanence: houses and tables and roads and hills and people. If we try however to fill out these closures, as science has attempted to do, we uncover a chaos of activity. It could not have been otherwise. How activity appears within any closure is decided empirically, but the presence of activity is a necessary consequence of the characteris-tics of closure. Brownian motion might not have applied to small particles buffeted by molecular movements, but as science added layers of closure in an attempt to describe matter each additional layer of closure inevitably generated activity in the form of change. So it is that on the most appar-ently empty and static of things micro-organisms are in turmoil. As we look closer, atoms are in constant movement and elementary particles move at immense speeds. Although activity is added with additional mate-rial, it is, as we have seen, present at any level. We do not have to resort to atomic vibration to find movement in the meteor. It is present without any further differentiation. Every further introduction of material, the colour of the meteor, its crystalline structure, its temperature, will introduce new activity. In each case there will be superficial closure, for otherwise there would not be new material at all, but if the closure is examined activity will be uncovered.

Most philosophers have sought to eradicate the failure of closure but some have recognised its inevitability. Derridian deconstruction is one such example and consists in the relentless playing out of the failure of linguistic closure, thereby bringing to the fore one aspect of the character of language and of closure in general.[13] In the context of the prevailing assumption that linguistic closure can succeed, that it can accurately describe the state of things outside of language, such a strategy has an important function. Deconstruction thus makes of the failure of closure a method, but by concentrating on the failure of closure deconstruction begins to make it look as if it is not possible to say anything, for even the terms used in the construction of a 'theory' of deconstruction must themselves fail. As a result the relentless pursuit of the failure of closure, as we have sought to demonstrate, threatens a collapse of meaning altogether.[14] If all terms are deconstructed in their moment of realisation, material is lost in a sea of texture, and finally we have nothing. A world full of possibility, but a world without any particularity. As we escape closure we lose the constraints of material and in that sense the world becomes more rich but at the same time the attempt to eliminate closure removes all that is provided through material. The demonstration of the failure of closure is significant in so far as it enables us to catch sight of a characteristic of closure and thus of openness, but pursued exclusively it is primarily destructive. In the aftermath of the deconstructive cultural wave, what is perhaps of more significance is not that deconstruction is possible but that it can be staved off sufficiently for us to get by. Which brings us to the third characteristic of closure: the requirement for stability.

In order to successfully realise material, closure must be sufficiently secure to allow for localised stability. While ultimately any closure fails, each closure is made possible by the provision of an environment in which the closure can be held as if it was secure. In simple systems of closure, as we shall later see, closures are secure because there are no alternatives available within the system. In more complex systems, like those that allow for human experience, the variety of possible closures is contained so that for the moment each closure can appear safe. In such circumstances, closure, as we have identified, is under threat either from alternative closures that might replace it, or because the texture generated by the material associated with the closure offers new closures which in turn potentially undermine the current closure. In order for closure to take place therefore these potential threats to closure must for the moment be contained. This is achieved by the assumption that closure is complete: that it could be carried through and thereby exhaust openness.

The assumption that closure is complete can be regarded as having two aspects. On the one hand alternative closures are not at the point of realisation entertained. On the other hand although texture is a consequence of the realisation of material, the possibilities for conflict between texture and the current closure are ignored. We have already come across an example of the assumption of completeness in the context of perceptual closure. At the point when we see a face in the dots we both refuse to entertain any other alternative, and assume that the texture generated will be compatible with the closure 'face'. Later we may see the face as something else, or we may abandon the closure on the basis that under examination the details of the face do not make sense, but at the time of realisation both of these possibilities are excluded.

There are occasions when we can contemplate two alternatives as being equally capable of closure, as in Wittgenstein's duck/rabbit,[15] or in Antoine de St Exupéry's hat/snake,[16] but even in such circumstances when one closure is adopted we are unable to realise the other. At the moment of perceptual closure the realisation of material is exclusive. Such an outcome is necessary to provide sufficient security to allow for material to be realised. Material is discrete, particular, self-same. If alternative closures were available at the same moment for the 'same thing', the material realised by either closure would no longer be discrete, particular and self-same, and as a consequence they would both collapse.

In the case of linguistic closure the assumption of completeness is at once apparent. On hearing or seeing a written word or sentence we assume that it means one thing. We know of course that words and sentences can be taken to mean different things, but we usually assume that the intention on the part of the speaker or writer is to provide a single meaning. In the event that alternative closures are available we can in the manner of the duck/rabbit swap from one to another, but even so we are likely to seek to abandon one closure in favour of a single meaning on this occasion. The assumption of closure becomes more evident when we are the author. When we say or write a word or sentence we have a unitary meaning in mind. Of course we may come to see that our formulation could be interpreted otherwise and this can on occasion be intentional, if rare, but at the moment of inception we provide one linguistic closure. There are instances in language, poetry being an example, when the potential to realise a single closure is deliberately avoided. This phenomenon will be examined in detail later, but even in such cases it is still only possible to provide a single closure at any particular moment. A poem is like a complex version of the duck/rabbit in which a plethora of alternatives are available, but as with the duck/rabbit only one of them can be understood at any one time.

In order to illustrate the assumption of completion examples of closure have been chosen that are unusually precarious. These examples have been used precisely because they demonstrate that even in circumstances where the closure is immediately threatened, we are capable of realisation simply by excluding the threat on a temporary basis. The majority of perceptual and linguistic closures are by comparison relatively stable. This stability is a deliberate and orchestrated effect of the combined web of closures in which we both personally and socially operate. The assumption of completion is thus made easier by the provision of an environment which enables the relative stability of the closure. This environment is acquired through childhood and through the social character of language is the outcome of many individual closures both past and present.

Since each individual closure requires relative stability, the same principle applies to the network of closure taken as a whole. The dogmatic philosopher and the religious fanatic share the notion that their system of closures is to some degree final and complete. However for those of us who operate on the basis that our closures taken as a whole are insecure, it remains the case that on any particular occasion any given closure requires the assumption that it is secure to enable realisation of material, and at the same time the network of closures on which it relies will be held as if it were complete. Although the assumption of closure need only be maintained temporarily it must nevertheless operate from moment to moment. We can regard ourselves as standing on shifting sand, but from moment to moment we must operate on the assumption that the current state of the sand is held still. As we shall see, the attempt to overcome the failure of closure through the provision of an environment in which our closures are secure is one of the primary motivations of human behaviour.

2

SYSTEMS OF CLOSURE

Body and mind

All forms of life, be they elementary organisms or human
beings, have this in common: they are closure machines.

So far an attempt has been made to provide an initial indication of the
character of closure. It is time to turn to the manner in which layers of
closure are combined to form systems of closure. These systems of closure,
as a result of the realisation of material, are able to intervene in openness.
In addition to describing the operation of elementary systems of closure
an attempt will be made to offer a preliminary outline of the operation of
biological systems of closure and the human system of closure in particular. The purpose of such a description is to indicate how in principle a
system of closure might be capable of providing experience and language;
it is not intended to offer a detailed account of the human system of
closure for such a description will require a dialogue between the framework offered here and extensive empirical observation of a kind that is not
currently available to us. Moreover, the description offered is itself a
linguistic closure or set of closures and is thus itself the outcome of the
system it itself describes.

Let me begin with a mechanistic metaphor and the following claim:
systems which are capable of intervening in openness to achieve specific
outcomes do so on the basis of closure. In this respect, all forms of life, be
they elementary organisms or human beings have this in common: they
are closure machines. The forms of closure employed are very different but
the principles underlying them are unchanged.

Most of us are inclined, from the outset, to oppose the notion of
ourselves as machines. There are many reasons that might influence us in
this respect but perhaps the underlying motivation comes from a sense that
there is more to us than a collection of mechanical parts – however
complex their alignment. There are some who might argue that this sense is
merely a desire to be more than a machine, a desire to assert the importance

27

of ourselves. In the context of closure however it can be seen to be more soundly rooted. It is through closure that we are able to intervene effectively and interact with other things and people. Closures are not however the same as openness and thus if pursued always fail. Although we rely on closures to make sense of openness and to provide us with a world we can understand and manipulate, we are at the same time able to become aware of the limited nature of closure and its incapacity to reflect the character of openness. Since we owe to closure our experience and our reality, the recognition of its failure is sensed only at the margins of our experience. Yet it is this inchoate sense of the failure of closure that is responsible for our opposition to the notion of ourselves as machine. We resist the idea of ourselves as machine because we sense that there is always more to our experience than we can express, and this 'more' to experience is not compatible with our notion of machine which has a 'what you see is what you get' character. The 'more' to experience that follows from the failure of closure plays an important role in our cultural pursuits and what we understand life to be about, and as we shall see is also that which in certain respects we most value.[1] It is not surprising therefore that most of us are uncomfortable with the description of ourselves as a machine for it thereby appears to reduce and belittle human experience: by removing the 'more' from experience it has apparently also removed that which we regard most highly.

By proposing that we consider all systems capable of intervening in openness as closure machines I am not thereby wishing to deny the 'more' to experience. The operation of a closure machine allows for unlimited potential and in sophisticated systems the identification of the failure of closure. Our notion of what it is possible for a machine to do is thus as much at risk by describing human beings as closure machines as our notion of what it is to be human. It is possible to refer to these systems capable of intervening in openness as closure machines because they could in principle be constructed: and moreover, I would argue, if we are to construct intelligent computers or robots they are likely to operate along these lines.

There is one further point that should be made before providing a schematic description of the closure machine. The description of ourselves as closure machines does not thereby assert the primacy of the physical. Science has made it look as if it is possible to reduce openness to a physical description, as if the physical metaphor takes primacy over all others. From this perspective if one were capable of describing the physical constituents of the world and the forces that governed them one would have described all that there was. One of the consequences of the theory of closure is that there can be no such privileged story. The value of the stories of science is that they enable interventions in openness that would not otherwise be possible. It is not that they uncover the real character of openness. The description of the closure machine is part of that set of stories and should be judged on the same basis that applies to scientific stories. 'The closure machine' is itself a closure. It does not, and could not, exclude alternative closures which might provide other insights into the nature of the human condition. A mechanistic metaphor perhaps, but not one that proposes a mechanistic universe.

A system for intervening in openness must have a basis for intervention. It will be argued that closure provides that basis, and that in the case of human behaviour many layers of closure combine to enable precise and successful intervention. The capacity to intervene is enhanced by the realisation of experience in the form of a perceptual world, and this experience is the outcome of the layers of closure currently available to the system. While all closures consist in the provision of particularity, different types and forms of closure can be distinguished because the nature of the material that is realised through closure varies according to the circumstances in which the closure takes place. It is not the character of closure that is different in these instances but the circumstances in which the closure takes place and the character of the material that is thereby realised. In the case of the human system of closures, it will be argued that the highest levels of closure – those most distant from openness and preliminary closure – are provided by linguistic closure and other types of intersensory closure and for the most part it is these types of closure which shall concern us. The case will be made that these closures have a special place because they enable the self-aware character of experience and influence the manner in which all closures are perceived. Despite the complexity and importance of language and other high-level closures the principle that underlies such closures is the same as that which applies to the simplest form of closure: namely the provision of particularity by holding that which is different as the same.

For a system of closure to provide a means of intervention in openness and thus to function as a closure machine, it requires a means of converting the flux of openness into an array of particularities. This initial layer of closure will be identified as 'preliminary closure'. As with closure generally, preliminary closure consists in the realisation of particularity as a consequence of holding that which is different as the same. This is achieved through the realisation of material in response to openness. The most minimal example of a system of closure consists of a single preliminary closure. Such a system requires two discrete states, or at least states that can be held as if they were discrete. It is not difficult to provide mechanical examples of such systems which allow for a single preliminary closure. A mousetrap for example, can be regarded as having two discrete states: it is either set, it is ready, or it has sprung, it has gone off. Many different causes may have led to it being in one state or another: it may have been sprung by a mouse, but it could also have been knocked by someone or something, or someone could have deliberately set it off. In the context of the mechanism all of these variations are of no consequence, it is either set or it has sprung. The diversity of the immediate environment is thereby reduced to single state and its absence: it is either set or it is not set. Any mechanical arrangement that enables a system to alternate between two or more discrete states is thereby capable of providing the basis for preliminary closure. For example, a bell or a gate could function as the basis for preliminary closure. The bell can either ring or not ring, the gate can be closed or not closed. The bell may ring as the result of the wind, or a person or animal shaking it, but the cause of the response is in the context of system of no consequence. The bell either rings or it doesn't. Similarly, the gate may be in one state or another because it has been deliberately moved, or because something or someone has dislodged it accidentally, but these variations are not relevant in the context of the state of system, which in this case is the position of the gate. In either case the cause of the bell ringing or the gate closing is infinitely varied, but in the context of the system the variety of inputs is not accessible to the system and thus of no consequence.

A preliminary closure is provided therefore by a mechanism that has a discrete response to its environment. The states of the mechanism might be described as open or closed, full or empty, on or off, up or down. The naming of the states is unimportant, what is important is the distinctness of the states. By this means the mechanism allows for the diversity of openness to be reduced to two alternatives, which is the same as one thing and its absence. The response of the mechanism provides a preliminary form of closure. It is preliminary because although the discrete response has the characteristic of material there is no texture that can be accessed, for the remainder that is left over from closure is inaccessible: the mousetrap, gate or bell, provides discreteness and therefore material, but there is no texture available, for that remains outside of the mechanism lost in openness.

In mechanical systems reliant on preliminary closure alone it will be readily agreed that there is nothing in common between the state of the mechanism and the circumstances which determined that state, other than that one has resulted in the other. The mouse is not the same as the sprung trap, nor is the wind the same as the bell. We cannot determine the character of one from the other nor is there anything inherently the same regarding the form or the content of these elements. This same principle applies to all closure. Closure does not consist in the identifying of something, or in picking something out. It can be seen that closure provides a new outcome, which is not the same as the circumstances from which it was realised. There is nothing in common between closure and the circumstances from which the closure is realised. In the case of an elementary mechanical system of closure this is self-evident, for we clearly cannot deduce from the state of the mechanism the state of affairs that caused this to occur since an unlimited number of different states will be capable of producing the same outcome.

Preliminary mechanical closures of this type can be combined to produce further outcomes. If for example the mousetrap was linked by string to a bell in an appropriate manner it would ring each time the trap was sprung. A linking of mechanical closures in this manner produces a series of responses but it remains a preliminary closure, because the relationship between the ringing of the bell and the initial circumstances is not different from that between the trap and those same circumstances. The combination of these preliminary closures is itself another preliminary closure. Preliminary closures can be linked therefore but no matter how complex the linkage the outcome is itself only a further type of preliminary closure.

The first level of non-preliminary closure does not consist in the linking of preliminary closures but requires one preliminary closure to be held as one with one or more other preliminary closures. The outcome of holding one preliminary closure as another preliminary closure is to realise a new form of thing, a new form of material, rather than the provision of a new type of preliminary closure. In the realisation of a new form of material a hierarchy of closure is generated. It can be seen therefore that unlike preliminary closure subsequent forms of closure in the system are realised from the internal state of the system itself, and thus from other prior closures, in such a way that the closure is not predictable from any individual prior closure alone. In the case of linked preliminary closures the state of any part of the system can be determined by the state of the initial preliminary closure. If the bell rings, the mousetrap has sprung. In the case of non-preliminary closure the closure is not linked to the state of any individual preliminary closure but to the state of some or all of them. For non-preliminary closure to occur the system requires more than a single preliminary closure in response to external input, and in a biological system this typically involves a great number of preliminary closures.

A mechanical example of non-preliminary closure can be generated from the preliminary closures offered by the previous example of a mousetrap. Suppose that instead of a single trap we have a hundred traps. A non-preliminary closure is realised if the system is capable of providing a discrete response to the state of at least two of the traps. Thus for example, if the system is organised so that a bell rings when two or more traps spring within a five-second interval. In such a system we cannot determine the state of the bell from the state of any one of the traps. We have in this example the first level of closure beyond preliminary closure. This new closure does not respond to the same inputs as the first nor is it predictable from any one of the preliminary closures. Yet as with preliminary closure it realises particularity by holding that which is different as the same. In this case the number of ways in which two or more traps, in an array of one hundred traps, can spring within a given interval is so great as to be unimaginable. Through the closure realised by the ringing of the bell all of these countless different things are held as the same.

Most biological systems can be seen to have many layers of closure beyond preliminary closure. The function of these closures is to provide effective intervention in openness. The principle by which such a system operates can be seen by analogy with a mechanical system. If the response of a machine is based on preliminary closure alone, its capacity for intervention is directly linked to the state of its preliminary closures. With the first level of closure beyond preliminary closure, a whole range of further responses becomes possible, which are not directly related to the state of any one preliminary closure. As a consequence the response of the machine to its circumstances can become more sophisticated and potentially more effective.

Consider a simple mechanical system: a pressure pad in a floor operates to open a door which closes again after a given period. Such a system consists of a linked set of preliminary closures. The state of any part of the system, assuming it is working normally, is predictable given the state of the pressure pad. The capacity for the machine to intervene is limited to one set of responses: the mechanism of opening and closing the door in a given time period. When first-level closures are introduced the capacity of the machine to intervene becomes more complex. One such first-level closure might consist in holding two operations of the pressure pad in close succession as a new closure. For this closure to be achieved new material would need to be provided. This might consist in the movement of a switch or some such mechanism. Through this new closure and the realisation of new material the machine could be designed to produce a different mechanical response: that, say, the door stayed open longer before closing, or perhaps did not open at all. The system can be made more complex with the provision of other first-level closures combining certain combinations of depressions of the pressure pad with different responses of the door opening and closing mechanism. Utilising a combination of such first-level closures the system as a whole could vary the length of time the door opened on each occasion the pressure pad was operated depending on the number of times the pressure pad had been depressed in the recent past.

In this mechanical system consisting of first-level closures it can be seen that the state of the machine is not predictable from the current state of the pressure pad. The provision of new closure has allowed for a more complex response to the inputs. The responses of the machine can be made more sophisticated if further closures are possible. If instead of a single pressure pad, and thus a single type of preliminary closure, the system involves other pads and thus a number of preliminary closures – for example at different entrances to a building and throughout its internal doorways – first-level closures could be arranged to instigate the opening and closing of doors dependent on the relative use of various doorways. Such a system could generate a complex response to what is an elementary action: the depressing of a pressure pad, and could perhaps be utilised to direct traffic flow into, out of, and through a building.

Although the mechanical system outlined is very primitive by comparison with biological systems it illustrates the basic principles underlying the operation of a system of closure in general. The depression of the pressure pad provides preliminary closure by operating as a switch, as a result the flux of openness is held as either on or off, up or down. It does not matter to the system what has caused the operation of the switch but the switch allows each of these different 'events' to be held as the same. The process of closure results in the provision of material – which in this case consists in the position of the pressure pad as either up or down – through which a response is made possible. At the next level of closure, combinations of preliminary closures – in this case combinations of positions of the pressure pads – are held as a new thing. The new 'thing' that is generated by first-level closure is in addition to the preliminary closures and requires new material – in this case the operation of a switch or set of switches which then triggers the different mechanical response of the system. It can be seen that the material that follows first-level closure relies upon preliminary closure but is in addition to preliminary closure. It contains preliminary closures, and thus contains the possibility that the first-level closure might be made differently. Moreover each preliminary closure remains the same as it was prior to the first-level closure, but it is now held in the context of the first-level closure and to this degree is different in the light of the first-level closure. It can be seen therefore that preliminary closure is not simply a set of data from which a pre-arranged set of consequences follow. The nature of the data is interpreted differently in the light of the character of the first-level closures. The preliminary closures are thus open to different types of first-level closure. Some of these types may prove, from the perspective of the inventors of the machine, to have a purpose, a great many others may have no viable function.

It is important to note some of the characteristics of this mechanical system. Firstly, there remains nothing in common between the closures of the system and the actions that instigated them: the position of the switches tells the system, or an intelligent observer of the system, nothing about the nature of the circumstances that led to the depressing of the pressure pads. Secondly, the first-level closures are not predictable from any one of the preliminary closures. And thirdly, alternative first-level closures could always be provided that could be used to generate wholly different outcomes. These characteristics are not only applicable to mechanical closure but it will be argued typify closure in general. It can be seen therefore that although preliminary closure provides material and particularity, an array of such material remains open since subsequent closure can combine the preliminary closures in any number of ways. As a result of preliminary closure a system realises material and is therefore able to intervene selectively. The provision of material does not however result in the eradication of openness for the array of preliminary closure is itself open. Preliminary closure can be conceived therefore as a mechanism for providing openness in a form that allows the system to function. Preliminary closure is in this sense preliminary to the real business of closure which begins with the first layer of non-preliminary closure.

Biological examples of closure operate on the same basis as the mechanical closures already described but there is an important additional characteristic, although one that could in principle be added to the mechanical system described. In the context of the elementary mechanical system described the responses of the machine are not relevant to the success of the system as seen from the system itself. Biological systems on the other hand typically have a self-referring loop, with the consequence that the outcome of the system of closure has an impact on the sustainability of the system. It is this circularity that allows the system to utilise different and competing closures which are then abandoned or retained according to their capacity to enable 'effective' intervention, as defined by the system itself and its own sustainability. It will be argued however that aside from this characteristic, biological systems, although hugely more complex, do not differ in principle from the elementary mechanical systems that have been described.

The detailed character of these biological systems has to be determined empirically – which is not to say that empirical observation leads to an accurate description of openness but that the process of using a theory to intervene in openness is the means by which our linguistic closures are made more effective. The current state of understanding of such systems however suggests three distinct levels or forms of closure each of which consists of many different sub-types. These three forms of closure will be identified as: preliminary closure, sensory closure and intersensory closure. I shall argue that in the case of sensory and intersensory closure these forms of closure in addition to containing many sub-types are organised into successive layers of closure that form hierarchies.

As with a mechanical system, it can be assumed that the interface between the biological system and openness provides preliminary closure. Human sense organs, for example, can be regarded as functioning to realise preliminary closures. Current theories of the operation of the sense would seem to support such an account. So in the context of vision, for example, the cells at the back of the eye, have two discrete states: polarised and unpolarised. When the cell, a neuron, is in a polarised state it generates a response: the release of neuro-transmitters. The neuron can be seen therefore to hold the diversity of openness as in one of two states: it either fires or it does not fire. The neuron provides a preliminary closure which is to hold openness in this particular way. In the place of infinite flux the system identifies a particularity. If we look at the preliminary closures of vision in more detail we can find further preliminary closures that provide links in the chain. The photopigment found on the outer layer of the photoreceptor has two states, in which retinal is either bound to opsin or it is separated. This preliminary closure in turn results in the opening and closing of the sodium channels – a further preliminary closure – which in turn results in the release of neurotransmitters.[2] The neuron can be regarded therefore as a linked set of preliminary closures that functions overall as a preliminary closure. In this respect, as with mechanical closures, the state of the neuron and its response has nothing in common with circumstances that caused this state to occur, nor can we deduce anything about those circumstances from the state of the neuron. The neuron simply fires or it doesn't, and from the point of view of the system at this level of closure nothing is shown or identified in the 'world'. It doesn't for example tell the system that a photoemic particle has hit it. Such an account, which is the part of the current state of the closures of science, is a conclusion that is only possible in the light of a whole history of culture and language which have been possible through higher levels of intersensory closure. The neuron on its own does no more than provide a particular preliminary closure: it polarises and releases neurotransmitters, or it does not. The response of the neuron does not provide the organism with a knowledge of openness, it simply provides a particularity through its firing or not firing. A particularity which in turn allows the organism to provide further closure. The mechanism offers a means of holding openness as something in particular, but that something is not 'out there' in openness, but a product of the closure.

Taken as a whole the combined output of neurons in the eye provides a huge array of preliminary closure which is then available for layers of subsequent closure. This array of preliminary closure does not in principle determine what further closure is possible and can be considered therefore to be open – although in practice biological systems are liable to constrain the ways in which first and subsequent level closures can be realised. The preliminary closures realised by the mechanism of the eye thus function to provide a basis for closure through the provision of particularity which is at the same time open.

Sense organs can be regarded therefore as providing for the possibility of a closure machine by realising preliminary closure. In doing so they realise the first layer of material in the complex and hierarchical system of closure that constitutes a biological system of closure. In the human system of closure I shall argue that the highest levels of closure are realised by linguistic closure, itself a type of intersensory closure, and that the system as a whole is responsible for self-aware experience. The way in which these closures function must be determined by empirical observation (an observation which inevitably takes place in the context of our own current systems of linguistic closure) but it can be seen that some form of system of closure is necessary if an organism is to be able to intervene in openness. Neurons do not as a result have to function as they do, or produce a response or trigger in a particular manner, but any sensory system however primitive or sophisticated will require a mechanism that provides preliminary closure and subsequent levels of closure.

The preliminary closures of a biological system taken as a whole thus produce an effect like bells placed in a tree. No matter what causes the bell or bells to move, the wind, or the movement of a branch, or a person or an animal, the output is the same, the sound of the bell. There is nothing in common between the cause of the bell ringing and the sound it produces. Yet with each sound of the bell there is a closure, for the flux of openness is held as one thing. Each sense functions in this manner, having its own response to openness. In this context we do not therefore see the colour of the world, or hear its sounds, but rather hold openness as coloured and as having sound. These ways of holding openness do not tell us what openness consists of, or how it is divided, but provide a means of making further closures which in turn enable us to vary our behaviour and intervene more effectively. A myriad of linked preliminary closures from each of the different senses thus provides the material for the system to function at all.

In the context of the human system of closure the first level of closure following preliminary closure can be identified as 'sensory closure'. This proposal is based on our current understanding of the operation of the human brain, which is to say on our current framework of linguistic closure, a framework which could in principle be radically different. In due course it may be that further forms of closure can be identified, even within the framework of our current understanding, which separate preliminary closure from sensory closure, but such an outcome would not alter the overall principle by which the system functions. It can be supposed therefore that each sense realises a different type of sensory closure, and each type of sensory closure itself consists of layers of closure the outcome of which is to provide material which holds as one a group of preliminary closures, thereby realising what might loosely be described as patterns. The character of patterning realised by sensory closure is accessed with difficulty for further higher-level closures contribute to perception thereby obscuring the nature of the sensory closures which precede them. The current state of research does not enable us to go into detail beyond preliminary closure but an attempt can be made to indicate the principles that are likely to underlie sensory closure. Different types of preliminary closure can be distinguished within each sense. The cones of the eye are an example of such difference, which in this case contain three further sub-types. Sensory closure it can be supposed realises patterns within each type of preliminary closure – ways of holding groups of each type of prelimi-nary closure as one. Then further levels of sensory closure will hold different types of preliminary closure from the same sense as one, or will hold patterns realised from different types of preliminary closure from the same sense as one. As a result of layers of sensory closure, which may run in parallel or may be hierarchical, patterns are realised from preliminary closure which can be used to influence behaviour. The particular type of material realised by the system, the particular type of pattern, may be prescribed in advance in which case the physiology of the organism can be regarded as being hard-wired. In such circumstances the openness within the system is limited by its construction. As we shall see, even in such cases the system can have a capacity to learn from experience and as a consequence to alter the character of its intervention, while in more sophisticated systems the type of material that is realised is itself capable of variation, with the result that the system has a greater range of possible interventions. The current scientific story regarding the processes involved in, for example, visual perception, suggests that the plasticity of the system in mammals at least is such that the capacity to adopt alternative closures, if current closures are unsuccessful, can extend to low-level sensory closures and possibly even to preliminary closures.[3]

In the context of this account of the operation of closure in biological systems, an account which is intended as no more than an initial sketch of the likely overall character of such a system of closure, each sense can be seen to provide many different types of its own form of preliminary closure. Subsequent closure also specific to each sense can be then taken to identify patterns within each type of preliminary closure and between types of preliminary closure. It can be supposed that further sensory closures group these patterns into what may later become the basis for material things. It can be seen that while each additional layer of closure provides particularity in the form of new closure and new material it also provides texture. Each new closure is in addition to the prior closures which remain available to the system. Texture is thus the consequence of the difference held within closure but which is available to the system in the lower-level closures. Returning to the example of the page of dots, having realised a line in a page of dots, the page is now held in this light. Texture is generated by the new material for the prior material will not be exhausted by this closure and further closure will be possible. Is there a break in the line? Is it crooked, or irregular? With each additional layer of closure new material is added, but the previous closures are not lost and are structured in the context of the new closure thereby generating texture. Preliminary closures are thus increasingly textured as new sensory closures are realised and additional layers of closure provided.

In the case of human sensation it would seem that the early years of life see a spectacular increase in the type and sophistication of sensory closure resulting from the interplay of various levels of closure in conjunction with the requirements and demands of the rest of the body. Our current understanding of the operation of the senses suggests that for each sense there is no limit to the number of closures that can be generated or new closures that can be found even though preliminary closures come in strictly limited types. Every part of our visual field for example can be regarded as a more complex version of the page of dots. As a result there are in principle no bounds to the number of new patterns that can be found or new distinctions that can be made. The stability of our perceptual world is in the face of this welter of possibility a remarkable feat. It is a stability that we can presume is hard won through layers of closure built up and refined over time. So it is that with the increasing provision of closure the flux of openness becomes increasingly textured so that eventually there can appear to be no space left for the unknown – as if closure has filled in all the gaps and explained everything.

It is apparent to us that some senses are textured more than others, with a resulting proliferation of things. Vision and sound are, for most of us, the most highly textured, so that the field of vision and sound are filled with material: with particularity. Touch is usually more impressionistic, offering fewer particularities, fewer things. The reason for this is presumably due to the potential of the system for intervention as the result of the material generated by the sense in question, which in the case of touch is for humans usually less than that of vision or sound. An analysis of sensation in terms of closure suggests however that there can in principle be no constraint on the extent to which any sense can be textured. For the large number of preliminary closures of each sense can be combined in an unlimited number of ways through first and higher-level closure. Initial evidence in support of this conclusion can be identified. In the case of touch, for example, the capacity to generate additional closure is notable in those for whom touch plays a more important role: as in the cases of those individuals who have impairments of sound or vision. Moreover experiments have suggested that it is possible to experience something akin to the detailed closures of sight with touch.[4] Such an outcome would be expected given the account that has been proposed, for the capacity of each sense to impose closure and realise material – with the resulting provision of particularity and the ability as a consequence for the system to intervene – can have no theoretical limit.

Following sensory closure, the next stage in the hierarchy of closure in biological systems will be identified as 'intersensory closure'. So far we have been concerned with intra-sensory closure: those closures that enable the identification of patterns and particularities within a particular sense by holding preliminary closures, or sets of preliminary closures as one. This happens firstly at the stage of preliminary closure, and then in successive layers of closure in which previous combinations of preliminary closures are held as the same through the provision of sensory material which can be regarded as patterns in preliminary closure. The higher-level closures of each sense provide particularities or things which include prior and lower-level closures. The closures of the human visual system realise for example shapes, which rely upon prior closures of colour, and edge.

Each level of closure could however in principle be held differently so that the closures which realised colour and edge are themselves the product of an earlier closure and as a consequence can themselves be abandoned in favour of new and alternative closures which in turn allow for different higher-level closures in the form of different shapes. All of these different closures can however be considered as being of the same form in that they are all the product of visual sensory closure. In contrast, intersensory closure holds closures realised from one sense as the same as closures from another. A smell – a pattern realised from the olfactory sense organs – can for example be held as the same thing as a visual pattern and a new closure realised. In this context a fox that scents a chicken and looks for the chicken in search of food, can be regarded as having held a scent and a visual pattern – and probably a taste – as one, and intervened accordingly. The system of closure allows not only for different closures within each sense to be held as one with other closures of that sense, but for different closures realised by different senses to be held as one with closures from another sense. Sounds and tastes are held as one with images, and images as smells and sensations of touch, and so forth. Intersensory closure is the means by which these diverse closures from different senses are held together as one. Long before language is employed intersensory closure is realised in order to intervene effectively in our environment.

The division of the mechanism of biological systems of closure into preliminary, sensory and intersensory closure is not intended to be definitive. Further examination of these systems may identify divisions within these forms of closure. The many layers of sensory closure may come to be regarded as containing different forms of closure, identifiable by a divergence in the form of material that is realised in each case. The same may turn out to be an appropriate account of intersensory closure. The categories of sensory and intersensory closure are to be regarded therefore as an initial starting point influenced by our current understanding of perception. In the context of this starting point sensory closure realises material in the form of sensation, intersensory closure realises material in the form of thought.[5]

Each layer of closure is a way of holding prior closure and one way of thinking about this way of holding prior closure is to regard it as an explanation of prior closure. The patterns of sensory closure can thus be regarded as explanations of the chaos of preliminary closure. It is as if the patterns of sensory closure serve to explain how the preliminary closures come to be as they are. In the same way intersensory closure holds divergent sensory closures as one and can in this manner be seen to provide an explanation for the variety of sensory closure. Intersensory closure provides a means of accounting for the complexity of sensory closure. It is in this way that complex biological systems are able to construct a reality through intersensory closure which can be regarded as the 'cause' of their sensory closures. In this manner intersensory closure realises closures which are in addition to sensory closure and are not therefore associated with any one sense. In order to hold closures from two different senses as one it is necessary to realise a new form of material, a new form of thing, through which the diversity of sensory closures are held as one. Thus it can be supposed that the fox, in the previous example, holds a smell – the scent of the chicken realised through a set of olfactory sensory closures – as one with a visual pattern realised through a set of visual sensory closures, through the provision of a new form of material which is non-sensory. This new form of material can be described as the thought of the chicken. The thought, a type of material, is neither the smell nor the visual pattern, but is that which allows these two entirely different prior closures to be held as one – to be held as the same – although they have nothing in common.

One likely objection to the account of closure that has been offered is to argue that the particularities which have been identified as the products of each layer of closure are in fact only identifiable as the result of the operation of language. The thoughts which it has been supposed enable the fox to combine visual and olfactory sensations have been identified using language. Similarly the products of preliminary closure, and sensory closure, are only identifiable as separate 'things', as separate states of the system, because the categories of language allow us to make these distinctions.

The account of the human machine in terms of closure that has been offered proposes that each layer of closure provides a way of holding the diversity of closures available to the system from prior and lower-level closures. In this respect each layer can be regarded as an explanation for the diversity of prior closure, as a postulation of something other than the prior closures by virtue of which they can be accounted for. In the human system of closure it is proposed that language provides the highest level of closure. In this respect it offers an explanation of each of the preceding layers. This does not however have the consequence that language need be regarded as being single handedly constitutive of experience or of an external reality. One of the consequences of the account of closure as it has been outlined is that while language plays an important role, and is itself, as I shall shortly indicate, a form of intersensory closure, language need not be regarded as a necessary requirement for the realisation of something akin to what is indicated in language by the phrase 'an external reality'. The Kantian legacy has encouraged us to suppose that it is the concepts of language that are responsible for the provision of experience.[6] In the context of closure however, linguistic concepts, although a vital and important type of intersensory closure, can be seen to be only one type, and there is no reason to suppose that non-linguistic intersensory closure should not be capable of enabling the realisation of physical objects, or more precisely things that function to explain sensory closure in the manner in which for us physical objects do – as the behaviour of young children and animals would trivially appear to suggest.

Since, as Kant argued, experience is possible only if we can identify it as mine, and since such an identification requires the ability to identify an external other as not-mine, it follows that in the context of closure, the prerequisite of experience is not language but rather intersensory closure. For intersensory closure postulates something by virtue of which the material products of different senses are capable of being held as one. This is however precisely the function of the framework of physical objects that make up external reality, and in this sense therefore intersensory closure can be regarded as enabling the realisation of an external world. So while the nature of a non-linguistic reality will inevitably be very different from a system that has the additional closures of language, it need not be considered to be so different as to be incomparable. It is for this reason therefore that the framework of what for us is identified as reality can be regarded as the outcome of intersensory closure and not some additional and separate form of closure that is linguistic. Furthermore it would appear that purely sensory closure will not allow for the provision of an external reality. For a biological closure machine that was restricted to a single sense, or to a set of senses that could not be held as one, would have no means of realising a closure which could be held as something other than the sense in question and would therefore be unable to realise material which functioned as an explanation for sensory closure in the way that physical objects do in our reality. The only manner in which this conclusion might prove to be unfounded, in the context of the general framework of the account that has been offered, is if the category of 'sensory closure' required further subdivisions of closure which themselves allowed for material of a different form.

The claim has been made that language is a type of intersensory closure. It is time to give an initial account of what is understood by this claim. A detailed account of the mechanism of linguistic closure will be provided in Part II, but an indication of the argument is offered here in order to allow readers to have an overview of the structure of closure being proposed. Language as a type of intersensory closure holds one sensory closure as one with a different type of sensory closure. So it is that a linguistic mark, for example a written or spoken word, itself the product of sensory closure, is held as one with a totally different sensory closure, or as one with another linguistic mark. As with all forms and types of closure, the process of linguistic closure is achieved through the realisation of material which allows that which is different to be held as the same. The material realised in the case of intersensory linguistic closure is meaning. The outcome of a particular type of intersensory closure, meaning can be regarded as being a type of thought. (It will be noted that there are types of thought other than meaning as the preceding account of intersensory closure implies.) Intersensory linguistic closure is a process that occurs in an individual and its outcome, meaning, is therefore also something that takes place in the context of the individual. The dictionary meaning of a word is distinct from meaning in this primary sense, meaning as the outcome of linguistic closure. The dictionary meaning of a word, as we shall examine in more detail later, seeks to identify those linguistic closures that competent speakers commonly and typically associate with a linguistic mark. The main concern here however will not be with the social or dictionary meaning but with meaning as the product of linguistic closure. This is not to suggest that meaning is a purely subjective phenomenon, if by this it is understood that meaning is simply created by the individual. Linguistic closure is inherently social[7] but the realisation of closure must take place in the context of the individual and before we are in a position to understand the social constraint on linguistic closure we must first examine the process of linguistic closure as it occurs in each individual.

As with other types of intersensory closure, linguistic closure relies on prior sensory closures. In the case of spoken language, for linguistic closure to take place individual sounds realised by sensory closure need to be identified as units of language.[8] The units of language are not words as such but those marks which enable us to realise linguistic closure. These linguistic marks are distinguished from other sounds, themselves the product of lower-level sensory closure, and identified as the same on different occasions even though they are not identical. Thus if we hear a language with which we are unfamiliar and which is very different to our own, we are initially unlikely even to be able to identify and thus repeat the units of sound in the language. As the aural closures become increasingly textured and it is possible as a result to identify finer details we are able to identify the sound of the same mark be it a word or phrase. Through aural sensory closure we are able to hold a sound as the same as another even though it is never in fact identical. It may be at a different

43

pitch or different volume. It may come from a different direction. Yet through aural closure all of these different sounds are held as the same through the identification of a pattern which allows us to identify a linguistic unit such as a word or phrase.

Amongst other uses, we think of words and phrases of language as referring to things in the world. This association of a word or phrase with a 'thing' in reality can be explained in the context of closure in the following manner. Linguistic closure in spoken language holds sounds, which have been identified as linguistic in character through aural sensory closure as the same as another sensory closure or closures, or intersensory closure or closures, through the provision of material in the form of meaning. For example, through intersensory linguistic closure we realise meaning which enables us to hold the sound 'blue' as the same thing as the 'blue' we perceive as a result of visual sensory closure. The new linguistic closure is not simply the combination of the preceding sensory closures, for in such a circumstance nothing new would be added by the closure: no material would be realised. Linguistic closure finds in these two diverse sensory closures that which is the same through the provision of material which allows these divergent things to be one thing.

Language as the product of linguistic closure has the characteristics of closure in general. Like all other forms and types of closure therefore it does not consist in the identifying or naming of something. Instead two or more different things are held as one and the same through the provision of new material in the form of meaning. In the case of the spoken word 'blue' the linguistic closure consists in holding two sensory closures as one, a pattern in sound as one with a pattern in vision; but there are instances in which linguistic closure involves the holding of many different sensory and intersensory closures as one. For example, the meaning realised by the word 'tree' can be regarded as the outcome of holding as one many different sensory and intersensory closures. On the one hand there are the sensory closures required for us to realise the written and spoken forms of the word 'tree'. Then on the other there is the set of sensory and intersensory closures that have enabled us to form the idea of a particular tree. This set of closures might include sensory closures such as 'its' visual shape and colour, the feel of 'its' surface, and the sound the wind makes in 'its' branches. The idea of the tree is the intersensory closure realised as a result of holding all of these different sensory closures as one. The meaning of the word 'tree' in this instance can then be realised by a further closure which holds as one these two sets of closures. If we have already come across the word the meaning will in addition be influenced by the meanings that have been previously realised when the word has been used. It can be seen therefore that the meaning of a single word may rely upon a larger number of prior closures in addition to linguistic closure. It can be seen that the meaning remains the product of closure and is not a description of openness or the identification or naming of something in the world.

One consequence of this account of meaning is that the marks of language, its words and terms, realised through sensory closure, can be seen to function in two distinct ways. On the one hand they serve as cues to linguistic closure; and on the other as tags of linguistic material that has already been realised. When a linguistic mark functions as a cue it serves to identify the possibility of linguistic closure. A child learning a language, or an adult coming across a new word, or a new use of a word, treats the mark as a signal that a linguistic closure is possible and could be realised. Attempts are then made to find an appropriate linguistic closure. In the case of the descriptive use of language this will involve the identification of a sensory closure, or the postulation of sensory closure, that is to be held as one with the linguistic mark thereby realising linguistic material in the form of meaning. From this point onwards the mark no longer functions as a cue, but is instead a tag to the meaning that has been realised through linguistic closure. For a proficient English speaker therefore the linguistic mark 'tree' functions as a tag to a linguistic closure – the meaning of the word – which itself is the outcome of the occasions on which the word tree has been used, and all of the sensory closures associated with these occasions. The meaning of the word 'tree' is thus that which is held to be the same between a great number of different sensory and intersensory closures.

Traditionally philosophers have been concerned about the existence of universals: does 'a tree' refer to a thing which exists, in the same way that we can seemingly point to the existence of a particular tree? In the context of the theory of closure it is not universals but the character of particulars which needs closer attention. For particulars and universals are both the outcome of closure, are both the product of the provision of particularity. In neither case do they exist if by that we are to understand that they are a part of openness, but in this sense there is no thing that exists. When we see a particular tree, its existence is the outcome of many layers of sensory and intersensory closure. The closures that realise the particular perceptual tree do not thereby identify an existent thing in openness. Similarly, when we refer to trees in general, as in 'a tree is a large plant with an inflexible trunk', the existence of the universal is an outcome of linguistic closure, it does not thereby identify an existent universal in openness. While particulars and universals are the product of closure and do not therefore indicate that something exists in openness, both particulars and universals exist in the sense that they are closures that have been realised.

As we acquire language, and our linguistic closures become detailed and complex, our sensory closures become increasingly textured by these linguistic closures with the outcome that our capacity to access sensory closure independently of language is limited. So much so that in the twentieth century where concerns with language have been predominant, it begins to look as if there is no experience at all without language. However, while it is a truism that we cannot describe sensory closure without language this does not have the consequence that language creates the world. Linguistic closure realises meaning which in turn influences the realisation of further sensory closure. There is no reason to suppose that this process is unique to language. It is likely that other types of intersensory closure have a similar outcome. Linguistic closure does not dictate sensory closure, but encourages certain sensory closures to be realised. Language thereby plays a vital contributory role in determining perception but experience does not consist of linguistic closure alone.

The precise relationship between closure and experience will be considered later, but for the moment experience can be loosely thought of as the outcome of all the closures available to the system at any one time. In this context it can be seen that while experience is directly affected by language because language involves high-level intersensory closures which play an important role in organising our lower-level closures, language is not responsible for providing experience nor is experience limited by language. In this sense it is neither the case as Wittgenstein put it that 'the limits of language are the limits of my world' nor as Derrida remarked 'there is nothing outside of the text'. It is the limits of closure that are the limits of my world and not language; and while there is no thing outside of closure, it is also the case that outside of closure there is potential for everything. Our experience is shaped in the context of linguistic closure, but experience is not exclusively linguistic. Language realises the highest-level intersensory closures and thereby provides the framework within which other closures are viewed but it does not thereby eradicate lower-level closures nor does it make it impossible to refer to them.

Every aspect of experience, be it sensation, perception, thought, imagination, or memory can be considered therefore as a product of the combined layers of closure. Each of these aspects of experience has as a result the character of closure. Perception can be thought of therefore as a realisation, a making real, of openness through a system of closure. Despite the complexity of the layers of closure required the underlying characteristics of closure remain. We perceive individual things, we do not perceive openness. In whatever way that 'something' is seen, it is seen as something. Yet there are always alternative possible ways of seeing the 'same' thing. Perception is unlimited. It is also stable and seemingly complete. Perception appears to be final. A person or a car is perceived; yet alternative closures and other perceptions could be realised. Instead of seeing the person we could see their clothes. Instead of the car we could see a reflection. The illusion of perception is that we see these different things at the same time, as if all possible closures were at once available to us, but they are not. We are able to 'explore' each closure separately, and provide further closures from the texture held within the material. If we see a person, we can look to see whether they are cheerful, whether they are walking with a spring in their step. If we are looking at the clothes we can notice that a button is missing, or the shoes are undone. We are able to make these different closures rapidly in succession, with the result that all of these closures are available to us even though each one is made separately. We can look at the shoes, and then at the laces, and then at the scuffs on the toecaps, and the worn down heels. And if we look closer we can see the stitching and the creases in the leather. And then we can look at the whole shoe again. We could not exhaust the possibilities of closure held within the shoe yet any particular closure appears to exhaust texture, if only temporarily. Perception is thus both unlimited and seemingly complete. Its failure, its inability to capture openness is found in the infinity of its task.[9]

When we look at the person, and the clothes, and the shoes, and the laces, we experience in succession a series of different closures each the outcome of a complex hierarchy of closure. There is no limit to the number of perceptions, the number of additional closures that might be realised. Each of these are ways of exhibiting openness, each the product of many layers of closure which realise material and exhibit texture. Each perception is presented as this thing and nothing else. We assume closure; we do not bother to check it out. We do not see those 'things' held within the closure until we carry out further closures. Yet the impression is that we have seen everything. And if we tried to look at everything we could not succeed. If we tried to look at all of the shoe we would fail, for there would always be more that we could look at: reflections in its surface, patterns in the creases, variations in the stitching, the grain and break of the leather.

An individual perceived thing is both particular, it is this thing and not anything else, and although it is taken to be complete it is in fact incomplete. The discreteness of perception is provided by the circle of material that is a consequence of closure. On its own the circle of material would however merely provide an empty form without any filling. The content of perception, that which makes it appear continuous, is texture. The texture of perception is the way preliminary sensory closures are held in the context of material. Texture is not openness, it is not as if we perceive openness; but it is open in that the residual preliminary closures held within higher-level closure are capable of being themselves closed in any number of ways – as we have seen in the example of the page of dots.

In summary therefore: any system capable of intervening in openness operates on the basis of closure, whether it is a mechanical or biological system. In the case of biological systems the hierarchy of closures can be divided into three basic forms: preliminary closure, sensory closure, and intersensory closure. Each of these forms of closure has many types, and within any type of sensory or intersensory closure there is a further hierarchy of closure. It is through the combined interaction of intersensory closures that we are able to realise an external physical world. Language is a type of intersensory closure which itself has different sub-types, each of which have themselves layers of closure within them. In combination, the human system of closures is responsible for our experience.

There is a further aspect of linguistic closure that has not been considered: through language closures are shared. The sharing of closure has far reaching consequences: for as a result the character of each individual's space owes to the closures of society as a whole and is the outcome of a history of closure. In addition to individual experience therefore, our beliefs and theories about the world and the structure of social organisation are also products of closure, with the result that their characteristics can also be traced to the structure of closure.

Inevitably, in an attempt to provide an overview of the operation of the human system of closure a great deal of ground has been covered. There are many objections to such an account to which no response has yet been made. The intention however has been to provide a broadbrush account so that the reader can have an overall story within which to place the more detailed consideration of the issues raised by this account in the remainder of this book.

A final caveat as a reminder of the partial nature of the present summary: the account or theory of closure so far provided is itself a linked set of linguistic closures, and illustrates as a consequence the characteristics of closure.

3

THE PURPOSE OF CLOSURE

Closure enables intervention in openness and is confirmed by that intervention, but the 'how' of intervention cannot be understood and lies outside of closure.

Closure has been described as a process which realises material and texture and which in successive layers provides the elements of experience, thereby determining our thought, and our social interaction. In the case of biological systems the function of closure however is not to realise material or provide experience but to intervene in openness. If, to use an earlier example, we have realised a face in the dots, the combined outcome of sensory and linguistic closure, we can find the face again, or cut it out, reproduce it, modify it. In the absence of the closure these interventions would not be possible. The interventions could occur in a physical sense, but if they were to occur they would do so as if by accident. A stone falling down a hillside can be considered to intervene in openness, but the manner in which it does so is passive: the stone is lost to the flux of openness. Moreover to imagine the stone intervening in openness is already to have imposed closure, for of course the stone in the absence of closure is not identifiable as a thing. Closure through the provision of material enables the system whether biological or mechanical to respond on the basis of the realised material and thus to intervene to achieve a particular outcome.

In complex systems of closure the interventions made possible through closure are chosen from a range of alternatives available to the system. This choice is made on the basis of the outcome that is required, and the intervention is in this sense directed. It is to be supposed that further sophistication enables the system itself to be aware of the choice of closure and the intended outcome, with the consequence that the directed intervention is self-aware and the intervention is purposeful. In systems where the availability of alternative closures allows for directed intervention, the

49

intervention is the means by which the closure in question is both confirmed and stabilised.[1] Without the capacity to intervene and modify closure as a consequence of the intervention the system of closure would be unstable, for there would be no grounds for maintaining any particular closure in the face of alternatives.

Complex systems of closure enable highly directed interventions in openness but even preliminary closure alone is capable of enabling intervention that can be considered directed, although not purposeful in the sense of being self-aware. Take the previous example of the preliminary closure offered by a mousetrap. Such a system can be made partially self-sustaining if the mouse is trapped unharmed, and only released when it has run on a wheel long enough to generate sufficient energy to reset the trap. If this hypothetical trap is then attached to an arm that moves fixed distances in a fixed period, say one metre at the end of each day, and if further it is supposed that the movement of the arm is a function of the trap, it is not difficult to envisage a circumstance in which the mechanism gets better at catching mice and therefore is able to be self-sustaining for a longer period. The relationship between the springing of the trap and the movement of the arm need not be complex for this to take place. For example if the springing of the trap had the consequence that after a given interval, say five days, the arm instead of moving the trap the given fixed distance in a fixed period, returns the trap to the location at which the mouse was caught for a period of time before resuming its previous movement. If it is assumed that a mouse is more likely to be caught in the same location – on account for example of the proximity of a nest or a run – it follows that the mechanism will get better at catching mice and therefore improve its own sustainability. Despite the fact that such a mechanism operates with a very limited system of preliminary closures, it is still capable of intervening in such a manner that it improves the capacity of the system to maintain itself. In addition any such mechanism would quickly have its own particular pattern of 'behaviour' developed as the result of its own 'history', and could be said to have learnt, in this limited sense, from its past actions. As a result any two such mechanisms are likely to operate differently even in roughly similar circumstances, and the movements of the machine would not be determinable without a knowledge of the principles under which it was operating and its history.

A relatively simple system of inanimate closures is capable therefore of surprisingly complicated interventions. In the system described however the closures themselves are fixed and are not influenced by intervention. While closure in general is unlimited within an individual system this need not be so. Although different mechanical systems can therefore employ an unlimited number of different preliminary closures, most mechanical systems operate with a fixed and small number of such closures. An important characteristic of many biological systems is that the system of closure allows for alternatives, so that the closures in question are maintained and reinforced, or abandoned, in response to the interventions that the system of closure has itself made possible. As a result the system can improve its capacity to intervene as well as improving its interventions. A fixed system of preliminary closure can result in behaviour that learns from its past interventions in the sense that the interventions are more successful in the completion of a particular task in future. Such learning might be said to be passive since there is no internal change to closures of the system. It can be seen that it is the ability to alter closure in order to improve intervention that allows the system to be capable of active learning.

While animate systems can allow for a variety of closures, this variety must nevertheless be constrained. Even a relatively small number of preliminary closures is capable of providing the basis for an almost limitless number of first-level closures. The number of patterns that can be generated in a page of one hundred dots has been already identified as being unimaginably great. By comparison the number of preliminary closures in the case of animate sensory systems is far greater. In the human visual system there are estimated to be hundreds of millions of photoreceptors, each of which functions to provide a preliminary closure. The number of possible first-level closures, even if these are restricted to the determination of various patterns of colour and shape, has therefore no conceivable limit. As a result the system must have some way of limiting this welter of possibility. One way of achieving this is to restrict the number of available closures by hard-wiring them. Instead of realising any pattern in the dots, the system seeks to realise only a particular closure, a particular given pattern of dots. More productively it can also be achieved by using the intervention and its outcome as a means of retaining the closure or seeking a new one. The 'outcome' is not a change that takes place in openness, for this is not identifiable to the system, but consists in the impact of the intervention on the subsequent closures that are available to the system.

This process can be identified in the context of a human baby. A baby realises sensory closure thereby providing material in the form of patterns of shape and colour, sound and so forth. The number of possible patterns has, as we have identified, no limit, but it is to be presumed that the physiology of the sensory organs and the brain is attuned to certain patterns rather than others. The ways of closing openness available to the baby remain very great. It is reasonable to suppose that initially interventions are made on the basis of these closures which appear at first almost random but as some of these interventions are successful – itself an outcome of other sensory closures – some closures are retained and repeated so that gradually the framework of closure becomes more effective.

The same process applies when as adults we realise linguistic closure. We can describe objects, or people, or circumstances, in any number of ways, but some of these descriptions will prove more useful to us than others – 'useful' in the context of the system of closure with which we are currently operating. As a result we are inclined to retain them until a more effective closure is realised. Closure can be seen therefore to provide the basis for our intervention in the world, and it is only through closure that we are able to intervene. Intervention however is not only made possible through closure but is the means by which the closure is maintained. Without intervention closure would still realise material but there would be no reason to realise and hold one closure over another. Intervention allows us not only to choose one closure over another but to retain it despite its inevitable failure in the face of openness. However we hold openness through closure, openness remains other than this. Everything is, in this sense, not what it is, for however we hold the world it is other than how it is being held. We are able to ignore this failure of closure because the advantages of holding openness as this thing are made apparent through intervention.

High-level closures may not directly enable intervention but allow for intervention indirectly through their impact on lower-level closures. As a consequence the relationship between intervention and the closure becomes less clear cut. A great deal of linguistic closure is of this form. If we hold an object as 'food' we are encouraged to eat it. If it turns out to be solid and inedible we either abandon the linguistic closure or modify it. In such cases the linguistic closure is directly related to intervention. In the sentence 'food is not only a necessity, it is a delight' the linguistic closure serves to organise other linguistic closures, and thereby alter the manner in which intervention takes place. In such cases, typical of a great part of linguistic closure, closure is not straightforwardly either retained or aban-doned on the basis of any particular intervention. While closure functions therefore to enable intervention, high-level linguistic closure often does so in an indirect manner. Furthermore as I shall argue in Part III, there are many circumstances in which high-level closures are pursued for their own sake and without direct concern for their capacity to allow interven-tion. One of the many consequences of this is that although the linguistic closure is relatively unstable in so far as it can be supplanted by an alterna-

tive linguistic closure without an obvious impact on the capacity to intervene, it is at the same time potentially more stable in so far as it is less
likely to be threatened by the failure of intervention.

The complex web of closures that provide human experience enable us
to intervene in openness to great effect, but we do not thereby gain access
to openness independently of closure. Nor could we do so for openness is
not any thing. What is experienced is not an intervention in openness but a
capacity to handle the activity of texture. The intervention itself lies outside
of our experience but we are aware of the consequences of intervention
through its effect on our preliminary closures and thereby on our system
of closures as a whole. Intervention has an impact on the material
provided by closure and on the activity shown in the texture associated
with higher-level closure. As a result it can be seen that intervention is not
understood, both in the sense that we are unable to access openness, but
also in that any intervention has consequential effects on texture which are
not fully accounted for by reference to our closures. We intervene and to
this degree therefore are not capable of knowing what has been done. We
know that things have changed, and can offer closures that seek to express
that change, but the changes that take place in our preliminary closures are
not captured by the high-level linguistic closures that seek to express that
change. It is in this sense that intervention can be regarded as the handling
of activity, the handling of the change in texture that is not capable of
being captured by higher-level closure.

It is only material that is known, only material that is realised by
linguistic closure. Once the page of dots is closed into a set of patterns, a
collection of things, these can be known, but without the patterns the dots
are open and unknowable. We partially know what happens as a result of
our intervention in openness through its effect on our closures. There is a
great deal that remains unknown for it is held in texture. If for example we
colour in one of the shapes formed by the patterns in the page of dots we
partially know what has happened – the pattern has been filled in. At the
same time there is much else that has happened that we do not know: new
patterns that have been created and old ones that we had not realised that
have been lost. The circle of material realised through closure exhibits
activity in texture and enables the handling of that activity through intervention. The manner in which the activity is handled is not fully known,
but it is through that which we can know that the closure on which the
intervention was based can be confirmed and reinforced.

While we are aware of what happens as the result of intervention in
openness, and can partly know its outcome, how we intervene is unavailable
to us. The 'how' of intervention lies outside of closure in openness. What is

intended by this claim can be illustrated with an example from ancient Egypt. It is said that the Pharaohs of Egypt were able to retain their power by predicting the date of the flood of the Nile.[2] For the populace this confirmed their belief that the Pharaohs were Gods. The Pharaohs, and their secret advisers, were able to produce this feat of prediction by what is now for us a familiar closure – the calendar. The Egyptian calendar applied a different symbol to each rising of the sun – often identified by the position of the stars – and after a number which we now know to be roughly 365 symbols, the Nile could be expected to flood. The Dog-days, that is to say the rising of the star Sirius, heralded the flooding of the Nile and sacrifices were timed to coincide with this event. The prediction is the outcome of a series of closures, only made possible by prior closures such as 'the sun', Sirius, and a system of naming days. As a consequence activity can be handled. The Egyptians could prepare for the flood, they could grow their crops more successfully. The Pharaohs could retain their power. But this does not mean that it was understood how activity was handled in order to achieve this outcome. For the ancient Egyptians the how of their intervention remained mysterious. From our perspective, from our closures, it appears that we are able to provide the explanation for that handling of activity, that we can understand that handling. Explanations in terms of the sun rotating around the earth, the forces of gravity and the masses of the bodies involved, provide further material which in turn results in an extended capacity to handle activity, but they do not result in the understanding of that activity or the means by which it was handled other than by a repetition of the material that enabled the handling. We have the impression that the timing of the flooding of the Nile is now understood, because the success of our current closures reinforces the notion that closure is complete. It is however in character no different from the ancient Egyptian story. Just as the ancient story could be successful without understanding why this was so, the same is true of our current tale. The current explanation provides an account in terms of the forces of nature, but it remains just as unclear how those forces achieve the outcome described. As a closure it enables a different handling of activity from the Egyptian closure, and in large part a more powerful handling – although not from the perspective of the Pharaohs for whom the closure would have been disastrous – but it has not succeeded in squeezing out openness.

The story of science has made it appear as if closures are complete, as if openness is fully known, but the closer we look the more the failure of closure becomes apparent and with it the 'how' of intervention becomes mysterious once again. This may become more apparent if everyday closures are considered. Material enables the handling of activity and because of the requirements of closure we often assume understanding when there is none. For example if we know that by walking to the end of the street we can walk in the park, we do not need to know how this is possible in order to be able to do it – the closure enables intervention and is confirmed by the successful handling of activity. How the intervention is possible is unknown, and is for us unimportant for we need to know merely that the closure is successful.

Although we cannot know the 'how' of intervention it is frequently assumed that we do so. It is for example sometimes imagined that in order to develop technology it is necessary to understand how it works, but in order to handle activity we do not need to know, indeed we cannot know, how we handle that activity. It is the capacity to intervene successfully that encourages the supposition that we understand the world and the mechanism of intervention. This supposition is sometimes maintained by the presumption that someone else can explain how the intervention is possible – how the closure works – even if we are not ourselves capable of doing so. We may not know, for example, how by turning the key in the ignition of my car it starts, but we can suppose that there is someone who does. In fact, at every point in the patchwork of the relevant material the how finds itself in openness. In an explanation, from the turning of the wrist to the firing of the cylinders, there are a series of closures none of which explains the handling of activity. By providing further material the closures give the impression that texture is exhausted, that closure is complete, but for every additional closure the question how will elicit the search for more material. The whole edifice of science will in the process be brought into play. To understand how turning the key in the ignition starts the car we will gradually be enmeshed in electricity, magnetism, atomic structure, and forces. Each new closure provides new material and new knowledge, but along with it new activity. At the end of the process we will be no closer to understanding how that activity is handled, although the new activity generated by the new closures will be more varied and offer more possibilities for other interventions. In answer to the question: 'how does it work?', we provide ever more material, but only succeed in generating further activity requiring further material. How and why closures work is not therefore itself capable of closure.

All material is in this sense magical. It enables intervention that cannot be understood. Ancient magicians were those who had access to closures that others did not know, in the same way that the Pharaohs had access to closures not available to their subjects. This gave them a supernatural character. It is now thought that their magic has been explained, as the knowledge of herbs, metals or the weather. No such thing has taken place. More powerful closures have been realised, more powerful magic that can subsume the feeble closures of those magicians. We have simply lost sight of its magical character. Anthropology has many accounts of tribes who on being observed by a Western scientist believe that the observer has access to some very powerful magic. Magic that produces sound and images from boxes, and makes travel swift. We are inclined to smile patronisingly believing that we merely have knowledge – the technology behind radio and television, and motor vehicles – and not magic. The closures behind the technology do indeed provide us with knowledge and understanding and enable us to handle activity, but they do not explain how the closures enable intervention. How the closures are successful remains incomprehensible and in this sense is our magic.

Science grew out of alchemy but was able to distance itself from the magical character of closures by having a method which allowed for an organised search that would enable the handling of activity. In abandoning the notion of magic science has made it look as if there is an endpoint of understanding, that complete closure is possible. As Wittgenstein noted in the *Tractatus* this assumption is mistaken,[3] but it is also one of the great strengths of science. To describe a closure as magic makes it look as if it is not possible to have a method by which to improve it. Science has instead encouraged a procedure by which every explanation requires further explanation, and every explanation is open to challenge. In so doing it has gradually amassed a collection of closures that appear to provide an account of how closure is successful. In denying its magical character science enabled its closures to be altered and changed which enabled them to be more successful. This was in marked contrast to previous systems of closure which had generally asserted the correctness of the closures – often to protect and maintain the social position of those who controlled them. We now operate in a predominantly scientific culture in which the method of science is largely accepted. As a result our capacity to intervene in the world has been greatly extended, but at the same time our understanding of the magical character of that intervention has been obscured.

It can be seen therefore that intervention is both the product of closure and the means by which closure is maintained, extended and reinforced. Nevertheless the mechanism of intervention must remain hidden from us and the attempt to provide a complete account of the means by which closure enables intervention will end in failure. There is, for example, a strain in philosophy that has consisted in seeking to demonstrate that a common-sense view of the world is justified despite apparent threats to its assumptions. Such an approach, common in empiricist philosophy, can be regarded as seeking to validate what non-philosophers take for granted. Yet from the common-sense perspective the completion or justifying of closure is unnecessary. The capacity to intervene to a purpose is taken as sufficient justification for the closure. A demonstration that closure is complete or can be made complete with the provision of further closures is not thought to be required. So long as a closure, for example 'chair', enables intervention – it can be sat on, moved, found, and generally used – this is all that is required. If a philosopher, with the backing of the Cartesian tradition, calls into question the existence of such an object on the basis that we do not have proof of its external reality, the common-sense response is to say that this object can be touched, felt, and moved. If the philosopher counters with dreams and devils, the likely response is to say, 'it doesn't matter, one can get by without solving all of these sophistical arguments'. If philosophy is taken to consist in an attempt to complete closure, when such closure is neither possible nor necessary, it is an inevitable response.

In this context the problems of epistemology can be seen to be largely a consequence of the attempt to provide complete closure when the completion of closure is not a possible strategy. Epistemology seeks to give an account of how knowledge is possible. In so far as this is an attempt to demonstrate how material can provide knowledge of openness it will fail, for there is nothing in common between closure and openness. This will be the case whether the epistemological problem is expressed in the context of meaning and the world, or sensation and reality. In the context of openness and closure it can be seen that the attempt to demonstrate knowledge of openness is a misguided goal. Language does not describe the world, if by the world we mean openness; and perception is not a version of reality, if by reality we mean openness. Material provides us with knowledge not because it describes openness but because it provides a basis for intervention.

In summary, and by way of a postscript to this initial attempt to provide an introductory overview of the framework of closure, I have sought to outline the principle structure of openness and closure; to describe the general characteristics of closure; to indicate in the broadest of terms how human experience is the outcome of layers of closure; and to suggest how the operation of a system of closure can through intervention result in further changes to that system. The nature of this overview has been driven by the desire to enable the reader to gain a broad impression of the overall character of the account that is being proposed so that subsequent and more detailed examination of particular aspects of closure can be placed in context. It has however also had the outcome that there has been little opportunity to provide examples and argument to support many elements of the account, and there has been little attempt to respond to likely criticisms. Succeeding chapters will attempt to address this balance.

For the moment it can be seen that in the context of the account of closure which has been offered, the nature of the human condition is to require closure in order to understand the world and intervene to effect. At the same time we have evidence of the failure of closure through the inability of closure to exhaust texture. We need closure but, however detailed and however effective, it remains unsatisfactory. Later this predicament will be examined by exploring the ceaseless search for closure that motivates so much of human behaviour, and the search for openness which is the corresponding outcome of the recognition of its failure.

Finally it should be noted that the account of closure that has been offered is itself the product of the system of closure described, is itself a linguistic closure and has the characteristics of closure in general. Before examining in more detail the search for closure and the search for openness, it is important therefore to provide a more detailed account of the closures of language if the status of the theory of closure itself is to be clarified. Moreover, it is linguistic closure which more than any other form of closure typifies human experience and it is linguistic closure that determines the character of knowledge and is largely responsible for the structure of social organisation.

Part II

LANGUAGE AS CLOSURE

Introduction: the question of language

The question of language has hijacked philosophy. Until the mid-to-late nineteenth century language was for the most part regarded as the transparent medium which enabled us to describe experience and the world. The descriptions provided by language might be more or less persuasive, more or less true, but language was the innocent messenger, the carrier of thoughts or concepts. In contrast philosophy for the last century has been preoccupied with the role of language. Not only has language ceased to be regarded as the neutral conveyor of our thoughts, it has become the pre-eminent philosophical issue. It has not even been uncommon for language to be held as the only legitimate concern of philosophy at all.

The philosophical preoccupation with language has often perplexed outsiders and a case can been made that it has contributed to the marginalising of philosophy as a discipline. Philosophers have been criticised for being concerned with grammar when they should have been concerned with matters of real importance. These criticisms however rely on the assumption that the medium through which we convey our thoughts and ideas, our theories and our beliefs, is incidental to the message which is thereby conveyed. Once the neutrality of that medium is in question, and moreover once it appears as though the message is only capable of being formulated as a result of the character of the medium, the central importance of language is seemingly unavoidable.

Philosophy has become mired in the question of language because although the importance of the relationship between language and the world cannot be evaded, no credible account of that relationship has been forthcoming. Realists have been unable to give an explanation as to how language refers to the world; while non-realists have found themselves trapped in language unable to account for how it has content, or how it is possible to say anything at all. So the puzzle has persisted and philosophy has been largely incapable of considering other issues, circling and re-circling the question of language. A case can be made therefore that almost all of the major philosophical figures of the last century from either the European or the English-speaking traditions – Russell or Wittgenstein, Heidegger or Derrida, Quine, Dummett, Davidson or Rorty – have all primarily been concerned to address the question of the role of language. No philosophy can be taken seriously unless it has a response to this question, yet no philosophy has been able to provide a satisfactory response.

Against this background in which for a century language has been philosophically centre stage, the account of closure that has been given so far has been more inclined to draw attention to what language does not do than to highlight its importance. So it has been argued that while it is trivially the case that we cannot describe our experience, or the world, without using language, language does not provide experience alone, and experience need not be linguistic. Language does not provide the limits either to our experience or to our world. There is plenty, not to say almost everything, outside of the text. It is closure as a whole, not language as one small sub-set of closure, that provides experience. It is closure that provides the limits to our experience and to our world. It is closure beyond which there is nothing, or more precisely no thing.

Yet these claims are of no value unless the account of closure offered here is also able to provide an account of the role of language and its relation to reality, and thus also an account of itself. Moreover the account of language will need to be given in a manner that does not fall to the problems of self-reference that were outlined in the Prologue. So it is to this central issue that I shall now turn. In the opening chapter of this part the primary question of the relation between language and reality is addressed. A description is given of the mechanism by which linguistic closure enables language to be held as reality, and therefore words to be held as things. In the next chapter the focus shifts to the relationship between elements of language, and to the mechanism by which one set of linguistic terms is held as one with another different set of terms. These two aspects of language, the relationship between language and reality, and the relationship between the terms of language itself, are shown to be the product of different types of linguistic closure. It then becomes possible in the third chapter of the part to provide an overview of the structure of our linguistic closures taken as a whole. The part concludes with a consideration of the implications that this description of language has for our notion of truth, and as a consequence proposes an account of how truth might be understood in a world which is held as not-thing.

In the context of closure, language is not the sole determinant of our world, but is itself the outcome of the process of closure, being the material realised as the result of a particular type of intersensory closure. How language is acquired; how it refers to reality; how the closures of language are themselves combined; what language enables us to achieve; what language is both capable of saying and what it is not capable of saying; all these will be shown to be explicable in the context of closure and will be seen as the outcome of the underlying structure of closure and openness. So it is that in later parts of the book it will become possible to demonstrate that through language the structure and characteristics of closure can be found to determine the shape of our theories and the character of knowledge; the organisation of society and the power relations between individuals. While therefore language may not be the sole determinant of the character of the world, as the outcome of closure it can be seen to influence every aspect of our experience and our social environment.

4

LANGUAGE AND THE WORLD
Practical closure

> Language is a system of marks that function initially as cues
> to the realisation of linguistic closure and then as tags to
> those closures. As a consequence language has a transparent
> relationship to that which we hold as the world.

A preliminary account of language was offered in Part I. Language was
described as the outcome of closure. More specifically, as the outcome of a
particular type of intersensory closure: 'linguistic closure'. As with all
closure, linguistic closure realises material, and in this case it is in the form
of meaning.

In saying that linguistic closure realises material in the form of
meaning, I wish to use the term 'meaning' in a precise and particular way,
and one which is different from its common use. The familiar use of
'meaning' identifies the socially agreed meaning. In the case of individual
words it is the meaning to which a dictionary seeks to approximate. It is
this sense of meaning, a sense that is publicly available to speakers of the
language, with which most philosophers and linguists have been primarily
concerned. However, I will refer to this sense as the 'proposed closure'.
Instead I will use 'meaning' simply to identify the material, the thought,
that is realised through linguistic closure. This internally generated
meaning realised by an individual, which has sometimes been pejoratively
described as subjective and psychological, has often been regarded as
subordinate to the socially agreed meaning and as a result perhaps has
received relatively little attention. In contrast the account of language that I
am about to put forward regards the internal realisation of meaning as
being the driving force behind language. In so doing it does not deny the
importance of the socially agreed meaning, and, as I shall argue much later
in Part V, socially agreed meanings leave little room for manoeuvre on the
part of individuals, but nevertheless it is by understanding the process

whereby an individual realises meaning in this subjective sense that we are able to give an account of the operation of language in general.

In order to bring light to the question as to how language can be understood to refer to the world, or more precisely what we take to be the world, I will further distinguish between two quite distinct ways in which this meaning, internal to the individual, can be realised. Both are examples of linguistic closure, but I shall refer to one as practical closure and the other as formal closure. In both cases, as with all closure, the closure consists in holding that which is different as the same. In the case of practical closure, the marks of language, its words and phrases, are held as one with an aspect of experience or 'reality'. While in the case of formal closure the marks of language are held as one with other linguistic marks.

This chapter is concerned with the central question how language is hooked onto the world, and it will therefore be primarily concerned to describe the first of these processes: the process of practical closure whereby we associate elements of language with elements in that which we take to be the world. The following chapter will be concerned with the question as to how elements of language are combined, and will therefore be engaged in a description of formal closure. The distinction between practical and formal linguistic closure does not of course imply that the nature of closure is different. In either case the closure consists in the holding of that which is different as the same, and in both cases the closure is achieved through the realisation of material in the form of meaning. It is for this reason that both are types of linguistic closure. The difference between practical and formal linguistic closure is solely due to the character of the 'things' that are being held as one.

In Part I an example of practical linguistic closure has already been considered. The example considered was how the word 'tree' might be used to refer to a tree. It was proposed that an individual could hold the sound of the word 'tree', itself an auditory closure, as one with the visual, auditory, and tactile sensory closures that in combination provide the experience of a tree. This holding of the sound of the word as one with a combination of other sensory closures is the process of practical linguistic closure. Practical closure is possible through the realisation of meaning, which takes its place alongside the other closures available to the system and which as a result contributes to the character of reality as experienced by the individual. The relationship between the marks of language and their meaning, and between meaning and the world, is not therefore the result of a grid superimposed on previously differentiated items, as if words were matched to their meaning and meaning to things in reality, but rather the relationship is the outcome of the process of linguistic closure which is responsible for realising meaning and thereby forming an indissoluble link between language and what in the context of the system of closure in question is the world.

It will be apparent that before linguistic closure can take place sensory closures need to have been realised which provide the starting point for linguistic practical closure. It will be necessary to have realised, on the one hand, closures that for example enable identification of the word 'tree' and, on the other, to have realised the material that provides the visual, auditory and tactile sensation of a tree. A very preliminary account has been given of how sensory closure provides patterns which provide the basis for the sensations which we associate with, for example, seeing a tree. For the present however our concern will not be with sensory closures or with the perceptual closures that follow further intersensory closure, but rather with the closures that allow for the identification of words and phrases, the marks of language.

Linguistic closure realises meaning, but in order for linguistic closure to be possible it is first necessary to identify the sound or visual pattern as an element of language. This process is itself the product of a compound set of closures. To realise a linguistic closure the sound of the word or phrase must have been identified as a term of language, and this in turn requires that it has been identified as a discrete sound. In order to identify the words and phrases of spoken and written language therefore they must themselves be the outcome of aural or visual closures. For example, in order to identify the sound of the word 'help' as a word, it will be necessary to hold this sound as one, and thereby separate it from other aspects of our aural sensory field. Only then is it possible to hold this auditory closure as a linguistic mark. In order to hold the auditory closure as a linguistic mark, this particular auditory closure will need to be held as one with many different sounds produced in different circum-stances. Thus to identify the word 'help' as a word, in addition to identifying the sound as a discrete unit, it is necessary to identify this sound as one with other sounds of language. Furthermore to identify this sound as the word 'help', it will be necessary to hold the auditory closure on this occasion as the same as other prior uses of the term, which will inevitably have been different. It can be seen therefore that the mere iden-tification of the terms of language is already the consequence of a complex set of closures. It is only having achieved these closures that it then becomes possible for linguistic practical closure to take place whereby this linguistic mark is itself held as one with another and different type of sensory closure, with the consequence that meaning is realised.

Any sensory closure can be a mark. It is not as if there is something special about words that enables them to act as marks, nor is language special in being based on marks. Through intersensory closure we can hold any one sensory closure as any other, and in this respect any sensory closure can be taken as a mark of another. A sentry on duty may take movement as a mark of the enemy. The feel of a surface in the dark can be the mark of a light switch. It is not necessary to provide linguistic closure before providing a behavioural response. The sentry does not have to provide a linguistic description of what is experienced as movement before taking appropriate action. Similarly following the contours of a wall in the dark does not require that we provide linguistic names for each of these contours in order to find the light-switch. The linguistic closures associated with the words 'enemy' and 'light-switch' have an impact on experience and will alter the perceptions of the individuals in these examples, but there is no reason to suppose that it is necessary to 'translate' experience into language in order to intervene appropriately, nor need we assume that it is necessary to operate with linguistic closures in order to have experience. Linguistic closure cannot merely consist therefore in the holding of a sensory closure as a mark, for all intersensory closure has this character and any sensory closure can be held as the mark of any other.

If we are to discern a distinguishing characteristic of the closures of language from other types of sensory closure it is not that they are based on marks, but that the marks are self-generated. The sentry who holds movement as the enemy through non-linguistic intersensory closure, has previously realised movement from preliminary closure but the sentry was not responsible for causing such a change in preliminary closure. In the case of the closures of language, the instigator of the linguistic closure is responsible for generating the mark. It is possible therefore to define language as a type of intersensory closure in which the mark is self-generated. This definition provides a rather wider account of language than might normally be employed. Such a definition extends for example beyond that which we typically associate with language and includes gestures such as pointing, or expressing an emotion such as anger. In such cases the individual produces a behaviour which is to be regarded by others as a mark of something else. In the context of this definition of language both animals and babies exhibit language both through gesture and sound, and possibly also through touch.

Language based on written or spoken words has a further and vital characteristic: in addition to being self-generated the marks are held not only as other sensory closures but as other linguistic marks. It is this additional characteristic that can be identified as responsible for the huge system of interconnected linguistic closure that is language. It is not necessary however to suppose that there is anything special about words and phrases that makes this additional characteristic possible. A similar system of interconnected closures could be generated from any type of mark that could be self-generated. One can surmise that the reason language has developed in the form that we commonly know it – in spoken or written words and phrases – is because humans are capable of providing with relative ease a huge number of different sounds, and pictorial shapes. It would be difficult for example for us to develop a language based on smell or taste, but it can be supposed that a closure machine capable of generating a variety of smells or tastes with ease could in principle use these as the basis for language. What is important therefore in the context of linguistic closure is not that the marks of language come in the form of words and phrases, a consequence simply of the physical workings of the human closure machine, but that the system of marks is self-generated and the marks are held, through formal linguistic closure, as one with other marks as well as other sensory closures.

It has been argued that linguistic closure, as with all closure, consists in the holding of that which is different as the same through the provision of material. A linguistic mark is held as the same as another sensory closure or as the same as another linguistic mark, and this is achieved through the provision of material in the form of meaning.[1] While sensory closure realises the marks of language through the provision of material in the form of the sound or sight of words or phrases, linguistic closure can be seen to realise meaning. While the product of sensory closure is sensation, with each type of sensory closure having its corresponding type of sensation, the product of intersensory closure is thought. In this context the product of intersensory linguistic closure, whether practical or formal, can be regarded as a particular type of thought, which consists in the realisation of meaning. It is because linguistic marks are self-generated and are themselves applicable to other linguistic marks that linguistic closure plays such a prominent role and is liable to obscure other types of intersensory closure for the closures of language allow for the realisation of an ever expanding system of closures which do not require external stimulus. It is for this reason that for us thinking largely consists in the manipulation of meanings. Linguistic meaning and thought are not however synonymous, for there are other types of intersensory closure and other types of thought, but unlike language they are not self-generating nor do they have the characteristic of linguistic marks that they can apply to other marks, as a result they do not form a system of intersensory closure and are instead transient and as a result are largely eclipsed by linguistic closures.

In order to understand the mechanism of practical linguistic closure consider an example where a proficient speaker of English comes across a new word. Suppose that we are visiting a zoo with a friend. We stand outside a cage and our friend says: 'An aasvogel'. Our response is the outcome of a sophisticated system of closures which in combination provides us with experience. In this case our sensory closures are likely to identify as one the sound of the utterance as a linguistic mark or combination of marks. This realisation separates the sound of the words from the other sounds that are contemporaneous. Further sensory closure associates this sound with other prior linguistic marks and realises it as a linguistic mark. Having realised a linguistic mark we are in a position to attempt to realise linguistic closure. We do so by looking for a sensory closure which we can hold as one with the mark. There is, of course, no limit to the number of sensory or linguistic closures that could be held as one with the mark. In the light of our current closures – themselves the product of previous experience – and perhaps a pointing gesture on the part of our friend we are perhaps most likely to hold the mark as one with the large bird at the back of the cage. There are however an indefinite number of other alternatives such as: the strange patterning on the wings of the bird, or the curved reflection of the artificial lighting in the pond, or the high pitched call of one of the birds, or the smell emanating from the cage. When offered a new linguistic mark we are likely therefore to briefly allow it to remain open in search of an appropriate linguistic closure. At this point the linguistic mark can be considered as a cue to a linguistic closure which has not yet been realised. The mark is left open as we look for a closure that can be realised as a meaning.

I have argued that linguistic closure is realised by settling on a particular sensory closure or set of closures and finding in it that which is to be held as one with the mark. In this case it has been supposed that the combined sensory closures consist in a large bird at the back of the cage. We therefore hold the sound 'an aasvogel' as one with the set of sensory closures which that make up a large bird – for the sake of simplicity the set of sensory closures will be assumed to consist simply of the visual image. The question to be faced is how the sound of the word 'aasvogel' is held as one with the visual image. For it is an image with which it in principle has nothing in common. The case that I shall make is that this is achieved by realising an intersensory closure in the form of a meaning which both prior closures can then be seen to have in common. The sound and the image are thus both connected to the same meaning, and in this respect can be held as one even though we know them to be wholly different. It can be seen that there is nothing actually in common between these two sensory closures, we simply choose to hold them as one and the same and in order to do so have to realise another layer of closure in the form of meaning. Since the sound of the word is not the same as the image, or indeed anything like it being different in every respect, it is only through the provision of a meaning, something else which both the sound and the image have in common, that these two different things can be held as the same. They are not thereby held as identical to each other for we remain aware of course that the visual image of the bird and the sound of the linguistic mark are entirely different, but they are held as the same in so far as they have something in common which enables them both to be part of a single whole. The meaning realised by practical linguistic closure thus allows these two different sensory closures to be held as the same in this respect. In the same way that sensory closures realise sensation by holding different preliminary closures as the same. The thing which is realised, whether a meaning or a sensation, is the product of closure and is in addition to that which preceded it.

It might appear at first from this example that nothing has been added by the realisation of a linguistic closure. The sound 'aasvogel' still sounds the same, the image of the bird still looks the same. So what has changed? The sensory closures on either side may not have changed, but a new closure has been realised. A new closure which is in addition to the prior available closures and which enables intervention which was not possible previously. For example, we now have a means of picking out this partic-ular bird in the zoo because the meaning that has been realised will have identified a something in virtue of which this bird is an aasvogel and which thus enables us to distinguish it from others. As a result there will be many consequences for how we might be able to intervene. It can be seen therefore that through the realisation of the meaning the sound and the image are not altered but they are associated with a meaning. The sound 'an aasvogel' is no longer a sound that functions as a cue to the real-isation of meaning, but is now associated with the meaning that has been realised and thus with the image which enables it to be held as one with the image in this respect.

In circumstances like this, where the sophistication of our current array of available closures already results in a highly differentiated perception, it is easy to overlook the role of linguistic closure and suppose that all that is taking place is the naming of perceived objects. The words of language do not name sensory closures, or things in reality, nor do they have anything in common with them, but through practical linguistic closure are held as one with them as the result of the realisation of material in the form of meaning. The closures of language thus echo the character of closure in general which neither describes openness nor has anything in common with openness but holds that which is different as the same through the realisation of an addition in the form of material.

The realisation of practical closure provides additional material but it leaves prior closures intact. It is for this reason that we are able to mistake the marks of language for labels attaching to perceptual things or objects in the world. It is because prior closures are undisturbed that practical closure does not usually result in any obvious perceptual change and as a consequence the marks of language can appear to be merely labels. In certain circumstances such as finding a face in a page of dots or choosing between ambiguous images (like the duck/rabbit) the consequence of practical linguistic closure becomes more apparent for reasons I shall consider shortly, but in large part the impact of practical closure on the rest of our experience is so minor as to go unnoticed. Nevertheless, practical closure always results in a new realisation and as a result has effects both for our capacity to intervene and for perception even though these may be sufficiently diffuse or minor to go unnoticed by us.

The mechanism of practical linguistic closure and its impact on our capacity to intervene may become more apparent if we consider a further example from a visit to a zoo. A collection of snakes is displayed so that each type of snake is housed in a separate glass fronted cage. Beside each cage a label identifies the particular snake in question. There are many cages and initially we are intrigued and scrutinise the labels and examine the snakes carefully. As a result we decide, at least to our own satisfaction, what makes each type of snake a such-an-such. After many snakes and many labels we become tired of the process and simply scan the remainder of the cages glancing at the labels and the snakes but making no underlying connection.

While examining the initial cages we realise material through linguistic closure by identifying some characteristic or set of characteristics that makes this type of snake a such and such. Each label is held as the same as the snake through the provision of a meaning. Assuming we are given no guidance this realised meaning may or may not be similar to the socially accepted meaning that would provide the dictionary definition of the term, but for the purposes of this example this is not of concern. To realise meaning is not effortless on our part, for we have to realise a meaning which will enable the snake and the label to be held as one. It is not sufficient simply to hold the perceptual image of the snake as the label because as the snake moved so the label would no longer be appropriate, nor would it apply to other snakes in the same cage. We have to realise something that enables us to hold different snakes, at different times, in different circumstances, as one thing, which is the same thing as the label. We do so through the realisation of meaning as the result of linguistic closure. The provision of a meaning does not alter the appearance of the snake, nor the appearance of the label, for it is in addition to both. Once the meaning is provided we are able to find this meaning in both the sensory closure and in the linguistic mark. We look at the snake and see a particular zigzag pattern, or a coloured collar. We see the linguistic mark and we associate it with the same pattern. The possibility of the pattern had always been there but was not previously identified, much in the same way that when we find a face in a page of dots the possibility of the face was always there but was not realised. Practical linguistic closure results therefore in the addition of material even when the marks are used for the first time and appear merely to be labels. The addition of material in turn enables intervention: we could return to the reptile house later and find a particular snake when asked to do so by someone using the relevant mark.[2] Moreover we can do this irrespective of whether our practical linguistic closures might be said to be correct in the context of the linguistic closures realised by proficient zoologists.

In those cages where no practical closure is realised, where the label and the snakes remain unconnected, no meaning is provided. We of course see the snakes and the labels, but there is no addition, there is no new closure, no material that enables these two different things to be held as one. As a result there are no consequential effects for intervention or perception. If linguistic closure has taken place we could find a particular type of snake, when asked to do so, even if it was moved to a different cage and all the snakes were jumbled up. While in the absence of linguistic closure it is unlikely that we would be able to find the snake, for the association between the label and the snake would have no basis. Without linguistic closure the mark is not associated with a meaning but only with that set of closures which make up the present experience. Aside from the fact that we would be unlikely to remember the mark because it would not have undergone linguistic closure and would thus have no function, even if we remember the label we would be unable to identify the snake for the label is not associated with any particular snake. This outcome would apply unless some other intersensory closure enabled an association, for example based on the position of the cage.

In passing it should be noted that the description that has been offered of the role of practical closure in the realisation of meaning on the part of an individual is quite distinct from a description that might be given of the dictionary meaning of a given mark. The dictionary meaning might be understood as a core set of practical closures associated with a given mark by a majority of proficient speakers or of speakers whose expertise places them in a position of authority within a given culture. I will later argue, for reasons that flow from the account of practical closure given here, that it is not possible to identify any unchangeable core meaning and that a looser definition as those closures commonly and typically associated with the tag by relevant individuals is more appropriate. Even so these associated closures will always be capable of revision and will not be universal across the speakers of the language.[3] While therefore we can attempt to provide a dictionary meaning for a given tag this is not the meaning realised by the individual as the result of practical closure, nor is it the focus of my concern here.

One of the consequences of this account is that in addition to enabling intervention the realisation of linguistic closure also alters perception as a direct result of the provision of material. It has been argued that perception is the outcome of the combined application of currently available closures to preliminary sensory closure. The closures currently available to us at any one time realise material which in combination make up reality. Any individual closure inevitably plays a small part in the realisation of experience, but each closure plays some part. In the case of the current example, the linguistic closure realised in response to the snake labels results in the provision of material which in turn has a consequential impact on perception. The change in perception is minor, since the contribution of the linguistic closures associated with the labels is small when compared with the system of closures currently available to the system. Nevertheless a change in perception must occur in each and every case since new material is available. The perception of the snakes will change so that they are seen not simply as snakes, but snakes with a pattern, or of a type. The possibility of finding a pattern was there in the prior sensory closures but it was not yet realised. So that in each case through the realisation of linguistic material the perception of the snake shifts in some respect.

The combined effect of a great many additions of material, each of which individually has little discernible impact on perception, can be considerable. Someone with a great knowledge of snakes for example will have a different perception of any individual snake than someone who has no knowledge of snakes at all. The expert on seeing the snake realises many closures not available to the novice, the result of which is to greatly enrich their experience in this respect. It might be argued that the expert is merely identifying aspects of the snake that the rest of us can see but cannot label, but it would seem that this is not the case for the expert is likely to be capable of making distinctions that the novice is not capable of discerning.

In the same way a shepherd will be able to distinguish between his sheep in a way at first unavailable to the rest of us. The shepherd's closures realise material which results in a different perception to those who do not know the sheep. As a result the shepherd can identify each individual animal in a manner that is not possible for the rest of us. The snake expert and the shepherd are not simply labelling characteristics that are already there, but are realising new closures and new material with consequential effects for perception and intervention. In the same way when we see a face in a page of dots we are not simply labelling a face that was already there. We frequently suppose that telling the difference between snakes is simply a matter of noticing a certain visual characteristic, but as I have tried to demonstrate the visual characteristic is not one that is simply lying there waiting for us to identify it but is one that we have to realise in the same way that we realise a face in the dots. To see a line along a snake we have to hold the markings on the snake as a line, in the same way that we have to hold the dots as a face. Until we take the trouble to realise such a practical closure the meanings associated with the terms will be largely empty. In one sense the snake expert and the shepherd see no more than the rest of us, assuming similar visual acuity and thus similar levels of preliminary closure, but their perception is different because of the subsequent and higher-level closures of the system. In the same way that someone who had studied the dots on a page would see countless shapes and patterns not available to those who had not seen the page of dots before.

It can be seen therefore that the realisation of new material through new closures has the double consequence of changing on the one hand our capacity to intervene and on the other our experience. The realisation of linguistic closure and the provision of material in the form of meaning is but one example of this mechanism. For the most part, each individual closure results in such a small change to the system of closure as a whole and thus to perception that its consequences are almost imperceptible, yet each and every closure through the realisation of material contributes to currently available closures and consequentially must change experience and the capacity to intervene in some respect.

So far examples have been chosen in which the realisation of linguistic closure is instigated by an unfamiliar linguistic mark. In such circumstances the mark functions as a cue to linguistic closure and the meaning that is realised is then associated with the mark. In most cases however the use of language depends on the use of familiar rather than unfamiliar marks. When we come across a familiar linguistic mark it is already associated by us with a meaning. In such circumstances instead of functioning as a cue to the provision of linguistic closure the mark operates as a tag to identify the previously realised meaning. Having identified the previous meaning, we then seek to apply this meaning in the particular circumstances. If practical closure is involved, we will seek to hold the meaning identified by the mark as one with some aspect of sensory closure. The circumstances of the use of the mark on this occasion will always differ from its use on previous occasions. So in order to achieve closure it will be necessary for us to realise that which can be held as the same between the prior meaning of the mark and the current circumstances. In the event that we are unable to find a similarity, linguistic closure is achieved by extending the meaning associated with the mark. In either case new material is realised which either results in a refining of the meaning associated with the mark or an extension of that meaning. Future uses of the mark will then function as a tag to this new meaning. Marks thus function in two distinct ways: as indicators of prior linguistic closure or closures, in which case they operate as tags, and as instigators of closure, in which case they operate as cues. When we first come across a mark it cannot operate as a tag for it has yet to be associated with a meaning, but subsequently it functions first as a tag to the current meaning associated with the mark which in turn then functions as a cue to the realisation of material in the current situation.

This mechanism can be followed in the case of the further use of the term 'aasvogel', incidentally a South African vulture, in the example identified previously. Suppose that following the initial use of the term we have realised linguistic closure which has provided us with a meaning based on a large bird that we identified at the back of the cage. When the term is used again we recall this material and seek to apply it in the new circumstances. In order to apply the term in the new circumstances we will need to hold the prior material associated with the mark and some aspect of our current experience as one and the same. If we are standing in front of another cage and our friend says 'another aasvogel' we will only be able to realise closure if we can hold some aspect of our current experience as the same as the bird we previously identified. Thus if it is applied to a different bird in a different cage closure will only be possible if we are able to hold this bird as the same in some respect as the previous closure. The bird will of course always be different. It will be of a different size and the colouring will not be identical, it will be behaving differently. The new closure will hold the current sensory closures as the same in some respect with the meaning already associated with the term. This is made possible through the realisation of new linguistic material in the form of a new meaning which now supersedes the previous linguistic closure. This can be achieved by identifying something in common between our current and our earlier experience as a result of which the meaning of 'aasvogel' will have been refined, or if in this instance there is apparently no similarity closure is achieved by extending the meaning we previously associated with the mark. In this manner the meaning we associate with a mark is gradually both extended and refined, so that in this example our understanding of the linguistic mark 'aasvogel' would evolve as we identified that which was the same in each of the instances in which it was employed.

The meaning we attach to any term is thus the combined outcome of the linguistic closures that it has realised in its previous uses. As the use of the term proliferates so the meaning, the material realised through linguistic closure, becomes a more complex amalgam of previous closure. The meaning of a word for an individual speaker is the combined outcome of all the ways in which the word has been used to realise closure. A mark of language is thus a tag to its current meaning which is a closure formed out of all of the prior realisations of the term. Any commonly used word in a language is therefore for a proficient speaker of the language associated with a meaning which has been realised from many different prior linguistic closures – the outcome of all the prior uses of the mark – which are held as one through its current meaning.

The marks of language can be confused for labels that refer to ready-made things in reality because individual things, and reality as a whole, are themselves the outcome of the combined interaction of the closures of the system including linguistic closure. Each linguistic closure has, as we have seen, a consequential impact on our perception by adding to the available material. Instead of the openness held in the array of preliminary closure, like a vast page of dots, we have instead a hugely complex set of patterns and things, the product of the closures of the system. Linguistic closures are an important part of all closures available to the system, and therefore play a major role in fashioning the character of reality. It is because the linguistic closure associated with the word 'table' has a place in our system of closures and thus already contributes to the manner in which we differentiate between things, that it begins to look as if the word refers to a set of things out there in the world which are already differentiated in this manner. In the context of the account of language outlined it can however be seen that linguistic marks never simply label things already present to us, for every time a mark is used it functions not only as a tag to prior closure but as a cue to new material. There will therefore be some way in which this particular table contributes to the meaning we associate with the mark for it will be different in some respects from all other tables that we have come across. Whenever a mark is realised new material is provided for that is what is involved in our understanding of the word or phrase. If new material was not provided we would not have realised linguistic closure and would not have been able to provide the mark with meaning. The meaning provided as a result of the realisation of the mark as cue is not a label of something already in reality but is the provision of a new closure with new material. Once that material has been realised it has a consequential effect on perceptual closure with the consequence that the word appears to label a thing in reality. As a result the familiar use of words appears to involve the naming of already differentiated things. When a word is used in an unexpected context, one which we often refer to as metaphorical, its impact through the provision of new material is often more evident. As a consequence we can become aware that we have something new, something additional to our previous experience. When we see a face in the dots we clearly have something we did not have previously. The material provided by the linguistic closure has an immediately identifiable effect on our perceptual experience. Subsequent references to the face will however appear merely to refer to something that was already there. In such circumstances it is easier to identify the process whereby the mark functions initially as a cue to indicate the possibility of closure, but once the closure is realised it functions as a tag and can appear to label something that we perceive in reality, since the new material realised has a hardly noticeable effect on perception. More commonly the role of the marks of language as cues is obscured and it appears that they are only tags. As a result the primary role of the marks of language as the instigator of linguistic closure and the realisation of meaning is hidden.

The relationship between language and the world can thus be seen to be the outcome of practical linguistic closure. The marks of language are not labels for things in openness, for there is nothing in common between a mark and openness and in any case openness is not differentiated and therefore cannot be labelled. Nor are the marks of language labels for things in reality. The things that make up reality are themselves the product of a hierarchy of closure: the combined interaction of all the closures of the system, but the marks of language do not simply label these things but in addition to being tags to previous linguistic closure, are cues to new linguistic closure. As tags they appear to label reality, for reality is itself the outcome of the very closures with which the tags are associated. As cues they initiate a search for new linguistic closures which if realised will have consequential effects on perceptual closure and our ability to intervene in openness.

Where the workings of language have been misunderstood it is usually because its role as a cue to new meaning has been overlooked and instead its passive function as label and tag has been identified as its primary task. In the process the relationship between meaning and reality becomes unfathomable. As a system of tags or labels some philosophers have sought to find that which connects the two, that which they have in common. Since there is however nothing in common in terms of their structure there appears an unbridgeable gap between language and the world. In the context of closure, we can see language as a system of marks that function initially as cues to the realisation of linguistic closure and then as tags to those closures. As a consequence language has a transparent relationship to the world. Language as a system of linguistic closure is in combination with our other closures responsible for the provision of reality, and as a set of tags labels that reality. Although there is no connection between language and openness, either in form or content, language and what we take to be the world are connected by an umbilical cord that cannot be broken. An umbilical cord that is created through the realisation of practical closure.

5

LANGUAGE AND ITSELF

Formal closure

Formal linguistic closure enables marks to be combined independently of sensory closure, and more generally provides the organisation of space.

Through the realisation of meaning practical linguistic closure enables the marks of language to be held as one with other parts of sensory closure, but it is formal linguistic closure that enables the marks of language to be combined. It differs from practical closure for instead of holding a word or phrase as some aspect of experience through the realisation of meaning which the word and the sensory closures share, formal closure realises meaning through which one linguistic mark can be held as another linguistic mark. Since most uses of language involve more than one linguistic mark formal closure is involved in the majority of language use. Formal closure is even present in the case of single linguistic marks and has already been implicit in the account of meaning that has been proposed of an individual word or phrase, for the meaning we associate with a word is itself a closure realised from the variety of previous, and inevitably different, practical linguistic closures with which the word has been associated. Formal linguistic closure is involved therefore in the realisation of a single meaning which holds as one the variety of different practical closures associated with the same linguistic mark. The primary concern in this chapter however will be the role of formal closure in combining different marks, since aside from the self-generated character of linguistic marks, it is the capacity to realise new material through the combination of linguistic marks that makes language such a powerful form of closure.

The consideration of practical closure in the previous chapter was restricted to the use of individual marks. This was done in order to clearly identify the separate and central role of intersensory closure. The majority of language however employs combinations of marks and requires in addition therefore some element of formal closure. In an overall sense, formal closure is subsequent upon practical closure, since marks cannot be combined until some marks have been associated with meanings through practical closure. In any particular instance however formal closure may precede practical closure. In passing it should be noted that marks are not equivalent to the words of language but rather to the units of language which function as cues and tags to closure. A mark can for this reason consist of a number of separate words since in its initial use these can function as a single mark. While it can be seen that language use is initiated by practical closure and while practical closure provides the link between language and the world, it will be argued that formal closure is not tied to practical closure and can take place independently of sensory closure. Through formal closure language is able to escape the immediate constraints of the present circumstances – currently realised sensory and non-linguistic intersensory closure. Further layers of formal closure allow meanings to be ordered and structured thereby providing a linked system which is itself then capable of yet further levels of closure.

The structure of language, its grammatical rules and principles, are the means by which speakers of the language know what formal closure is proposed. Whether formal closure is realised will depend on the individual circumstances, and in the event that formal closure is realised practical closure may or may not follow. Take for example marks that are used to identify material things such as trees and houses, chairs and tables. As it has been argued the meaning of each of these words, for any given individual, is the outcome of the history of linguistic closure associated with the mark. New meanings however can be realised by holding any one of these marks as the same as another mark through formal closure. The sentence 'a chair is a table' proposes that we hold the linguistic closures associated with the word 'chair' as one with the linguistic closures associated with the word 'table'. The sentence proposes therefore that we find a means to hold the meaning associated with the linguistic mark 'chair' as one with the meaning associated with the linguistic mark 'table' through the provision of a new linguistic closure. Knowing what linguistic closure is proposed does not however mean that closure has been realised. If we fail to realise closure we are unable to provide a meaning for the combination of marks. In the event that formal closure is realised new material is provided which enables the different meanings to be held as one and the same. Closure might be realised by finding in the meaning we attach to the mark – our idea of a chair – a flat surface on which things could be placed. In doing so material is realised in the form of a new meaning, which is now associated with the combination of marks 'a chair is a table'. This in turn will result in a slight shift in the meanings associated with the individual marks 'chair' and 'table'. If a means of holding a chair as a table was not found closure would not be realised and the sentence would have no meaning. It can be seen therefore that the way words are combined shows what linguistic closure is proposed, through the meanings associated with these words and the rules of grammar that govern their combination, but the sentence does not thereby ensure that closure is in fact realised. In the event that closure is realised additional material is provided which has consequential effects on experience and the capacity to intervene. As a result chairs may be seen differently on account of the additional closure that is available. The additional patterning of sensory closure that may result from the new closure may then in turn encourage a different and possibly new use of a particular chair.

There is in practice no limit to the combination of marks. Some combinations are made with ease and can be used to indicate the meaning of a term that we have not previously come across even though practical closure is not presently available. We do not for example need to be able to provide practical closure to realise a meaning for the sentence 'An aasvogel is a South African vulture.' Formal closure will suffice to provide us with a meaning for the sentence. The ease of the formal closure in such circumstances can however obscure the realisation of new material which is the inevitable outcome of the linguistic closure. The realisation of new material becomes more evident when we consider formal closures that combine seemingly unlikely combinations of linguistic marks. Formal closure allows us to provide a meaning for such sentences as 'a tree is a house' (it is a place for all types of insects and birds to live) or 'a chair is a tree' (it is made from wood). Sometimes a combination of marks may appear too difficult for us to realise a new closure: 'a tree is a rhinoceros'. In such cases although we know what would be required to form the new closure, namely the holding of the linguistic closure associated with the term 'tree' as one with the linguistic closure associated with the term 'rhinoceros', we are unable to do so and as a result a new closure is not realised. Yet a new closure is possible. We could for example imagine a lecture given by a botanist which examines the protective role of the bark on a tree beginning with the claim: 'A tree is a rhinoceros. The thick skin has been central to evolutionary success'. Not only is it possible to realise meanings from the combination of any appropriate marks, but there are often many possible meanings that might be realised. 'A chair is a table' could be realised as 'these two objects have four legs' or 'both of these objects are household items'. There are many possible meanings that will allow 'a chair is a table' to be realised each of which holds linguistic closures associated with 'a chair' and 'a table', two different meanings, as one and the same in some respect. The closures that each individual will choose to realise will depend on the current state of that individual's system of closures, their space. The social character of language however ensures that there is relatively little room for individual differences, the mechanism of which I shall examine later in Part V.[1]

The description of the operation of language given so far has made no distinction between different types of mark and different combinations and uses of those marks. No distinction has been made between subjects and predicates, nor between nouns, verbs, adjectives, and so forth, nor has a distinction been made between different types of sentences such as questions, statements, and commands. In addition we have not distinguished between individual marks and sentences other than to give a preliminary account of how marks are combined. The reason no distinctions have been made between types of mark or their combination is because in the context of the account of language proposed there is no primary unit of language, other than linguistic closure. It follows therefore that the marks of language whether offered individually or in combination are an invitation to closure and function in the same manner. It is one of the characteristics of social interaction that we assume linguistic marks offered by others are capable of realisation; an assumption that we are only rarely prepared to abandon. The assumption that closure is possible applies as a consequence to all types and combinations of mark providing they are identified as such. When a mark, or marks, are offered the possibility of closure is assumed and we usually seek to realise such a closure through the provision of material in the form of meaning. Since all types and combinations of mark function to initiate closure, there can be no elementary type of mark that provides the basis of meaning. The division of words into different types, nouns, verbs, and so forth, has the consequence of making it look as if the world was already differentiated into things, properties, activities, and so on. Such distinctions can in the context of the account of language offered be seen to be the product of closure not the instigator of closure. The vehicle of language, the provider of meaning, is not therefore a sentence or a word, or any particular combination of marks, but linguistic closure; and linguistic closure can be initiated by different types of mark singly or in combination. There are no limits to the sort of mark that can initiate closure, and as a result there is no primary unit of language.

In seeking to demonstrate the mechanism by which combinations of mark are capable of initiating closure the examples chosen have been the marks of familiar physical objects. In deliberately choosing unusual, and at first sight impossible, combinations of mark the intention has been to highlight the underlying function of formal closure and its potentially unlimited capacity to combine marks. The majority of language use involves of course a more familiar application of marks and in such circumstances the function of closure is obscured because the marks appear to function as labels. In a sentence such as 'the tree is in leaf' the impression is given that the marks simply describe what is already perceived. The provision of linguistic closure however follows the same mechanism as that which applied in the example 'a tree is a house'. The sentence 'the tree is in leaf' proposes that the mark 'tree' and 'in leaf' are held as one and the same through the realisation of formal closure. In the event that we are standing in front of a tree whose leaves are out the formal and practical closure are readily realised. The resulting meaning appears simply to describe what is seen because the character of our currently available system of closures, our space, already differentiates openness in this manner and so we are unaware of the additional material provided by linguistic closure. As we move away from the routine uses of linguistic marks we are more likely to notice the provision of new material with its consequential effect on perception and intervention. The familiar use of linguistic marks largely replicates our currently available system of closures while the unfamiliar more evidently brings with it new material. As with practical closure, the familiar use of a mark is associated with its so-called literal meaning, while the unfamiliar use is often referred to as a metaphorical application. From the preceding examples it can be seen that the literal and metaphorical uses of a mark function in the same way and it is, in contrast to our everyday understanding, the metaphorical use that more clearly demonstrates the principle underlying both formal and practical linguistic closure. Furthermore the metaphorical use of a mark is not subsidiary upon its literal use but the reverse applies. The literal use of a mark is only possible in the light of a preceding metaphorical use which through repetition and over time acquires the character of a literal meaning. It is for this reason that examples like 'a tree is a house' which employ a metaphorical use of the words, and which have a consequentially greater impact on our system of closures, provide a better indication of the underlying function of linguistic closure than more familiar literal uses such as 'the tree is in leaf' where the character of closure is easily overlooked.

One of the consequences of this account of language is that marks or combinations of marks do not offer a description of the world which we can identify as being true or false but indicate the possibility of a closure which either is or is not realised. The same mechanism applies whatever type of word is employed or whether a single mark is offered or a combination of marks in the form of a sentence. In order to illustrate this point consider the sentence 'the sky is blue'. This sentence proposes that we hold as one the meaning associated with the term 'sky', and the meaning we associate with the term 'blue'. In principle the sentence could be replaced with a single word 'skyblue' which could be used to elicit a similar linguistic closure. The former sentence incorporates four different types of word, the latter appears to be a single noun. This is an instance of the general case that has been argued namely that the role played by the mark is not relevant to the overall operation of language, which is the provision of closure. There has been a tendency in the philosophy of language to identify sentences rather than words as the primary carrier of meaning,[2] with the result that individual words are seen to derive their meaning from their use in sentences and can properly be said only to have meaning in a sentence. The account offered here gives no such primacy to the sentence nor does it identify the sentence as being engaged in a different role than an individual word. The meanings we associate with individual words are the outcome of the prior occasions on which the words have been used to initiate closure. Often this will involve the combination of the word with other words in a phrase or sentence, but it may also involve the solitary use of the word.

The standard objection to an account of language in which meaning is identified with words rather than sentences is that truth and falsity appear to apply to sentences and not to words. For in a framework in which language is regarded as referring to the world, truth is the means by which meaning is determined. For it is truth that provides the connection between language and the world. It has thus been argued that sentences are hooked onto the world because they are true, and as a result meaning is seen as a product of truth function.[3] In the context of closure it can be seen that the marks of language do not either individually or collectively refer to the world but instead function as cues and tags to closure. As a consequence meaning is not a function of truth but the outcome of linguistic closure, and sentences are no longer the sole carriers of meaning. It will not be necessary to argue therefore that the use of individual words disguises an implicit sentence, indeed the reverse might appear to be the case namely that sentences function as individual words as in the previous example of 'the sky is blue' and 'skyblue'. It is not however that words are primary rather than sentences but that meaning is not a function of any particular mark or combination of marks but is the outcome of linguistic closure which can be instigated by any type of mark in any combination.

Linguistic closure, whether practical or formal, and whether instigated by a mark or a combination of marks, realises material in the form of meaning, but it also carries with it texture; and it is texture that enables formal closure to combine marks. As it has been argued in the case of an individual mark, the meaning associated with the mark is a closure realised from the prior occasions on which the mark has been used. For each individual therefore the meaning of the word 'tree' is not something that can be linked to any particular tree, or even to a combination of all trees that have been realised, but is the material which enables all of these separate things to be held as the same. While linguistic closure realises material it also correspondingly generates texture. The material realised by the closure enables the prior closures to be held as the same, but it does not thereby make them the same for they are different. While therefore we associate the word 'tree' with a meaning which enables us to hold all trees as one, it remains the case that each tree is different. Texture is thus the difference that is held within the sameness of material. Each linguistic closure provides material in the form of meaning, the nugget of sameness that provides us with something that we did not have before, but at the same time contains texture. Unlike material texture is open, but it is not the same openness that is the other of preliminary closure. Texture is the outcome of prior, lower-level, preliminary or sensory closures, which are open in so far as there are any number of higher-level closures that can be realised from them. The texture of the face in the dots is thus found in the dots, the texture of a tree in the sensory and intersensory closures which are held within the material realised by the linguistic closure we associate with the tag 'tree'.

It is texture within prior linguistic closures that makes the combination of marks possible through formal closure. We are able to hold the tree as a house, because the meaning we associate with the mark 'tree' contains texture which makes it possible to find within this thing something other which can be held as one with the meaning we associate with 'house'. If linguistic closure realised material without any residual texture it would not be possible to hold one meaning as another through the provision of a new linguistic closure for they would simply be different. In order to hold them as the same a new level of closure would be required in the same way that in intersensory closure two different things are held as the same by the provision of a new level of closure in the form of meaning or thought. It is because meaning is not a nugget of self-same material, because it includes texture, that we are able to look for closure, scanning the texture associated with the marks in order to find some aspect that enables them to be held as the same. Formal closure is therefore able to combine marks in sentences because the meaning associated with any individual mark contains texture which enables us to find in different marks that which is the same.

If marks were precise labels for specific things or properties, the combination of marks would either result in a tautology or a contradiction. Marks, as tags, appear to offer a single meaning, a single thing. Yet this 'thing' contains difference. Linguistic closure provides the illusion of a single meaning in the same way that we have the illusion that we see individual physical objects. It is the residue in the mark other than material, that is the means by which a mark can be held as another mark without being either vacuous or contradictory. The sentence 'the sky is blue' has content because the texture held in both 'sky' and 'blue' enables new material to be realised by holding them both as one. The material we associate with the linguistic mark 'sky' and 'blue' allows us to hold each of these meanings as one thing. We thus have the impression that the meaning in each case is singular. Yet if the meanings associated with these marks were in fact singular we would either be unable to realise the closure 'the sky is blue' or it would be tautological for the meaning of sky would incorporate its being blue. It is not possible to squeeze texture from the meanings associated with these marks but if we sought to do so we would have to seek to provide an exact and unitary meaning for each mark. We might attempt to specify precisely what was meant by 'blue' with the use of a colour chart. However, the more precisely the meaning of 'blue' was defined the less likely that it would be applicable to this particular sky, for the colour of the sky would differ from it. The same applies in the case of the mark 'the sky'. If the meaning of 'sky' was restricted to the linguistic closure achieved on the initial use of the term it could not be applied in any other circumstances. The meaning an individual associates with 'the sky' is a closure provided from the combination of all prior closures employing the mark and contains therefore the residual texture that stems from their difference. Only as a result of the residue in the closures associated with the marks can the sentence 'the sky is blue' be realised, and as with the individual marks the material realised will on each occasion that the sentence is employed be different although we have the impression that the meaning remains the same. As with all closure therefore, linguistic closure provides particularity through the realisation of material, but the particularity incorporates texture and thus difference.[4]

While formal closure relies on texture held within the meaning associated with marks, practical closure is achieved independently of texture. Practical closure holds a mark or combination of marks as one with an aspect of current experience. This is not achieved by seeking to find what is the same in the sensory closure that realised the mark, and whatever aspect of experience with which it is connected, for there is nothing in common between these two things. There is, except in special and rare circumstances, no aspect of a mark, as an aural or visual pattern, which can be held as the same as any aspect of the sensory closure with which it is associated. These two quite separate and different things are held as one through the provision of material which is not a sound, or a visual pattern or a sensory closure of any type, but which is a thought or meaning. The texture held within the sound of the word is thus unimportant. We do not look to the texture of the sound to find something similar between it and sensory closure. Instead we provide a wholly different thing in the form of meaning. The attachment of language to the world through practical closure is achieved through the provision of material which is of a different type to that which preceded the closure. This mechanism applies to other types of intersensory closure. Intersensory closure holds different forms of sensory closure as the same not by searching the texture held within the sensory material but by providing new, non-sensory material in the form of meaning or thought which enables these two different forms of sensory closure to have something in common with each other.

The way marks are attached to the world is different therefore from the way they are attached to each other. Marks are attached to the world through practical linguistic closure which as a type of intersensory closure provides a new form of material from the closures which preceded it. In order for words to be attached to each other formal linguistic closure, a higher level of intersensory closure that relies on prior practical linguistic closures, realises a new meaning by finding in the texture of prior meanings that which can be held as the same. In both cases the closure realises material and texture but practical closure does not rely on the texture held within prior sensory closure to realise meaning, while formal closure relies on the texture held within prior meanings to realise new meaning. While practical and formal closure are therefore types of linguistic closure and types of intersensory closure, for they both realise meaning based on linguistic marks, the manner in which the realisation takes place is different.

It is the combination of practical and formal closure that is one of the contributing factors to the importance and power of language as a type of intersensory closure. Aside from our ability to provide linguistic marks ourselves with ease, what makes language powerful is that in addition to realising meaning through practical closure we are able to manipulate those meanings and realise new meaning through formal closure. Without practical linguistic closure however formal closure would not be possible.

Language begins with practical closure, for it is only through practical closure that the first marks can be associated with a meaning. It is only once meanings have been established that formal closure becomes possible. Meaning realised through formal closure takes its place in our currently available closures, our space, which will have potential consequential effects on our perception and our capacity to intervene, but these consequences may not actually take place because they have no application to our current sensory closures. Formal closure has the consequence that the realisation of meaning is dissociated from the circumstances in which that meaning is held as one with sensory closure. We are able through formal closure to provide a meaning for 'the sky is blue' or 'the tree is in leaf' without applying the sentences to a particular sky or a particular tree. The meaning realised is thought but it does not change what we currently perceive because it has no application to our currently available sensory closures. However, if the words in the sentences had not already undergone practical linguistic closure or were not connected through formal closure to practical linguistic closures it would not be possible to realise a meaning at all. Most readers of the preceding section will have realised a meaning for 'aasvogel' through formal closure. As a consequence it is now possible to understand sentences in which the mark is used. Assuming that readers had not previously been acquainted with aasvogels, the meaning associated with the word will not yet have undergone practical closure but through its links to other words that have undergone practical closure, such as 'vulture', a meaning can be realised for sentences in which the word is employed. Formal closure is thus dependent upon practical closure but its capacity to provide new material without requiring an application to current sensory closures has a profound and wide-reaching effect on our ability to manipulate space and thus to intervene effectively in openness.

Formal linguistic closure can be seen therefore to enable marks to be combined independently of sensory closure, and more generally provides the organisation of space. Formal closures enable us to refer to those things which are not present, both in time and in place. It also makes possible imaginary worlds in the form of fictional accounts. The great majority of our use of language operates therefore at the level of formal rather than practical closure. Yet with the possible exception of the fictional use of formal closure, the purpose of formal closure is to enable practical closure. Formal closure is therefore usually realised on the assumption that in the appropriate circumstances it can be realised as practical closure, or that it organises our space so that practical closure can be realised. If practical closure is not realisable in the appropriate circumstances or our space fails to enable practical closure we will be encouraged to abandon the formal closure and seek an alternative realisation. Suppose we are told by someone 'the tree is in leaf' but when we go to look we find it bare of any leaves. The formal closure initially realised will have been based on the meanings we already associate with the marks. On finding the tree has no leaves we are forced to reconsider our realisation. We might think a mistake had been made or a different tree was intended, but suppose this was not the case and that standing in front of the tree with no leaves we are once again told 'the tree is in leaf'. At this point we will be forced to look for an alternative closure. How can we hold this combination of marks along with our current sensory closures? We might realise closure as 'at this time of year the tree is in leaf, but there are no leaves, so there is something the matter' or 'a tiny shoot that we had previously overlooked has just appeared – the first sign of spring', but there will be other possible means of realising closure, each of which will have consequential effects on our perception, on our capacity to intervene, and on the future material we associate with the marks.

It is because linguistic closure involves both formal and practical closure that we are able to separate the meaning we associate with a word or a sentence and the application of that meaning to the world. One of the consequences of this that has already been identified is that it can appear that language functions by our determining meaning and then seeing whether the meaning applies to the world. Such an approach can be seen to have motivated truth-functional accounts of meaning and language.[5] In the context of this account of language however it can be seen that although formal closure enables the realisation of meaning independently of sensory closure, our system of linguistic closures as a whole must in any particular instance be capable of being held as one with the sensory closures immediately available to us. In so far as formal closures contribute to our space they are inextricably linked to practical closure. As a result, although formal closure is capable of realising meaning independently of sensory closure it necessarily has an impact on our space and thus, if not in the current circumstances, there will be potential circumstances in which any formal closure will have an impact on our perception and our capacity to intervene. While we can determine meaning independently of sensory closure in any particular instance, taken as a whole our space must allow for practical closure at all times. Formal closure does not therefore realise meaning which is then applied to the world, but it temporarily realises meaning which contributes to our space. In due course, and in the relevant circumstances, this meaning may be confirmed by practical closure, but it may equally be abandoned or modified. We think we know what 'the tree is in leaf' means and until this is challenged by sensory closure it therefore takes its place in our space. When it is challenged the meaning is abandoned and replaced with one capable of practical closure.[6]

The account given here has concentrated on seeking to describe the mechanism by which an individual is able to realise a linguistic closure. It may at this stage therefore appear that I am proposing a purely subjectivist account of meaning. This is not however the case as it will later become apparent. Each individual acquires language not in a vacuum but in the context of other speakers who share linguistic marks. Each individual is therefore encouraged to adopt linguistic closures which are not at odds with the closures of others. As a result the meanings an individual associates with any linguistic mark are heavily dependent on the network of marks currently available within the linguistic community as a whole. It is for this reason therefore that any one individual can adopt new linguistic closure only at the margins of language, and it is because the capacity for new closure only happens at the margins that it is possible to confuse the habitual use of linguistic terms for the underlying mechanism of linguistic closure.

Philosophers have sometimes been misled by the structure of formal closure into imagining that this is how language functions in general, while it can be seen that formal closure is derivative of practical closure and does not account for the relationship between language and the world. While formal closure is derivative upon practical closure it liberates language from the immediate circumstances, thus allowing for the indefinite extension of closure and the incorporation of the experience of others and their closures into the modification of our own space. Much of the remainder of this book is therefore concerned with formal closure and its impact on practical closure, our perception, and our capacity to intervene in openness.

6

THE ORGANISATION OF SPACE

> The character of space determines our currently available
> closures and potential closures. Over time, through practical
> and formal closure its character changes in ways that are
> both profound and intangible.

Space, the organisation of prior linguistic closures, plays an important role
in determining the character of experience. It does so because the
linguistic closures that can be realised at any given point are dependent on
the nature of space; and because linguistic closures provide the highest
level of closure within the human system of closure and thus provide the
organising framework within which preliminary and sensory closures take
place.[1] Since experience is the combined outcome of all of our currently
realised closures, the nature of space through its influence on these
closures has immediate consequences for the character of perception and
experience as a whole.

An individual's space determines which linguistic closures can be
realised in any given circumstances because in general prior linguistic
closures provide the cues to current linguistic material. Once a mark is
associated with a meaning it takes its place alongside other meanings
which are then available for future use as cues to new material. With the
realisation of our first linguistic closure, for example, we not only have
material in the form of meaning but we are able through memory to recall
this meaning so that the same mark can be used in future as a cue to a
further realisation. The process whereby the meaning associated with a
mark evolves has already been described. The meaning develops as it incorporates new meanings which have been realised through linguistic closure
initiated by the further uses of the same mark. Since space is the product
of all prior linguistic closures it evolves along with changes in each of the
marks and through changes in combinations of marks. The product of
linguistic closure, space can also be seen as the current organisation of

tags. Tags are marks associated with a given meaning which are now there-fore available for use as cues to future closure.

It is possible to indicate in a preliminary manner how the character of space develops. In childhood we realise a huge number of new linguistic closures from marks that initially function as cues but which as tags take their place in our space: a space which does not so much grow in size as become more dense. It can be supposed that an individual's first linguistic closure distinguishes between this particular thing and everything else, providing a differentiation between a 'this' and 'not-this'.[2] Once realised, the closure can be retained for future use so that any future sensory closure can be held as 'this' or 'not-this'. With each additional closure further divisions within this space of 'this' and 'not-this' take place. Later through formal closure links are made between the closures of the current space realising new material. Then further layers of formal closure in turn provide an organising structure for the meanings already realised through practical or formal closure, the function of this organising structure being to preserve and sustain the closures already realised. For although each closure provides a particularity it is at once threatened by the difference held within the particularity and by the possibility of alternative closures. Tags are therefore organised so as to be able to maintain current closures and realise future closures successfully. It will be argued however that such a goal is not independent of closure, but is itself the outcome of the pursuit of closure in general.

In this framework it can be seen that the relationship between tags, and thus between all prior linguistic closures, is a consequence of formal closure. It is thus through formal closure that it is possible to build a stable picture of reality about which we can communicate. Formal closure is not a pre-requisite for a reality – linguistic intersensory closure as I have argued previously is sufficient – but it is formal closure that allows us to order, refine, and communicate that reality. Although we have the impression that we experience reality directly, the experience of reality is not achieved all at once but is the result of placing currently realised closures in the context of the organisation of previous closures, an individual's space. As we develop from a baby to a child and then to an adult our web of closures becomes more dense by the accretion of closure but it is also increasingly structured so that each new closure is placed in the context of other prior closures. We no longer see a tree, but a type of tree, and then not just a type of tree but a type of tree with a particular character. The tree is an oak, or a holly, which is young or old, healthy or weak. Such closures are only possible in the light of previous closures which not only identify a tree, an oak and a holly as different things, but which has linked these closures into a hierarchy thereby providing an organisation within which the individual closures are placed.

The structure of closure and the mechanism of linguistic closure in particular has the consequence that the formal closures of personal space are interdependent and in turn influence each other. One way of picturing this interrelation is to suppose that the formal linguistic closures that provide personal space form a hierarchy of layers which influence each other. Such an account makes it possible to distinguish between the roles of different types of formal linguistic closure. The first layer can be taken to consist of elementary linguistic closures which are not themselves composed of other linguistic closures. Since a linguistic closure can be associated with a tag or tags, first layer linguistic closures would sometimes be associated with individual words, but they could equally be associated with sentences. First-level, or elementary, linguistic closure can then be distinguished from higher levels in so far as the closures are not realised from a composite of prior linguistic closures. All further layers of linguistic closure are the product of compound closures realised through the amalgamation of elementary linguistic closure.

In order to describe the operation of space a further distinction can be made in the layers found within linguistic closure. Compound linguistic closure requires the realisation of a single closure from a combination of elementary linguistic closures. This form of compound linguistic closure is most commonly associated with phrases or sentences. It is similar in kind to elementary closure in that the resulting closure is unitary. Although the product of prior linguistic closure the outcome of the realisation is a single thought. It can then be supposed that further levels of linguistic closure result from linking together either elementary or compound closures sequentially thus providing groups of closures which will be referred to as 'stories'. These stories are in turn themselves nested thereby ordering the whole of space. It can be seen that each level of linguistic closure has an impact on every other level. A change in the realisation of an elementary closure will have consequential effects on those compound closures of which it is a part and in turn on those stories which contain or are related to such closures. Similarly if a story changes the ordering of closures within the story changes, which will in turn have an impact on the individual closures, both compound and elementary, which make up the story.

Space is a product of closure and develops through the acquisition of new closures. These new closures can occur at any level within the system of closure. They can therefore be elementary or compound or consist in the ordering of those closures through stories. The stories can themselves be regarded as being layered so that each new layer of stories seeks to embrace those in preceding layers and if the process of closure could come to a completion – which it cannot since all closure fails in the limit – it would realise a story which successfully embraced all others without residue thereby eradicating texture. It is the gradual accretion of closure and the organisation of space through the realisation of stories that enables the realisation of an account which allows the individual to have a story of what is happening from moment to moment. Through the character of space it becomes possible for the individual to carry a story of what is taking place, or more typically many stories that operate at different levels. This overall story or combination of stories can be seen to operate both at a subjective level by providing an account of what is happening to us, and at an objective level in providing an account of what is happening 'in the world'. In both cases it is the product of the organisation of our previous closures. It is for this reason that the character of space is central to our capacity to 'make sense' of current circumstances by providing the framework in which current closures are realised, and by providing the cues which initiate those closures. Space therefore not only influences the character of our current experience, but the character of what it is possible for us to experience.

The importance of an individual's space to perception and to the capacity to intervene in openness becomes clear if we consider the difference between someone who has considerable experience in a particular field and a novice. The experienced individual is able to intervene more effectively because the space with which they are operating enables them to realise linguistic closures unavailable to the novice and therefore to perceive 'things' which otherwise go unnoticed. On the basis of this perception the experienced individual is then able to intervene by calling upon the closures available from current space. It can be seen therefore that the ability of the experienced individual to intervene is not the outcome of increased knowledge about a reality which is the same for all of us, but is the result of the reality experienced by the individual being richer on account of the character of their personal space. This outcome was noted earlier in the example of an individual who was expert in their knowledge of snakes and a novice. The closures available to the expert depend on previously realised practical and formal linguistic closures. The capacity to be able to intervene effectively however stems not only from the acquisition of appropriate closures but from their organisation. The accretion of closures alone does not enable effective intervention as can be witnessed by a supposed expert whose breadth of prior closure is so great and so haphazard that he or she is no better able to choose an appropriate current closure than a novice. The accretion of closure increases the number of potential cues and thus the range of linguistic closures that might be realised in a particular instance. It thus influences perception but it need not result in a greater capacity to intervene. The organisation of our previous closures through formal closure can be seen therefore not only to influence how the world is perceived but to influence the capacity to intervene effectively.

It might be supposed that the organisation of space, loosely speaking the theory by which we understand the world, is driven by a logical constraint such as consistency, or perhaps by an aesthetic principle such as simplicity. However, it is not necessary to import a principle in addition to closure to account for the mechanism which drives the organisation of an individual's space, and which drives our stories or our theories in general. For the process of closure delivers its own form of consistency and coherence. The ordering of the closures of space can be regarded as solely the product of layers of formal closure which realise stories and which seek like all other closure to hold that which is different as the same. The adherence to closure alone through the continued provision of formal closure would, if it was possible to carry through, result in the provision of a single story. Such an outcome is indeed often consciously desired if sometimes tempered by a recognition of the impossibility of such a task. So it is that science seeks a unified theory to encompass all other closures, and so it is that each of us in our own way seeks to provide a coherent story of our own life and our own place in the world. For as we shall see in later chapters when we unpack what is meant by 'coherent story' we find a desire for closure, a desire to hold in as unified a form as possible the disparateness of our individual experiences.

Despite the mechanism of closure and the consequential attempts through formal closure to hold still the diversity of closure that is realised from moment to moment, space is in a continual state of flux. Not only does each linguistic closure result in meaning, but each linguistic closure has consequential effects on our space as a whole. For as well as realising material in the form of meaning, each new linguistic closure is linked through formal closure to other marks. If, for example, through formal linguistic closure we hold one mark as another the meanings we associate with each of the individual marks shifts for each now incorporates the other. These shifts in meaning have further effects on all the other marks with which those marks have in turn been combined. Similarly, if through practical linguistic closure we hold a mark, or combination of marks, as one with elements of current sensory closure, we not only realise material in the form of meaning, but that meaning has a consequential effect on the meaning we associate with the mark as tag. The tag is now associated with different material and this in turn has a further consequential effect on all those marks with which it has been previously combined.

Each new linguistic closure, whether practical or formal, has therefore a ripple effect on all the other closures available to the system. The complexity of these effects are such that we cannot be aware of all of the consequences of any particular linguistic closure and we do not attempt to complete formal closure nor could we do so if we tried. In this respect the function of much discussion and argument can be seen as an attempt to draw attention to the consequences of a linguistic closure or set of closures which have not been realised and thereby to encourage the closures to be reinforced, modified, or changed. Since the character of space contributes directly to the closures realised at any particular moment and thus to what we experience at that moment each linguistic closure not only plays a part in shaping our perception, but in ways we are not currently aware contributes to our future perception. The shifting character of space makes the past inaccessible and the future unpredictable. Not in the sense that we cannot predict what events will befall us, but that we cannot predict, nor in principle imagine, how those events will be regarded or who we will be.

The impact of the shifting character of space can be shown in the changing nature of our perception of reality over time. From an early age we constantly acquire new linguistic closures which contribute to the character of our space. Each new linguistic closure contributes to our space both by the addition of material and by its consequential impact on other closures. Over time we come to see the world differently, although the way the world is different is hidden from us because our previous set of closures is no longer available to us. Even though we may use the same mark we did as a child it no longer realises the same practical closure, because the meaning of the mark has changed as the result of being associated with a wider variety of practical and formal closures. It is perhaps for this reason that on returning to places that have not been visited since childhood we can have the impression that they are different in some unspecified and unspecifiable way. Of course, if a house is no longer there or a road has moved this can be expressed. Aside from such overt changes however there can also be a sense of difference that we have difficulty in accounting for. We may be tempted to explain the difference away by our change of perspective − we were smaller, objects and houses looked different. Such explanations are not sufficient, for even when we choose the same perspective there remains a residual sense of difference. In the context of the closure this change in perception can be accounted for as the result of the change to character of personal space. The school we knew as a child had links to other material, homework, play, our friends, lessons, which have since lapsed, and the school we see today is set in the context of new closures about education, teaching, architecture, unavailable to us as a child. Our closures and their inter-relations have changed, and as a consequence when we visit the school we realise a different set of closures. In the context of closure, it can be seen that we do not perceive an independent reality but hold openness in the manner of our current space with the consequence that our perception changes even in situations where it might be supposed the stimulus was the same. This is because there is a different personal space and consequently a different set of realised closures, with the result that perception is different and the capacity to intervene is also different.

It is perhaps because the addition of new closure usually increases our ability to intervene in openness that we are inclined to be dismissive of our earlier experience. We are, for example, inclined to explain away the difference between our current perception and prior perception as due to the limitation of our earlier experience. So we are inclined to argue that we now have a more accurate picture of the school. We suppose that we are more perceptive: we notice and are drawn to details we did not pay attention to as a child. Now we can see what it is really like. However, we are no more capable of seeing what the school is 'really like' now than we were as a child. We simply realise a different set of closures. In time to come it will change again. Although we are unable to access our childhood space and thus our earlier experience, through memory we are able to catch sight of the shift in the character of space, and perhaps of particular changes in material and texture. The door handles are not quite how they seemed to be, the corridor has a different feel. As we catch sight of this change, and in our inability to account for it, we have through a sense of the fragility and limitation of closure an indication of that which is other than closure. What might be described as a glimpse of openness, so long as it is not supposed that openness is thereby captured.

One indication of the changing character of personal space is that although we are able to notice that our current perception differs from our perception as a child we are unable to describe this change. We cannot point to any specific thing which has changed. The door handles may seem different but how they are different remains elusive. We notice the style of the handle perhaps when previously it had been taken for granted but we are not able to point to any specific sensory closure that was not available to us as a child. The handle may look different but the elements that make it up are apparently the same. A similar phenomenon applies to the example of the face found in a page of dots, or when we can see two different things in a single ambiguous drawing – such as Wittgenstein's example of a duck/rabbit.[3] In each of these cases our perception changes but how it has changed is elusive. The dots have not changed, nor has the pencil drawing, yet what we perceive is different. Once again in the context of closure this phenomenon is easily explained. What has changed in these instances is the realisation of new closures. We hold the dots as a face by the provision of new material made possible because of the prior realisation of a face which has then been incorporated into our space and is available as a cue. Someone who had never seen a rabbit, nor had any notion of what a rabbit looked like, would not be in a position to see the pencil drawing as a rabbit. The provision of new closures does not however alter the preceding closures. The page of dots is still a page of dots, the pencil drawing still a pencil drawing, but a new closure has been added which in combination with preceding closures results in a change to perception.

We are unable to point to what is different when we visit a school from our childhood or when the pencil drawing is seen first as a duck and then as a rabbit because our experience is the combined product of closure and the way that it is different lies outside of the currently realised closures. Closure provides us with experience but the way that it does so and how it differs from alternative closures is, as we have argued, not accessible for we have knowledge only of closures and not of what it is that makes them possible. Similarly it is through closure that we are able to intervene in the world, but what has changed in order for us to have been able to do so is hidden. So it is that when we cannot see something, or can't intervene successfully, we are baffled and perplexed, and when we have succeeded it seems as if it was straightforward all along. Yet we cannot say what it is that we have now seen, or how we have been able to intervene, other than by repeating the closures that made this possible. We now see the character of the school buildings, the shape of the classroom, the style of the interiors, but we cannot point to any particular thing and describe what it is about it that is different other than to repeat the closures that have made it different. And how it was as a child is altogether inaccessible other than through our hazy memory of closures made then. On occasion, as in the examples discussed, we are able to catch sight of the process of closure and the role it plays in determining our perception and thus in our capacity to intervene, although we are only able to become aware of its limitation and not of the openness that lies beyond closure. For the most part however the role of closure is obscured altogether as we are lost in our current space which we usually take to be the world.

The shifting character of our space has the consequence that our memories of the past are also in flux, with the result that we can have no direct access to our past experiences. As it has been argued, each new linguistic closure, whether the outcome of practical or formal closure, has a potential impact on all the closures that share the same marks, and they in turn on other closures. The meaning associated with any word is thus continually changing depending on the character of space as a whole, although often the change is so piecemeal that we are unaware of the shift. While therefore we can recognise that what we understand by 'school' as a child is not what we understand by 'school' as an adult, we are less likely to be aware of the shift in the meaning of a word from day to day. Our lack of awareness is due to the inability to access our prior closures and thus the inability to compare what we experienced previously with what is experienced now. All of our past can be seen in this sense to be inaccessible. No memory, no matter how vivid, gives us direct access to our past, for what we hold instead is the past closure, or set of closures, in the context of current space. The practical closures of the past are gradually lost as the character of space and the organisation of tags shifts as a consequence of formal closure. Over a short period the change is so imperceptible that our recollection of the past is usually easily accommodated within our current space. As the time interval increases it is not just our memory that suffers, but the shift in the character of space makes it increasingly difficult to access the memory which instead becomes infused with closures of the present. We are, on account of what we have been, but the route by which we have become so is, and must remain, obscure to us.

The inaccessibility of our past is what perhaps encourages us to be critical of our own previous behaviour. We not only forget how it was then but the character of our experience has changed so that we can no longer access how it was then. We view the past from our current space and our behaviour and decisions are judged by our current concerns. As a result we are inclined to patronise the past – we were innocent as a child, headstrong as a young adult. The account offered of the development of space proposes however a more complex relationship. Life is not the gradual accumulation of closure which is simply added to prior closure as if we become ever more experienced. Instead each addition results in a shift in the organisation of space and may result in the abandonment of prior closures. While therefore we never cease to acquire new closures their assimilation into space and their consequential effect on the organisation of space changes our experience and our capacity to intervene in the world: a change which may or may not prove more effective in any given circumstances.

The changing character of space brought about by new practical and formal closures over a long period of time can be seen therefore to be profound but intangible. The changed character of space that results from seeing a face in a page of dots is minor but easily identified. In these cases there is only access to our currently realised closures, but in the case of face in the dots the shift in closure is more apparent on account of the proximity of the prior realised closure. In between the long-term shift in the character of space and the immediate shift identified in the face in the page of dots, there are many intermediary examples of the shifting character of space. We know for example that a painting viewed for the hundredth time, or a piece of music repeated, or the journey to work, are different from the first occasion that we saw that painting, heard that piece of music, or took that route to work. These changes are often accounted for by saying that we concentrate on different aspects: we listen to different combinations of notes, we pick out new pieces of detail. Yet we have no means of returning to the initial closure. A painting with which we have become accustomed has little in common with the painting seen at first sight. Nor is it simply that we have explored it in more detail, for we cannot see it in the fashion that we did originally even if we wished to do so. With music it is even more apparent – the phrases we once found expressive have now lost their force, or perhaps the reverse: a piece that once sounded tuneless now has a pattern we desire to hear. The painting, the music, the street we know well, are no longer the same thing we first encountered. They are no longer the same because additional closures have been realised which are now woven into our initial closures and have thereby changed our space. So it is with all aspects of our experience. As we accrete new closures the character of our space changes and with it our perception of each and every thing changes also. As a result how we might intervene, what we might do, and what we might wish to do, are themselves in flux.

Underlying the shifting character of space is the precarious nature of closure. Since closure is unlimited, any closure can be replaced, and even if it is not its character is dependent on links to other closures which themselves are equally precarious. If closures were added to space in a random fashion the resulting flux would be so great that the whole system of closure would be in jeopardy. Formal closure is required therefore to ensure the relative stability of personal space. The framework of a fixed reality within which physical objects of an unchanging character interact can be regarded therefore as an important part of the stability provided by formal closure. This framework is of course itself part of personal space but it provides an organising structure within which other closures can be placed. Although therefore the pencil drawing shifts between rabbit and duck, and although our immediate environment changes with habit, so that the painting we see now is not the one we originally saw, and our journey to work is not the same as it was, our attachment to closures that hold openness as a fixed reality is such that these evidences to the contrary are accounted for within our current space. Thus the painting, the music, the street, our past, the pencil drawing, are held as if they remained the same, and instead we account for the change in our experience by putting it down to a change in our perception of a fixed external reality. It is through the provision of a reality which is relatively stable that we are able to maintain our system of closures even though that system is constantly changing. In this respect, Kant's attempt to provide a transcendental argument for the external world[4] can be seen as an outcome of the constraints on a system of closure. For a system of closure to be effective it must be possible to maintain any individual closure sufficiently to allow for intervention. Given that the closures available to the system are constantly changing stability is only possible by the assumption that there is something other, something beyond closure, which is fixed. This principle applies to the most elementary sensory closures as well as to high-level linguistic closure. The holding of that which is different as the same, the essential characteristic of all closure, is to hold the world as one thing when there is evidence to the contrary. This is achieved by ignoring the evidence to the contrary in favour of the assumption of closure – even though it is not in practice maintained. The particular character which as humans we suppose of the external world, that it has dimension and temporality and contains physical objects, is probably an accident of our system of closure, but it must be the case that any system of closure will require a structure which provides relative stability; a reality of sorts, even if it is not our type of reality.[5]

7

LANGUAGE, TRUTH, AND THE
FAILURE OF CLOSURE

While subjective existential truth applies to all closures at the
point of realisation, there are no closures which live up to
the notion of ideal truth. All closures are thus both existentially true and ideally false.

There is one obvious objection to the account of language that has been
offered to which so far no response has been given. The objection being
that the description of linguistic closure seems to leave no room for a
distinction between true and false sentences. Yet this cannot be the case. It
is to this objection and the matter of the relation between meaning and
truth that I shall now turn.

It can be agreed that an important consequence of the account that has
been given of linguistic closure is that linguistic closures are not true or
false in the sense that they correspond to things in openness. For language
does not refer to, or map, the world outside of closure. This does not
however have the consequence that the distinction between truth and
falsity has to be abandoned. As it has been argued, linguistic closure
realises meaning, but truth is not a property that might or might not apply
to these realised meanings, as if we could check the world to see if they
were accurate. Since a closure cannot be compared with an independent
reality, reality being the product of closure, it follows that at the point of
realisation all meanings are true for the individual concerned. Whether
meaning is realised by holding a mark as one with some aspect of experience or by holding a mark as one with other marks, at the point of
realisation it is for the individual true, for this is how the individual holds
the world. The truth of the closure for the individual is affirmed simply by
its presence. By realising the closure the individual holds the world in the
manner of the closure. The closure is reality for the individual at this juncture and the meaning realised is therefore necessarily true.

In the same way that truth follows the realisation of closure, falsehood is a consequence of the failure to realise closure. Failure of closure can arise for the individual because a closure is abandoned that was previously realised, or at a social level because others fail to realise closure. Truth and falsity do not therefore describe the relationship between a closure and the world but the occurrence, or the absence, of the realisation of closure either in the case of an individual or across a range of individuals. The case will be made that this simple account whereby truth is a consequence of closure and falsity a consequence of its failure is the basis for all of the ways in which the notions of truth and falsity are employed.

The claim has been made that all closures are true at the point of realisation. The most likely first objection to such a claim is to argue that we frequently entertain false thoughts. For example, we can realise the closure 'London is the capital of France' or 'Snow is black' and yet these claims are surely false? In response it will be argued that in the case of these counterexamples, contrary to our initial impressions, so long as the combination of tags 'London is the capital of France' and 'Snow is black' are realised they, like all other closures, are indeed true for the individual at the point of realisation.

'London is the capital of England' is identified as true because its realisation is not in question, while 'London is the capital of France' is less easily realised and we are correspondingly inclined to describe it as false. However like any combination of tags, closure is possible and if realised the sentence then becomes true at the point of realisation. If for example, the closure 'London is the capital of France' is achieved in the context of the opening lines of a contentious speech to French bankers, with the implication that the power of the city of London is sufficient for it to be regarded as the capital of France, its status for the speaker is no different from 'London is the capital of England'. Both have been realised and have thus generated material, and are therefore true. For members of the audience the closure may fail and for them it would be false.

Such a response is unlikely to have satisfied the hypothetical critic, for it does not address the central thrust of the objection that it is possible to think something that we know to be false. Is it not the case that the members of the audience who fail to realise the closure 'London is the capital of France' are still able to think it? And could the speaker not realise the formal closure 'London is the capital of France' in the sense that the town of London situated on the Thames, is the seat of government of France, and in doing so would it not therefore be false? The force of such an argument rests on the notion that we determine what a sentence means and then ascertain whether it is true, thereby obscuring the difficulty involved in providing realisation and making opaque how it could be that formal closures cannot be doubted. Members of the audience who failed to realise a closure would have known from the way the tags are combined what needs to take place in order for the closure to be realised – 'London' is to be held as one with 'the capital of France' – but they do not need to realise closure in order to understand what needs to take place in order for closure to be achieved. Similarly for the speaker to realise the closure, 'London is the capital of France', in its so-called literal sense, a means must be found by which it is possible to hold 'London' as the city in England as one with 'the capital of France' as the government of the country of France. To do so it is not sufficient simply to formulate the sentence. Formal closures are not immediately consequent upon a combination of current tags, any more than closure for a monoglot English speaker follows the repetition of a Finnish sentence. For closure to occur material and texture must be provided. In the event that closure is realised the individual is then inevitably also able to realise the same combination of tags prefaced by 'It is true that ...' The difference between knowing what closure is proposed and realising the closure is thus the difference between saying something and meaning it.[1]

In passing it can be noted that a major part of linguistic communication, on the part of competent speakers of a language, consists in the identification of proposed closures rather than the realisation of formal or practical closures. In accounts given of language both by philosophers and linguists, meaning is frequently associated simply with the identification of proposed closure rather than the realisation of closure. It is perhaps the concentration on this aspect of linguistic use that has obscured a recognition of the importance of closure for our understanding of the operation of language in general.

The difficulty of realising the closure 'London is the capital of France' is not due to the lack of a supposed relationship with reality which makes it true, but the impact such a closure would have on its constituent tags. The closure associated with 'London' will differ widely between speakers, not least between those who live in the city, those who have seen images of the city in photographs or films, and those who have no acquaintance with the city. However a proficient speaker is likely to have realised formal closures that characterise 'London' as a city in England. To retain these closures and realise 'London is the capital of France' will therefore require shifts in the closures associated with 'capital' or 'France'. Such shifts are always possible, the issue is whether they undermine previous closures which we are not prepared to abandon. In this instance we can for example entertain the notion that the capital of a country is not to be found in the country: the capital as the administrative centre need not entail a geographical relationship. Thus it might be said that the capital of Tibet is Beijing. The closure of 'capital' in this form could be broadened to the notion of the ultimate seat of power: thus we could imagine closures such as 'Washington is the capital of Panama' even if we choose to reject them. Such manoeuvres will have further consequential effects. The closures associated with the tags 'country' and 'province' will be affected. If the capital of a country need not be found in the country, the closure 'country' is more likely to be associated with cultural identity rather than administrative identity. Such consequential effects, and in turn the impact on other closures in which these tags take part, restrain the realisation of formal closure.

So in the case of the sentence 'London is the capital of France' the claim will not usually be realised because its impact on 'London', 'capital', and 'France' would be too damaging for a proficient speaker. The possibility of realising the formal closure is however always present, and can take place if the consequential impact on other closures is either desired or tolerable. The impression that we determine the meaning of the sentence 'London is the capital of France' and then determine its truth value is a result of the sentence containing tags with which we are familiar, in a grammatical form which indicates how a possible closure might be realised. The failure to realise closure is overlooked and we are left with the impression that we know what the sentence means but that it is false. In the context of the account given of linguistic closure it can be seen that until closure has been realised it is more correct to say that we cannot know what a sentence means. Thus we cannot know what 'London is the capital of France' means until we can realise a closure. If and when we are in a position to realise such a claim it will then, and only then, be apparent what is meant by the claim. Given the most frequent meanings associated with the terms the sentence 'London is the capital of France' appears to be false, but in fact if we seek to realise the sentence with these meanings we find ourselves unable to provide a closure and as a result the sentence is not so much false as without realisable meaning.

Our capacity to realise formal closure is dependent on our being able to hold a mark or marks as one with other marks. This can always be achieved by changing the material associated with the marks but in turn such a shift may result in the abandonment of other closures. It is to be presumed that closure is realised providing there are no costs to doing so. In turn whether we choose to realise closure at the cost of prior closures depends on the extent to which the closures in question enable intervention. Such a conclusion however is not to argue that closure is driven by its capacity to enable the handling of activity but rather that closure is confirmed and maintained by that handling. Since it has been argued that we can regard ourselves as a system of closure, it follows that the operation of that system is concerned to pursue closure. In this sense therefore the pursuit of closure is independent of the consequences. Yet the attachment to a particular closure will be reinforced by the capacity of the closure to enable intervention of the type desired by the system. Our capacity to realise and maintain closure is therefore a product both of the nature of our space and our ability to intervene. It is possible to hold London as the capital of France, or snow as black, but these closures will have consequential effects on the rest of personal space and as a result we will only choose to realise these closures if we can find a way to avoid conflict with prior closures or if the resulting capacity to intervene is sufficiently important for prior closures to be abandoned.

A further counter to this argument might be to argue that when an individual lies it is known that what is being said is not true and thus here is an example of a formal closure which is false. In the context of linguistic closure however an alternative account can be offered. When an individual lies it is not that a formal closure is realised that is known to be false but that a formal closure is pretended. Thus, if in a playground an older child malevolently tries to trick a younger child by saying 'London is the capital of France', the older child does not realise the closure 'London is the capital of France' which is known to be false, but rather the combination of tags is uttered without a formal closure being realised. As with all linguistic closure the falsehood we associate with lying stems from the failure to realise closure, only in the case of the lie the individual gives the impression to others that the realisation of meaning has occurred. Indeed, it is one of the difficulties of lying that it requires one to pretend to experience something – the realisation of linguistic closure as meaning – that is not in fact experienced.

The possibility of lying relies on the distinction between the utterance of a combination of tags and the realisation of closure, between saying and meaning. The same is also true of acting and other forms of pretence. The possibility of communication, and thus of language, however requires the assumption that the closure associated with the words uttered has in fact been realised. Lying, acting, and pretence are thus parasitic on the truthful use of language. The ancient Greek liar paradox was based on the statement 'All Cretans are liars' uttered by a Cretan. However, aside from the paradoxical character of the sentence, it is not possible that all Cretans are liars, for in order to communicate they would need to acquire closures associated with particular tags. If they were all liars no closures could be maintained for there would be no consistency in the application of the closure and therefore communication could not occur. It only appears to be possible because we assume that language has already been acquired, at which point it then becomes possible to pretend closure. We can only lie having acquired a set of linguistic closures, and the acquisition of those closures relies on the assumption that the words uttered indicate that closure has been realised. It can be seen therefore that the assumption that closure has been realised is a prerequisite of communication.

In the context of the account of truth and falsity that has been outlined it is now possible to consider how the terms true and false are applied in an everyday context. While all closures are true at the point of realisation for the individual concerned and closures are in this respect true by virtue of their existence, the primary meaning we attach to 'true' and 'false' is not concerned with this existential, and subjective truth. We are in general not concerned with existential truth precisely because we take it for granted as the framework within which communication takes place. Instead our common use of 'truth' and 'falsity' seeks to move beyond truth for the individual at the point of realisation, and seeks to generalise the notion first to other individuals and ultimately to all 'ideal' individuals in all 'ideal' circumstances.

We can distinguish a number of stages in the progress of our use of the tag 'truth'. There is the extension from individual existential truth to a truth which involves the predominant majority within a culture, thereby generating a social truth, a sense of what is commonly believed. Social truths can be regarded as those linguistic closures which are generally realised by members of a society, and are based on the notion of individual existential truth but extended to the society as a whole. A closure is thus socially true if it is realised by most individuals within a group or culture. This notion of social truth is then taken further so that it refers not only to currently held realisations but to future realisation, thereby generating supposedly timeless social truths. A further and final extension of the notion of truth is to postulate a hypothetical ideal truth which consists of those linguistic closures which would be realised by any ideal individual in ideal circumstances. Such a notion of truth not only includes closures currently realised but allows for the future realisation of closures which are not at present available. The meaning we associate with 'truth' incorporates all of these levels although its initial starting point comes from realisation alone. While subjective existential truth applies to all closures at the point of realisation, there are no closures which live up to the notion of ideal truth. Hence the apparent paradox that all closures are both true and false, true in the sense that they are existentially true at the point of realisation and false in the sense that they are not ideal.

All closures are precarious. They are at risk of being abandoned or supplanted. In such a circumstance they become false. One way of understanding the hierarchy of truth from existential subjective truth, through social truth to ideal truth, is to see these as seeking to identify the extent to which a closure is safe from abandonment. Thus existential truth is not at all safe, while ideal truth proposes a set of hypothetical closures which are entirely safe. The scientific or religious notion of truth is an example of ideal truth, so that we can speak of 'the truth' as an ultimate goal, which postulates a set of linguistic closures which will never be supplanted and would always be realised given ideal circumstances. It is the notion of ideal truth, a framework of linguistic closure which in ideal circumstances will always be realised, which is responsible for our everyday sense of truth as 'what the world is really like'. For what we take to be reality is the outcome of our currently available closures which are themselves the product of a history of social interaction. This reality changes along with our closures, but for an ideal set of closures reality would not change. The everyday meaning of truth as 'what the world is like' is therefore the supposition of a set of closures which is necessarily realised in ideal circumstances and which will as a result not be abandoned. This set of closures thus provides a definitive account of reality against which all other closures can be measured.

All closures will fail if pursued because they are a way of holding openness as something, and since openness is not some thing, closures cannot be completed but are a temporary residing point from which we can intervene. Which is to say that closures fail because they have nothing in common with openness. In any particular instance the failure of closure is shown by its being supplanted by an alternative closure, or by its being abandoned on account of its failure to enable intervention. At the point of failure the combination of tags associated with the linguistic closure becomes false, in the first instance at the level of the individual but in due course across a community as a whole. The failure of closure has the consequence that we cannot give an example of an ideal or realist truth, and the goal of ideal truth is an impossible one.

It might appear that if closures are retained on account of their capacity to enable intervention and abandoned due to their inability to enable intervention that there is after all something by virtue of which intervention is made possible. If this were the case the notion of an independent reality would have been introduced by stealth, and there would be truth in a realist sense after all. Intervention, however, does not require that we have a knowledge of openness. Closure can be more or less successful in enabling intervention without requiring that there is something in common between closure and openness. We are used to the realist idea that our capacity to intervene in the world is based on a similarity between language and the world. The force of closure does not however result from its capacity to describe or map openness. As it has been argued, no such description is possible since openness does not have the character of a thing that can be described. Linguistic closure gains its force by the provision of material and texture. The circle of material, the meaning realised through closure, does not tell us anything about the character of openness – but it does generate activity in texture and as a consequence enables the directed handling of that activity.

Some may not yet be convinced by this argument. It might for example still be asked why one closure is more successful than another – is it not because there is something in the manner of the particular web of closures that echoes openness? I shall argue that such responses are a result of a desire to validate our closures by giving them a spurious solidity. Closures are not successful because they are more or less approximate to openness, because they are more or less true in a realist sense. Closures are successful because they enable intervention. Intervention in openness can take place without knowledge of openness or how that intervention takes place. What is known is the consequence of the intervention which is shown in our closures. Different closures thus enable us to intervene in different ways. If a closure enables us to intervene in the world to achieve what was wanted, in the context of the system of closure as a whole, it is reinforced and the realisation is retained. If it fails to enable a desired intervention we are encouraged to abandon the closure. Every closure has consequences for intervention, and whether we retain or abandon a linguistic closure is a complex function of the intervention made possible through the closure, and the character of personal space in general. Closure is pursued for its own sake but successful intervention strengthens our attachment to the closure in question. It is of course the case that what is taken for successful intervention is in turn a product of personal current space. A successful closure is thus successful in a particular context, from a particular space, for there can be no account of success independently of closure. One reason for abandoning closures is that our desires change as the result of the shifting character of space. Successful intervention as the result of closure does not therefore require an underlying similarity, agreement, or identity, either in form or content, between closure and the world.

All statements are thus both true and false. There are circumstances in which they can be realised, at which point they are true at the point of realisation for the individual concerned. Yet there is no closure that is safe and which will not fail. This outcome stems from the initial character of closure. It is through closure that there is something and in so far as there is something it is true and cannot be doubted. Yet the something that is realised is not openness and has nothing in common with openness, but it is held as if it was openness and it must therefore fail. Instead of Wittgenstein's argument in the *Tractatus* that 'the requirement that simple signs be possible is the requirement that sense be determinate',[2] it turns out that there are no occasions when a linguistic closure has a determinate truth value and therefore that there are no simple signs in the Wittgensteinian sense.

The case has been made therefore that all closures are capable of being true existentially and are false ideally. In order to illustrate how these distinctions operate in practice a number of examples will be considered. Some examples have already been used which serve to illustrate the principle that any closure can be existentially true. We have seen that apparently false sentences such as 'London is the capital of France' or 'a tree is a rhinoceros' are capable of being realised. A similar case can be made for any well formed sentence. Even 'snow is black' is realisable in the context, for example, of the aftermath of an avalanche. The possibility of subjective existential truth is no more therefore than the possibility of realisation. We shall therefore dwell no further on the capacity for closures to be existentially true and shall turn instead to the failure of closure.

In order to illustrate the failure of closure both an empirical and an analytic claim will be considered. In both cases the claims will have the appearance of being ideally true. As an example of an empirical claim that has the appearance of being true in a timeless ideal sense consider the statement 'The sun rises every day.' The failure of closure in this instance, and generally, can be demonstrated by pursuing the closure in question relentlessly. In the process the elementary linguistic closures that make up the compound formal closure are also shown to fail. In this instance we can begin by noting that from our current astronomical standpoint it is not the sun that is moving but the earth. The sun does not therefore rise every day so much as the earth rotates once a day. Then again in the polar winter the sun doesn't rise daily; on the moon the sun rises only monthly and in space it doesn't rise at all. For these reasons the closure 'the sun rises every day' will not on all occasions be capable of realisation. From a realist perspective the truth of 'the sun rises every day' is dependent on something in the world. It therefore appears that by refining the phrase one could provide an ideally true version. In this case this might approximate to: 'from the perspective of those on the planet Earth, the sun rises every day in temperate latitudes'. Such refining however requires ever more refining. Each aspect of the new closure can provide the source of failure. The sun does not rise from every perspective. Visually it does not appear to rise for those indoors, or who are asleep, or blind, or if it is cloudy. It does not rise from the perspective of infants, the poorly educated or the mentally retarded, nor does it do so for animals. The sentence might therefore be refined further 'for intelligent, adult, humans capable of seeing the sun from the surface of the planet Earth, the sun rises every day in temperate latitudes if it is not cloudy'. Further questions follow. Is it the sun which rises every day or the appearance of the sun? How are we to understand 'rises' since on observing the sun its movement is only rarely vertical? And how are we to define a day? If it is defined in terms of the length of time between the sun rising on one occasion and the next, the claim has become analytic and is no longer empirical. If it is defined independently of the sun the claim will no longer be realisable in all instances. Attempts can of course be made to resolve these challenges by refining the closure still further. However with each revision new closures will be added which can in turn be challenged. In due course the single closure will involve a whole system of closures. In which case what has been achieved is no more than to claim that given a certain space the closure in question is realisable. This is however hardly ideal truth, and in any case it remains the case that even from this space there is no difficulty in realising claims that in virtue of the identity of the linguistic marks employed they have the appearance of being contradictory such as: 'the sun never rises' as, for example, an emotional or as a religious claim; 'the sun rises once in a generation' as a political or historical remark. Realism assumes a primary meaning which enables such examples to be placed to one side on the grounds that they are metaphorical uses of the word. No such primary meaning can be elicited however despite ever more careful attempts to define the terms. The closures associated with the tags 'sun', 'rise' and 'day' are no more capable than any other of being completed. Does the closure we associate with the tag 'sun' include the light from the sun for example? If it does include sunlight then the sun extends as far as sunlight reaches in which case it is not clear what would be involved in saying that it rises every day; and if does not include sunlight then we do not see the sun and so it doesn't rise at all. We may be inclined to dismiss such responses as sophistical puzzles but there is no sophistry involved. Closure cannot be completed hence the puzzle. Our inclination to dismiss the puzzle is due to our capacity to realise closure which assumes that closure can be completed. It is as if we want to say 'look I know what "the sun" means, stop messing around'. We are able to realise a closure from the tag 'sun' which provides us with material and as a result we know that it has meaning. When asked however to elaborate this meaning the closure fails as it will with all closures. As a consequence we are inclined to dismiss the challenge as subterfuge since for the time being we can realise a closure and it enables us to intervene. However, we are in fact no more capable of determining the closure associated with any of the individual constituents of the sentence 'the sun rises every day' than we are of the whole. We cannot complete the closures because the closure is not openness and thus if pursued can be seen to fail.

In the same way that empirical truth fails in its ideal or realist sense, the same is also true of analytic truth. Since analytic truths seek to define the relationship between formal closures without reference to practical closure the possibility of ideal truth looks more attainable. However in the case of everyday language formal closures cannot be isolated from practical closure and as result the residual texture associated with the material makes precise definition impossible. In a system of formal closures that have no connection with practical closure, as in mathematics and logic, it is possible to define the relationships between closures so that they are ideally true, a consequence that has many important consequences and which has encouraged some to seek to use logic as a means of describing language. However, as will be shown in later chapters, the parallel between mathematics and language is misleading.

As an example of the failure of analytic truth, consider the sentence 'a bachelor is an unmarried man'. As with examples of empirical truth the failure of closure that can be identified in this sentence is the consequence of our inability to define sufficiently tightly the elementary closures on which it relies so that all texture is excluded. As a result it becomes possible to find circumstances in which the sentence would not be realised and would therefore be false. 'A bachelor is an unmarried man' appears to be ideally true, but it is not difficult to generate circumstances when the meaning of 'bachelor' and 'unmarried' are different. For example, we would be unlikely to describe a man who has lived with a single partner for twenty years as a bachelor but we could well described him as unmarried. As a result we can imagine circumstances in which the following sentence could be realised: 'he is unmarried, but he is by no means a bachelor'. It can be seen therefore that there are circumstances in which 'a bachelor is an unmarried man' is regarded as false. The possibility of circumstances in which the sentence is not realised is due therefore to the closures associated with the tags carrying texture which is not identical – the marks are thus not in fact synonyms. It is possible of course to seek to define the two terms as equivalent but because they will be combined with other tags and be incorporated into space in different ways there will be points at which the terms differ. Simply asserting the terms as identical is not sufficient, for the meaning of a term is the outcome of realisation and cannot be legislated through definition. This circumstance does not apply to mathematics or logic.

The same arguments apply equally to apparently logical sentences such as 'something either is the case or it is not the case but cannot be both', the so-called law of the excluded middle. This 'law' is constantly broken. A chair is a table and it is not a table. A bachelor is an unmarried man, and a bachelor is not an unmarried man. Or, as we have just sought to demonstrate: sentences are both true and not true. In each case the challenge to the claim that 'something is either the case or not the case' is possible because the closure cannot be defined stringently enough to exclude texture, and this is not possible because the world is not a thing but is open. When expressed in a logical form as 'a v ~a' the failure of closure is avoided because the term 'a' already presupposes the possibility of closure. Namely that there could be a thing which was 'a'. If such closure was possible then ideally true sentences could be generated in the same way that logically true formulations can be written. By avoiding the specific and referring to the possibility of closure as an 'x', mathematics and logic make it look as if complete closure is possible and in such circumstance a system of ideal truth can be generated. If it were the case that there could be something that was 'a' then it would be the case that either 'a v ~a', but since we can give no example of a thing and therefore no example of something that could be 'a', it is not ideally true that 'something either is or is not the case but cannot be both'. Logic and mathematics appear to deal with closures. Instead they deal with imagined closures, but the imagined closures are impossible in the context of the characteristics of closure and openness. As a result we are forced to conclude that when the truths of mathematics or logic are translated into everyday language they are no longer capable of being ideally true and as a consequence will be seen to fail.

The argument put forward so far has aimed to demonstrate the claim that all closures are capable of being existentially true and are ideally false. What has yet to be shown is how notions of truth and falsity can be applied to distinguish between sentences. In such circumstances the notions of truth and falsity have necessarily to be applied between the extremes of existential truth and ideal falsity. Take the sentence 'Casablanca is the capital of Morocco'. Such a sentence is presumably capable of being realised by an individual and is therefore at that point existentially true. Later it may be abandoned by the individual perhaps because someone else they trust tells them that Rabat is the capital of Morocco, or because on visiting the country and seeking the parliament building it is found in

Rabat, or for a myriad other reasons, but so long as it is realised its existential truth is assured. It is possible to surmise that it may even be the case that a majority of individuals in England or the United States are able to realise the closure 'Casablanca is the capital of Morocco' and that as consequence the sentence can be considered socially true as well as being existentially true. Nevertheless, it would normally be the case that we would wish to assert its falsity. We do so either on the grounds that those with knowledge of the country do not realise the closure, or that in circumstances where the formal closure had consequences for practical closure, for example entry into the parliament building, it would be abandoned. The description of the sentence as false can therefore be regarded as being based on some notion of ideal truth. It is however a weak sense of ideal truth, for there will be circumstances in which the ideal subject and ideal conditions might not choose to realise the closure. Thus while the parliament building and associated ministries may be in Rabat, it could be hypothetically imagined that someone with detailed inside knowledge of the workings of the country might argue that it was not in fact a democracy but an autocracy run by the king; that his primary palace was based in Casablanca and that the court and advisers that surrounded him there effectively ran the country. As a consequence the apparent falsehood 'Casablanca is the capital of Morocco' could be realised and the ideal truth of the statement 'Rabat is the capital of Morocco' undermined.

In asserting the truth of a sentence therefore we are in practice usually asserting either its social truth, or a weak notion of its ideal truth. It is however the strong notion of ideal truth that is the notion of truth associated with the idea of an independent reality that remains the same independently of our observations of it. Ideal subjects in ideal conditions observe the essential character of reality. Or so a realist notion of truth supposes. This strong form of ideal truth, as we have sought to demonstrate is however an impossible goal – even for a single sentence, let alone for a complete system. The realisation associated with any combination of tags cannot be fixed, either for an individual speaker, or across a society. Nor can a closure be safeguarded from future closures which will force its abandonment, and since there is no limit to the form of future closures there is no closure that is safe, or could be safe.

While it may with relative equanimity be accepted that 'Casablanca is the capital of Morocco' may not be capable of being definitively false, and that in such circumstances ideal truth is an impossible goal, there are other statements about which we will have greater hesitation. If neo-Nazis deny the existence of concentration camps under Hitler are we forced to accept that this might be true despite our desire to declare it an insidious falsehood? And if we are not able to deny such claims is not the framework of closure itself a dangerous one? The context of closure does not however constrain our response in a way which limits our capacity to respond with sufficient vigour to such claims.

In the context of an account of language in terms of closure, faced with someone who makes the claim: 'The events commonly described as the holocaust did not happen' it is not possible simply to assert the ideal falsity of the claim. Yet this does not have the consequence that we are left defenceless. It would first need to be determined whether the closure was in fact realised, or whether it is pretended. In other words is it a lie, or is it genuinely believed? If it is a lie, it can be said to be false, in the sense of being existentially false. The pretence of closure may be demonstrated in many ways: the overhearing of a conversation with another individual with a similar political perspective for example. In such circumstances the claim can be said to be false in a definitive sense.

The force of such an example however only to a limited extent relies on cases of lying, and supposes instances where the claim is believed by the person who asserts it. In these cases the claim is existentially true. What is at issue is how it is possible to realise such a claim. It may be that the socially agreed facts are unknown to the individual, or that they are disputed, or that new facts are believed. Once it is apparent what allows the closure to be realised it is possible to combat such a closure as false. In the event that the individual denies the existence of Auschwitz, and on being taken to the current site still denies its existence, we may be unable to comprehend that such a realisation is in fact possible and dismiss it as existentially false. If however it is argued that the present watch-towers and railways lines were built by the victors of the war to humiliate the Germans further evidence would need to be addressed. Given the framework of linguistic closure there is in principle no limit to such a line of argument. No matter how much evidence is produced it will always be possible to defend the original closure: documents produced are false, or inaccurate, witnesses are themselves involved in a conspiracy, and so forth. Such a defence may require increasingly preposterous claims, and increasing damage to the space of the individual concerned, but it remains possible.

Is this sufficient? Is it not the case that Auschwitz did exist and millions of people were deliberately killed? Can we allow such claims to be only socially true? To declare that a statement is ideally true is to want to be able to draw a line beneath a closure so that no further discussion is necessary. Certainly in the framework of closure no such line can be drawn. For the completion of closure is never possible, and ideal truth requires the completion of closure. The SS guard responsible for unloading people from the wagons may account for his own behaviour on the grounds that these individuals were not human, and that murder is not involved. That such a closure has hideous consequences does not thereby mean that it cannot be realised. Furthermore it is precisely because it is possible to realise such a closure that we feel so strongly about its denial. If the SS guard was merely stating something false we could regard it as an error which could be corrected. Instead the realisation of closure is an endorsement of a means of intervening, a way of handling the activity shown in texture. We reject it because of the implications it has for personal and social behaviour, not because it is unrealisable or false in a realist sense. We can as a consequence describe the closure as abhorrent but, even though some countries have sought in this instance to outlaw the claim that the holocaust did not happen, we cannot outlaw its realisation, as if we could thereby make the closure impossible.

In recognising the inability to complete closure, and thus the illusory character of ideal truth, it has not been made impossible to pursue a less stringent notion of truth, nor has the possibility of error been denied, and with it the possibility of combating error with new realisations. Moreover, the adoption of a realist truth even if it were possible is not the solution it can appear. For faced with someone who denies our realisations, and instead asserts alternatives, the assertion of truth in an ideal or realist sense is likely to be of little help in ending the argument. A neo-Nazi, unaware of the evidence, may be capable of realisations which later will be abandoned faced with socially agreed facts. In this sense the initial realisation was in error. It is possible however to deny the agreed facts and to retain the initial closure by proposing alternative closures, and for this reason to say that Hitler was in error would be to sanitise his behaviour: as if had he only been told the true situation he would have abandoned his closures, his existential truths. The case being made is therefore that such issues are not about, nor are they decided by, ideal truth. If we try to extend a limited notion of error to that of ideal truth so that there are claims incapable of being abandoned, and claims incapable of realisation, we will succeed only in resurrecting a reality that is forever distant from us and thus unknowable. There is no need to appeal to such an unattainable notion to combat opinions with which we disagree. It is in any case a weak rhetorical strategy since an appeal to truth will have little impact on the holder of such views, since for them the realisations are existentially true.

One of the points that may have become apparent as a result of the discussion of this particular example is that our use of the notion of truth is often directed towards encouraging the realisation of closure. In doing so the notion of truth is employed along a spectrum from subjective existential truth to an unattainable ideal truth. At one end of the spectrum, it may for example be claimed that a statement is true in order to counter a suspicion of lying. In such cases, 'true' is used in its subjective existential sense to identify the realisation of closure. When an accused says to the jury 'I am telling you the truth when I tell you that I didn't commit this murder', the assertion of truth seeks to assert the realisation of closure. Normally this would be taken for granted for, as it has been argued, the assumption that closure is realised is a necessary assumption if communication is to get off the ground. In these circumstances however the jury may remain sceptical, for the phrase 'I am telling you the truth' is equally capable of not being realised as 'I didn't commit this murder'. The jury will seek to decide whether these utterances are likely to have been made by someone who is or is not realising the closures in question. The tag 'truth' is employed by the accused in the hope that the assertion of truth will encourage the jury to the view that closure has been realised.

More commonly truth is used to encourage the adoption of closures which are being resisted for some reason, or to parry challenges to a closure that we wish to retain. The force of the ideal or realist notion of truth is that it proposes that the closure is always realised in ideal circumstances and thus that there is no alternative but to realise the closure. It seeks therefore to be a means of persuasion. Social truth, although stronger than individual existential truth, still allows for the response that although the closure is widely realised by others it need not be realised by the individual in question. Ideal truth is often therefore used to provide a means imposing, defending, and justifying closures which we happen to realise. Moreover it thereby promotes an interconnected system of closure and the institutions and procedures which have been used to realise it.

So far the argument put forward has concentrated on the individual realisation of closures, with relatively little discussion of the social framework within which these closures are realised. However as it will be argued in later chapters closures realised by any individual are largely those realised by the society as a whole. Children are taught to use words 'correctly' and thus to adopt the closures common in the culture as a whole. In this respect the notion of truth plays an important role in providing stability within the system of closures available to a society. To say to a child that such and such is true is to encourage the child to adopt the closure. Through language and the common adherence to closure each individual is linked to the closures made by others and furthermore to the history of closure up until this point. As we grow up and acquire the system of closures of our culture we adopt the hard won realisations of generations of previous individuals who have in small ways contributed to the system of closure that now operates. One of the reasons therefore for the maintenance of the notion of ideal or realist truth is that it encourages social cohesion which has consequential benefits for each individual and the capacity to intervene. Instead of saying 'this is how I, or we, hold the world', ideal truth says 'this is how the world is', and in doing so provides a shorthand for the imposition of closure. The detailed mechanism by which this takes place is examined later in Part V, 'The politics of closure'. Indeed it will be argued further that some notion of ideal truth is required to enable language to retain sufficient stability for it to be useful. Without an agreed framework it would not be possible to realise any combination of tags in a sufficiently predictable manner to enable the effective operation of language. The handling of activity often, although not exclusively, requires communication. In turn communication requires that to a degree the same tags are used to effect the same closures, for if the use of a tag on one occasion had no relevance to its use on subsequent occasions there would be no basis for using it to realise a particular closure. Ideal truth in postulating closures that will not be abandoned requires tags that realise the same closures. Ideal truth therefore functions to unify the use of particular tags within a language. The defining of terms is thus part of an attempt to approach ideal closures. Yet, as has been shown, definitive definitions are not possible, nor are closures possible that are ultimately safe and while the force of ideal truth is that it encourages social cohesion it can also have the consequence of denying new and different closures. As with all closures therefore its impact on our capacity to intervene is complex, enabling some interventions and denying others, which in turn varies according to character of space as a whole.

In summary therefore, the certainty associated with truth stems from its use in the context of existential truth to indicate the realisation of closure, while the social and contentious aspect of truth, the sense in which we argue over the truth, stems from the character of ideal truth. Ideal truth is an extension of the notion of existential truth so that the closure is not only certain now, but certain for ever. If it was possible to complete closure ideal truth would be possible and it is because it cannot be completed that no closure is safe and that as a consequence ideal truth is illusory. The fundamental character of existential and ideal truth therefore stems from the character of closure. Namely that there is something only as the result of closure, and in this respect the closure is certain; on the other hand, closure must fail for it is not openness.

116

Part III

THE SEARCH FOR CLOSURE
Science, logic, and mathematics

Introduction: our stories of the world

There are many stories we use to make sense of the world. We have
personal stories about what we are doing and where we are. Some of these
stories operate momentarily and consist of apparently simple closures: we
are reading a book, we are eating a meal; or they may provide a longer
term framework: we are pursuing a career, we are looking after our family.
These personal stories consist of a linked network of closures and are
themselves connected to each other. They also take their place alongside
culturally agreed stories. These culturally agreed stories like those of our
personal space are also constructed from individual closures, and operate
at successive levels of generality. At an elementary level they are built
around closures providing information about the relation between things
and people. Times Square is a location in New York city. The Prime Minister
of Britain lives at 10, Downing Street. At a higher level of closure these
stories offer an account of the operation of many parts of our reality: the
workings of the human eye, the mechanism of a watch, the growth
pattern of a certain type of tree. At a higher level still they offer a broader
and more general account of our circumstances: the history of our culture,
the nature of reality, the laws that govern the behaviour of things. All of
these stories, be they the stories of an individual or the stories of a culture,
are the product of closure.

As the outcome of the process of closure our stories about the world
also carry the characteristics of closure. So it is that our personal understanding of the world, and therefore the mode and operation of our
lives, can be seen to flow from the character of closure. Similarly our
cultural understanding of the world, from the way we divide up the
world and its constituents, to the way these elements are combined, to
the character of our overall stories that seek ultimate explanations for the
nature of things, is the outcome of the process of closure and exhibits

117

therefore its characteristics. As we have already seen, through closure a system is able to intervene in openness, and through successive layers of closure we are able to realise sensation, language, and a reality which we take to be the world. In determining the structure of language and in enabling the provision of reality the character of closure at once influences knowledge – and it is to the form of these stories about the nature of the world, and the way they are influenced by the process of closure, that I shall now turn.

There are some areas of knowledge, the physical sciences being the most obvious example, that allow not only for seemingly precise description of the world but enable accurate prediction of some future event. If an account of knowledge in terms of closure is to have any conviction therefore one question that has to be addressed is how such closures are capable of providing us with what appear to be precise and accurate accounts of the world. In particular if the descriptions of science and the laws of physics are closures and not true descriptions of an independent reality, how is it that we can with their aid predict with remarkable exactness the movements of physical objects? In order to answer this question it will in turn be necessary to give a preliminary account of how logic and mathematics, on which these descriptions and laws depend, are themselves the product of closure.

The issues concerning the effectiveness of the stories of science, and the framework of mathematics on which they rely, are so central to an appreciation of the role of closure that these will provide the primary focus of this part. I shall begin however with an initial description of how the process of closure influences both personal and cultural space and thereby determines the character of knowledge. This description in providing an account of the process of closure might be thought to suggest a vantage point outside of that process. It is of course central to the overall approach that has been outlined that such a vantage point is not possible. If such a vantage point were attainable a true description of an independent reality would after all be available to us. What it is possible to do however is to operate within the closures of our present cultural space, and offer in the context of these closures a description of the relation between the process of closure and cultural space. Such a description is not to be taken as a description of an independent reality but is itself a closure with the characteristics of closure. It adopts the familiar framework of materiality for this is the primary story with which in the context of our current space we seek to explicate our circumstances. In doing so it should not be imagined that the material framework has been endorsed as a true description of the world, nor that it is necessarily primary.

8

THE INDIVIDUAL AND THE
SEARCH FOR CLOSURE

*From blind accretion to denial, individuals are at the mercy
of the search for closure and are themselves its product.*

In this chapter an attempt will be made to offer a description of the operation of the process of closure throughout an individual's life. Its aim is to
propose an overall story which the reader can use in order to understand
how closure influences our understanding of the world. It is intended as a
starting point. Assumptions are made which, although plausible, are no
more than reasonable guesses. In the light of further observation and
research elements of the account offered here may therefore benefit from
modification, but the provision of this initial story aims to illustrate the
general character of such an account and provide an indication of what a
more detailed version might look like.

Although closure, in all instances, consists in holding that which is
different as the same, we can distinguish different phases of the process of
closure in the development of the system of closure that operates amongst
humans. For although we can presume that the process of closure is operative from a point prior to birth until death, the system of closures that
results from this process is continually developing. So, for example, it can
be presumed that at the point of birth the system of closure does not yet
incorporate those closures, self-reference among them, which allow the
individual to be self-aware and thus to have self-consciousness and experience. At this point in the development of the system it is reasonable to
suppose that many closures are realised without an immediate purpose
other than the realisation of closure itself. The provision of experience
allows the individual, in due course, to consciously pursue closure, both as
an end in itself and as a means to achieve other goals. It will be argued
that these changes in the way closure is realised can be witnessed in an
identifiable shift in the pattern of human behaviour at various points
throughout life.

It has been proposed that the sensory and intersensory closures realised by a baby are not based on actual identities in the world but in the presupposition of identity through the realisation of material. The baby holds as one thing, two or more 'instances' which are themselves the consequence of closure. These initial sensory and intersensory closures need not be directed towards an intervention, but may simply be the product of the mechanism which operates to realise closure. No doubt the basis on which such early closures are adopted is in large part due to the hardwiring of the system which will encourage certain closures and discourage others. The extent to which such closures are hard-wired and the extent to which they are plastic has yet to be agreed. However it will be proposed that the realisation of such closures is blind in the sense that the closures are adopted by the individual independently of their success in enabling intervention. Evolution may have encouraged the adoption of sensory and intersensory closures that are likely to be effective in enabling successful intervention but since it need not be supposed that they are initially realised by the baby on the grounds of success for the system of closure to develop, it is reasonable to suppose that they are realised for their own sake as the consequence of the process of closure. Many of the closures realised will fail to enable any intervention and as a result are gradually forgotten. Others, may for example enable food to be eaten by identifying a colour, a taste, a texture, or a smell, and are therefore more likely to be retained. Although initially realised blindly, such closures are then maintained either because the physiology of the system provides for no alternative, or because the closures realised prove to enable intervention.

In a similar manner initial attempts at linguistic closure can be regarded as the outcome of the blind adoption of closure, rather than requiring a motivation for realising closure, such as the desire to achieve a particular end, or the attempt to label things in the world.[1] Nor need it be supposed that an infant in realising linguistic closure is replicating seemingly similar linguistic closures of adult speakers. In order to realise practical linguistic closure, the infant requires only to hold as one a sound or sounds of language with some aspect of the infant's own current sensory or intersensory space. The character of the infant's space as a whole, being a product of all closures available to the system, will as a consequence be very different from that realised by adults. As a result the initial linguistic closures realised by an infant may have little in common with the conventional meaning associated with the tags involved.

For example, while the infant may appear to correctly apply the terms 'mama' or 'dada' to the parents, this does not imply that the infant's space has thereby identified particular people, or even three-dimensional objects.

A closure has been realised that holds as one the sound of the word with an aspect of the infant's current closures. This may consist of a smell, or colour, or even simply the level of visual activity; and even these simple sensory closures are likely to be of a different character to adult sensory closures since they have not been organised by further layers of closure. Although therefore the parents suppose that the expression of such linguistic marks identifies them, no such identification has occurred, even though the infant may appear to apply the term in approximately those circumstances in which it is appropriate. The linguistic closure realised by the infant from the same tag takes place in a space so different to that of the parents or any other adult speakers that its meaning is not merely different from the meaning attached to it by adult speakers but is so distinct that its meaning cannot even be imagined, for any attempt to understand its meaning from the adult space will at once employ closures unavailable to the infant.

It need not be supposed therefore that initial linguistic closure is driven by the requirement to intervene; the realisation of closure for its own sake, the consequence of blind accretion alone, will suffice. As a result the linguistic closures realised through this process need not immediately have a role in enabling intervention. Nor need they immediately function to organise other sensory or intersensory closures or to enable effective communication. They are simply the product of the process of closure. The apparently random sounds generated by a baby can be interpreted in this context. In due course, again it can be presumed by chance alone, such accretion of linguistic closure results in the realisation of a linguistic closure that proves in some respect to be useful as judged by the closures of the system, and it is these closures that are then retained and refined. It is likely therefore that many linguistic closures will be realised before one happens to echo an adult closure. Through positive feedback the baby is encouraged to repeat and maintain such closures. A baby therefore that realises linguistic closures that echo adult language by seeming to identify parents or objects, need not be presumed to do so either because these individuals or objects are already found differentiated in openness or because the closures immediately function to serve a purpose.[2] The process of closure alone will suffice to account for the realisation of such linguistic closures. Only later as a result of further closures will some of these closures prove to be useful.

There is a further reason for proposing that the process of closure alone is sufficient to explain the realisation of the initial linguistic closures of the system. It was shown in earlier chapters that a closure can function to allow interventions which maintain the system, without its having been chosen to achieve such a purpose. In the case of biological systems such intervention can be regarded as the outcome of the process of evolution which ensures that unsuccessful systems of closure are not replicated. As a result of the evolutionary process, systems of closure that result in sensory and intersensory closures that enable intervention beneficial to the maintenance of the system itself are promoted. Although therefore these systems of closure are not chosen purposefully they can in a passive sense of directedness be seen to have been adopted because of their capacity to intervene. It is possible that linguistic closure operates in a similar fashion and that we are therefore programmed to form certain linguistic closures. While the claim that there is evidence of a universal linguistic grammar,[3] would support such a notion at the level of linguistic structure, the variety of languages and the sounds involved would appear to argue against this being the case in the formation of individual words. As a result there is a good argument for proposing that linguistic closure in so far as it involves the provision of individual linguistic sounds is not directed even in this passive evolutionary sense.

Moreover, not only can the pursuit of closure alone account for the adoption of linguistic closure, but the proposal that the linguistic closure is driven by a specific purpose faces an immediate problem. For although linguistic closure alters the manner in which activity can be handled, if linguistic closures were only introduced with a specific handling of activity consciously in mind language could not get off the ground, since it is only in a context that is itself the product of linguistic closure that it makes sense to speak of purposeful activity.

While as an infant closure is unconsciously pursued, in later life there are many instances in which closure is consciously sought or desired. The conscious desire for closure, examples of which would include the desire to make sense of the immediate circumstances, or the desire for understanding or knowledge, follows the acquisition of subjectivity and the provision of conscious experience. I have argued that these are themselves the outcome of the process of closure. An attempt will now be made to give a preliminary account of this development. It is an account which can be seen to follow in some respects Kant's description of the necessary conditions for experience, but translates this description into the framework of closure.

For a system to be able to identify its own experience as experience, it must be possible to identify that experience as belonging to the system.[4] If we were unaware of experience being ours we would be lost to it and in that lostness there would be no consciousness. To have experience therefore the system of closure must be capable of identifying the closures of the system as being a product of the system. Thus in order for an individual to be self-aware the closures of the system need to be identified as my closures. This in turn requires that a distinction is made between what is me and what is not me. While however experience requires subjectivity, closure does not. If it were otherwise we would face an impossible bootstrap problem.[5] It is proposed therefore that the search for closure begins as a process of which we have as yet no knowledge or experience. Through this process the realisation of closure proceeds to the point where the system realises the closure 'self'. Through this, and other related closures, self-awareness is made possible with the result that the closures can themselves be experienced. Closure is not driven therefore by the desire of a purposeful subject, as if realisations took place because they were wanted, rather the process of closure is in its provision of subjectivity also the basis for desire and purposeful activity. Subjectivity, experience, and purpose, rather than being the initiator of closure, can be seen to be the product of closure.

In this overall context a schematic description of the general mechanism by which closure can allow for the appearance of subjectivity can be proposed. For a baby closures occur, as food is eaten and air is breathed, it is part of the machinery of the body. The baby does not breathe because it desires to do so, but because that is how the system operates. From the outset closures multiply, as first sensory and then intersensory closures are realised and refined through failure. At some point the infant realises the closure that is his or herself. At this stage the closure is realised from the system of closure as material but is not experienced by the system as such, for at this point the system is not capable of experience. It is possible to go further and postulate a mechanism by which the closure of self could be formulated. It would seem likely that the closure 'self' is realised first in the context of physical identity since the infant's own body is part of its own visual field. It can then be supposed that the closure of subjectivity, the realisation of all closures as being 'mine', follows from this prior closure of physical identity.[6] The closure of physical identity is important for without it the realisation of the closure of subjectivity looks impossible. The identification of subjectivity is the holding of all closures available to the system as 'mine'. The problem with this closure is that there is seemingly no means of identifying a closure as 'not-mine' – for all closures are 'mine'. Yet if there were no means of identifying a closure as not-mine what would be involved in realising closures as 'mine'? It is for this reason that it can be supposed that the closure of subjectivity, the closure that all closures are mine, needs to be linked to the closure of physical identity. For in the physical identity which is the body it is possible to identify that which is other than this physical thing – which later in the context of subjectivity can be described as 'me'. It can be supposed therefore that the closure of subjectivity is the outcome of a parallel with physical closure where it is possible to distinguish between this body and that which is not this body. Having realised the physical body, it can be supposed that the system is then capable of realising relationships between certain closures and the behaviour of the body, and thus by extension that of realising the closure of subjectivity, namely that all closures available to the system belong to the body. It is at the point when all closures are realised as in this sense mine, that the infant can be regarded as capable of experience. For in the realisation of closures as mine, there is the corresponding realisation that there is something other which is not me. A something that the closures available to the system, my closures, might be held to describe.

This summary account suggests a mechanism by which the process of closure can lead to the realisation of subjectivity. It is proposed therefore that the self-aware character of experience, the subjectivity of experience, is an extension of a continuous process of closure. A vital extension perhaps, but an extension nevertheless.

The account given may have suggested a mechanism for the realisation of subjectivity but it makes no pretence to having solved the question of how the process of closure realises experience. Not least because there has been no description of a mechanism by which the system is able to access the state of its own closures. The provision of an intersensory closure identifying the system of closures as 'mine' does not have the consequence that the system thereby experiences for example the preliminary closures that are operational in the human system of vision. The system would appear to require something in addition, something that would allow the system to monitor its own state of closure. Yet such a monitoring would not achieve anything if it were another system of closure within the system. The schematic description of the process does however indicate an approach that might lead to a more complete account.

With the arrival of subjectivity the process of closure can on occasion lead to the experience of a conscious desire for closure. In a similar manner we desire water when we experience a thirst, but the desire is not the reason that we choose to drink. The human body requires to drink in order to function. Once subjectivity is established this mechanical process is, on occasion, associated with desire but the desire need not be regarded as the cause of the thirst nor does it explain the thirst. The function of the desire and the experience of thirst is the means by which the requirements of the machinery of the body are integrated into a complex system of closure which allows the system to 'choose' when to drink. It can be supposed therefore that desire is not the reason for drinking, although it is experienced by the individual as such. In a similar manner the search for closure which determines how individuals think and thus how individuals act, and which socially is responsible for the structure of knowledge and the ordering of society, is the manner in which as individuals we experience the process of closure. The desire for closure is not however the reason that closure takes place.

As an individual's space develops so the blind accretion that characterises early realisations of closure can be seen to be modified. There are two reasons for this. Firstly, with the emergence of conscious desire the process of closure is increasingly driven by the desire of the individual to intervene to achieve a particular end. As a consequence the process of closure becomes an intentional search for closure. Secondly, as the system of closure becomes larger and more complex, the realisation of new closure is potentially more damaging for any new closure may threaten prior closures and with it whole regions of current space. As a result the accretion of closure increasingly functions to maintain, develop and defend current and prior closures, with the consequence that the process of closure operates differently as personal space develops.

This process can be illustrated by considering again the initial example given of closure – the identification of a shape in a page of dots. The identification of the shape in the formerly random page of dots realises a new thing. It was supposed that the shape of a face was found in the place of what until then had been no more than a set of dots. There is little constraint on the realisation of this closure for although it obscures openness, by masking the potential for the dots to be held in other ways, it provides us with some-thing when previously there was no-thing. Let it be presumed that this closure is not chosen to achieve any particular intervention and is therefore blind. Over time we can further suppose that to this shape are added a host of others, so that the page of dots is gradually filled with shapes and patterns around which we are able to realise further layers of closure in the form of stories and relationships which link the images. It can be now be seen that while initially there was little constraint on the adoption of a new closure, if at a later point, a new closure is proposed it may not be realised even though in principle we are capable of its realisation. For in order to realise the closure we may have to give up other closures to which we have become attached and which are embedded in the system of closures which are now used to 'understand' the page of dots.

It is because the simple accretion of closure results in a profusion of competing closures, that the search for closure requires that this profusion itself undergoes closure, providing singularity in what has become diversity. Aside therefore from the constraint offered by the requirement to intervene effectively in openness, the realisation of closure is constrained by the requirement to contain the diversity generated by the realisation of closure itself. In this manner the realisation of closure which is at first unconstrained and blind, is increasingly constrained by prior closures of the individual and the desire to achieve particular goals through intervention. The search for closure thus undergoes a shift in character in course of an individual's life. Infancy and youth are typified by the relatively unconstrained extension of closure, while in adult life the search for closure more typically consists in the maintenance and development of currently available space to achieve particular ends. If as a result we tend to get less adventurous in our closures as we get older, it is because we have more to lose.

The theoretical endpoint of the search for closure would be the provision of a single closure within which all other closures are contained, and which taken together exhaust texture. Such an outcome would provide the completion of closure, and although impossible, is nevertheless an end towards which the search for closure can be seen to be directed. The search for closure can be conceived therefore as a lifelong pursuit that attempts on the one hand to eradicate openness by 'filling-in' the texture provided by current closures, and to contain the profusion of closure within a higher-level closure which enables them all to be held as one. The accretion of closure is in this respect not random or blind, but is influenced by prior closures and the organisation of space. Such constraints apply to some degree even in infancy, however the accretion of closure in infancy is nevertheless relatively unconstrained because the realisation of new closures, that seek to fill in the texture of prior closures or to provide organising sets of closures or stories within which to contain the diversity of lower-level closure, are less likely to undermine already realised closure.

The shift from accretion to defence can be illustrated in the case of the overall closure which enables us to account for our present experience. In one sense of course the understanding of the present is a function of the system of closure as a whole, but there is a narrower closure which can be regarded as the story of what is happening to us, which enables us to hold our current sensory and intersensory closures in a manner that provides them with some general coherence. This story realises a state of affairs, a reality, which we regard as responsible for the diversity of our closures. The story is maintained so that from moment to moment we have an organising closure which 'explains' what is happening. The search for closure continues with the provision of a wider story within which any momentary experience can be placed in a continuous history of such moments. In due course we can be seen to desire an account not only of what is happening to us at any point, but a more general story which provides a framework for all of our experiences. The pursuit of closure is further extended to include a story not only of our lives, but of the world as a whole. As we realise these higher levels of linguistic closure there is an increasing suspicion of new lower-level closures if they do not accord with the framework of the higher-level stories. For any new closure has the potential not only to threaten other presently held alternative closures, but the hierarchy of closure within which the whole system of closure is held. As a result there is an increasing desire to defend current space from the addition of closures which may prove destabilising.

While the defence of current space and the provision of higher-level closures to contain the multiplicity of lower-level closure becomes more prevalent as the system of closure develops, it would appear that blind accretion of closure is important throughout childhood and is not altogether abandoned at any stage in life. Even the desire for closure can paradoxically result in blind accretion, if that desire is for closure itself. Examples of such accretion might include an obsessive knowledge of sporting results, or the spotting of train numbers. While aspects of these closures may have other motivations it can be supposed that they are largely driven by the desire for closure in itself. In this context, the attempt to educate a child can be seen to seek to harness this process and thus initially operates on the basis of a child's desire for closure. As adults we continue to seek closure for its own sake in a plenitude of ways be it as the acquisition of knowledge of some arcane or specialised subject, or merely by listening to the news or solving a crossword. Increasingly however the realisation of closure is no longer blind but adopted in the context of currently available closure to achieve ends which are themselves the outcome of the prevailing character of space. How we hold openness as the world is a function of our prior closures but the way we add to these is, over time, more likely therefore to be driven by a specific intent. We identify things which we want and which we don't want, and we realise stories to encourage the one and discourage the other. It is in this sense that children are more innocent than adults, at play, rather than at work, with their closures.

It can be supposed that a point comes in the development of personal space when the currently held framework of stories and the elementary and compound closures on which they rely are largely self-sustaining. The organisation of personal space from this point may not be immutable but it would appear that its central facets are largely decided. Such a point it could be argued has usually been reached by the early stages of adult life. In place of the incessant 'why?' of childhood which serves to extend and fill out current closure, there is more often an attempt to adhere to the organisation of closure that we have already realised. It is through this hierarchy of linguistic closure and its interwoven and interdependent closures that we understand the world and make our way in it. The adherence to the closures of this space can be accounted for on the grounds that the stories and the closures will have proved useful and we are therefore reluctant to abandon them. Instead of seeking new closure there is likely therefore to be more concern to apply current closures and to support and defend them where required. As a consequence it is possible for individuals to come to fear the abandonment of any significant part of their system of closure, and, as they grow older and more attached to the successes of the system of closure, actively to avoid the possibility of realising new closure on the grounds that this might upset closures already held.

Childhood is often associated with learning and play, and adult life with work and conversation. This pattern of development seen in the light of the process of closure can be seen as the outcome of the shift in the character of closure from blind accretion to the maintenance and development of current closures. Childhood play can therefore be interpreted as the exploration of new worlds made possible through the realisation of new closure. In a personal space which is constantly growing through simple accretion, there is a continuous supply of new closures and thus new things. Play is desirable to the child because closure is desirable. Each new closure is in this sense a present, the provision of something new that was not previously available. Children at play thus delight in new closures and the new worlds they offer. The relative loss of our ability to engage in play which gradually takes place in our transition to adulthood can in the context of closure be seen to be the product of the shift in the character of the search for closure from the extension of space to its defence.

At the risk of repetition it is worth pointing out that the changes that take place in the search for closure are not the result of the shifting character of the desire of an individual. It is not as if as we get older the nature of our desire shifts from the uncovering of the new to a defence of the known. Rather it is a consequence of the process of closure itself. At the outset the search for closure takes place in the context of openness and is satisfied by the realisation of closures. Once however the body of current closures is sufficiently extensive, further extension of closure is at risk of undermining present closures. The emphasis of the search for closure therefore shifts to the development of space and in due course to the defence of that space. The more however personal space is defended the more difficult it is to receive and incorporate new closures, and so the defence tends to become ever more determined.

The system of closure realised over time by any individual can be seen therefore to be a consequence of accident and design. It is accidental since the process of closure begins through blind accretion. It is designed by the individual because the closures are retained on the basis of their capacity to enable intervention, and new closures are sought purposefully. Moreover through the acquisition of language we inherit the closures of the past which have themselves been realised as the consequence of a history of purposeful activity. Through language and culture therefore each individual is party to countless trials of past closure and its capacity to enable intervention. Although the process of closure is thus at root random, there is a history of purpose and design in each individual's space. As we shall see[7] this also has the consequence that we carry with us the desires of the past, and with it the institutional hierarchy and power relations of the past also.

The development of personal space as a result of the search for closure is thus, in some ways, analogous to a game of chess. Early moves do not determine the character of the rest of the game nor are they entirely driven by the need to serve a particular purpose, but they do make certain outcomes more likely and serve to open up the game as a whole. Early moves do not usually preclude any future moves, but over time the possibilities become increasingly restricted. Later on the moves are precisely targeted to some particular end. In a similar manner the realisation of early closures does not determine the character of our eventual space, nor do they rule out the realisation of any future closures. As the character of space becomes more precisely defined so closures are increasingly realised with the specific intention to enable a specific purpose, and the need to defend our prior closures gradually serves to restrict what is possible to the point where there may appear to be no alternatives.

Although the character of our space is the outcome of purposeful intent both as a consequence of our own desires, and as a consequence of the inherited desires of others through the acquisition of language, we remain caught in the accidental character of its initial composition. Closures or stories that we realised as a child through blind accretion or because of their immediate capacity to enable intervention become embedded in our personal space and can then sometimes only be discarded with great damage to the remainder of space that has grown up around them. As a result, although at each stage the capacity to handle activity influences our retention or abandonment of any particular closure, as an adult the structure of personal space may be such that it is not well suited to our current desires or the task in hand. Nor is this necessarily easily rectified, for the closures we choose to realise in a given circumstance are themselves the product of our current space. We may thus be unable to realise closures which might better satisfy our desires or enable more effective intervention. What we are capable of is itself the outcome of the accidental pattern of closures realised in our early years and throughout our life.

We are therefore all in thrall to the closures and the framework of stories that make up our personal space, but it is only when we are unable to satisfy our desires or intervene successfully that the restrictive character of space is made apparent. From moment to moment we are lost to our currently realised closures, since closures at the point of realisation are existentially true. We thus believe whatever we think at the time. We may know that in other circumstances we will think and feel differently but at the particular moment it does not seem like that for our closures are reality. Normally we negotiate the abandonment of unsuccessful closures by allowing the accident of circumstance to alter our currently realised closures, or by actively seeking alternative realisations. Our room for manoeuvre however remains circumscribed by our personal space, and it is when we are unable to satisfy our desires or intervene successfully that we are more inclined to become aware of the extent to which we are trapped by our own system of closures. Since closure cannot be completed and in the limit each and every closure fails there is the potential to escape the strictures of our current space. Our capacity to do so will depend on the character of our personal space and on our ability to carry out closure. In so far as our emotional state is a product of our capacity to satisfy desires, the search for closure in addition to largely determining the pattern and character of our thought can therefore also be seen to be responsible for our psychological well-being.

Psychological disturbance or illness can on occasion thus be understood as the outcome of an accidental pattern of closure which over the passage of time has generated a space that is incapable of satisfying the desires of the individual or enabling effective intervention in the current circumstances. Thus the individual operates on the basis of a system of closure that has grown over time, and which incorporates closures that are not successful and which may bring distress to the individual. These closures are not discarded because they are so embedded into space that to abandon them would entail the abandonment of a whole swathe of other closures that may be important in enabling intervention. Furthermore, once the overall shape of personal space has been determined, the search for closure will seek to maintain and shore up weakness in the system, thereby making it difficult to give up closures that would at one time have enabled successful intervention. The interest shown by psychoanalysis and psychotherapy in an individual's past and childhood is explicable in this context.

Relations with our parents and family are likely to be amongst our earliest closures and thus may be responsible for influencing the structure of our personal space in a profound manner. The importance of the relationship with our parents along with other early closures in the determination of the structure of personal space would suggest one reason for supposing that the attempted disinterring of such closures can on occasion prove a valuable exercise. The purpose presumably being the removal of closures that generate conflicts of desire and which result therefore in discomfort to the individual. Whether in practice space can be reorganised successfully by the uncovering of the structure of early closures is a matter of empirical evidence. The successful psychoanalyst or psychotherapist can in this context be regarded as one who more often than not enables the individual to rebuild their personal space in such a fashion that there is less conflict of desire. However there would seem to be no reason to suppose that the simple uncovering of early closures may in itself guarantee that conflicts that flow from them are thereby excised. Moreover since closures once realised cannot be denied, but only be forgotten and abandoned, there is a case for believing that in so far as there is a solution to psychological distress there are circumstances in which it is possibly more likely to be achieved by the adoption of new closures and new stories rather than the attempt to uncover old closures whose pattern will be difficult to throw off since they will be sustained by the network of closure operating throughout personal space.

In this context we can, as some have suggested,[8] regard mental illness as the adoption of closures that are not accepted or acceptable by the culture as a whole. Usually these closures will be less capable of enabling successful intervention – for otherwise they are likely to have been adopted more widely. Nevertheless the system of closure adopted by those traditionally deemed mad may have its own internal logic, sustained by the interwoven closures of the individual's space – leaving aside those cases where there is damage to the physiology of the brain. Assuming that a 'solution' is desired and is desirable, it may need to consist in the realisation of a set of closures that will provide an alternative 'reality' and thereby an escape from the current space.[9] It is of course one thing to offer such closures it is another for the individual to realise them and come to recognise their value and thus desire to retain them.

In summary therefore, it has been argued that the realisation of closure can be regarded as an end in itself, which in due course serves to define ourselves and our place in the world, and which as a consequence enables us to intervene in openness. The shift from the provision of new closures to the development of current space and thence to the defence of that space, and the shift from innocent accretion to purposeful adoption of closure, is the consequence of the process of closure. A restructuring of space is always possible to accommodate new closure, but over time the scale of the restructuring required becomes greater so that it is more rarely attempted. The conscious search for closure, itself the outcome of the process of closure, can be pursued in the aid of blind accretion, or in the pursuit of some purpose which is itself the product of the individual's system of closure. As the closures of the system become more extensive and more effective, the risks of introducing new closure increase and as a consequence the blind accretion of closure has a lesser role.

Such is the broad brush character of the search for closure as it applies to the individual. In the fine detail is found a whole array of human activity. The search for a purpose to our lives, and the need for an explanation of the world is the consequence of the desire for closure. We are lost to our own space, acquiring and defending it, and putting it to use. From the framework of play to the framework of work; from the blind accretion of closure to closure as a means to enable a particular end; from the desire for the new and the unexpected to the desire for the safe and the known; from birth until death what we are and what we do is the outcome of the search for closure.

9

THE STRUCTURE OF KNOWLEDGE

Knowledge is hierarchical and subdivided, unlimited and pyramidal. Its closures are interdependent, and have the appearance of being largely complete; and the future structure and use of the closures of cultural space is unpredictable. Each of these characteristics flows from the character of closure.

Individual systems of closure, not groups or collections, realise material. It is individual human beings therefore that realise closure, not cultures or societies. The capacity to convey a closure from one individual to another through the use of language has however far-reaching consequences for each individual's space. As we have seen the individual does not realise linguistic closure in a vacuum, as if each infant had to work out a system of language for themselves. Instead, the marks of language enable individuals to acquire the closures not merely of those with whom they come into contact, but to acquire an organised framework of closure that has been the outcome of the history of a culture. Although language consists of marks and not closures, these marks, and the way they are organised, carry with them the legacy of closures of the past, and more particularly those closures that provided a means of intervention that was in some manner held to be desirable.

Individuals therefore realise their individual space in the context of the marks of language and the current organisation of those marks prevalent in the culture. To some degree, although limited and often seemingly insignificant, each individual also contributes to that language and the current organisation of marks. The term 'cultural space' will be used to identify the language and the organisation of its marks prevalent at a given point in the culture. Each individual's space is thus a product of cultural space and contributes to that space.

Knowledge at the level of the individual is provided by each linguistic closure which offers a description of the world, since all realised closures are existentially true. Knowledge at the level of society however will be defined as those sets of marks of cultural space that carry 'authority'. The nature of authority will be considered in some detail in a later chapter[1] but for the present 'marks of authority' can be taken to mean those marks or combinations of marks that are realised by those individuals within the culture who are deemed competent in a relevant sphere. On occasion the marks of authority of cultural space will also be referred to as 'the closures of cultural space', for they are the marks which in the context of a culture are deemed to be realisable in all circumstances. Strictly speaking there can be no closures in cultural space but only marks, since closures can only be realised by individuals and not groups or cultures, and even those marks that are deemed realisable in all circumstances will nevertheless result in slightly different closures by each individual, for the structure of each individual's space will alter the manner of the realisation. Nevertheless the unifying role of language in imposing similar closures across a culture is great and the shorthand use of the description 'the closures of cultural space' can therefore be justified.

The case will be made that the structure of knowledge reflects the character of the search for closure on the part of individuals. This is because the closures of cultural space are the product of a history of individual realisations of closure. Three ways have already been identified by which an individual can realise new closure. New closure can be realised by 'filling-in' the texture held within prior closure; it can be realised by a higher-level closure that serves to organise lower-level closures; or it can be the result of simple accretion. Each of these ways of realising closure has consequences for the structure of knowledge.

The provision of new closure from the texture of prior closure has the consequence of defending current closure at the same time as realising new closure. At each stage the closures can superficially appear for the time being complete, but texture is always present. The search for closure involves the attempt to eradicate the gaps and squeeze out openness, as if it were possible to provide a world of material alone. As a result, knowledge as a whole and each sub-division of knowledge exhibits this character. For example, in the context of the description of our physical space, each individual realises closures which divide up the physical environment using the marks of language. Any inhabited physical space is thus broken into parts which are then further subdivided. We divide physical space into towns and villages, fields and rivers, hills and valleys, each of which has a precise name and location. Each of the closures realised from these marks is however precarious. There is the impression that each mark or set of marks identifies this one thing. Yet the closures realised from the marks that divide our landscape can only be retained by ignoring threats to closure or by engaging in further division.

In order to illustrate this point examples of some commonly used geographical linguistic marks will be considered. Linguistic marks such as 'the Matterhorn', or 'the city of New York', or 'the Thames' indicate closures that most us are able to realise. In doing so we have the impression that we know what is meant by these marks, and that in doing so we are able to complete closure. Yet the more carefully the closures are examined the less it appears possible to give an account of each of the closures, or set of closures, associated with these linguistic tags, and the more the inherent texture becomes apparent. The Matterhorn, for example, is held to be a mountain in the Alps, but standing at the location there would be difficulty identifying where the mountain begins and where it ends, whether it includes the rock underground or the snow on its peak. New York is a city in the United States, but does 'New York' refer only to the buildings or also its people? Are the closures we associate with these words purely geographical, or do they also involve a style of architecture and behaviour? The Thames is a river that runs through London, we assume the river includes the water that it carries, but if so what happens when the water passes into the sea, and if not in what does the river consist? Such challenges to the completeness of the closures of cultural space are met in defence with the supposition that closure could be completed with the aid of new closures and thus more precisely defined terms. Closures proliferate and then require defence and further closures, which in turn find their uses and direct the acquisition of new closures. Physical space is segmented on the assumption that in principle we could be so precise that all parts of the world are identified even if in practice this is neither attempted nor thought to be achievable. Our geography is thus organised into a hierarchy in which each thing and each aspect of each thing has its place. It is not however that openness is divided into a hierarchy but that the search for closure is responsible for the nested character of description because texture is at once both a threat to closure and is always capable of further closure. In this manner the hierarchical character of knowledge can be seen to stem from the in-filling of texture. It is also a hierarchy without end. Although we sometimes have the illusion that knowledge could be completed, if not as a whole at least in some respect, the closures associated with the marks of authority of cultural space carry within them texture. The search for closure cannot be brought to a close, and as a consequence knowledge can have no limit.

In response it might be argued that there is a confusion here between naming and knowing. The identification of the name 'New York' does not have the consequence that we know something about the world as a consequence. Knowledge consists in the identification of facts, not in the identification of names. Such a response would be undeniable in the context of an analysis of language based on predication. In the context of closure however knowledge is not the product of the application of a predicate to a subject and a determination that this application is an accurate description of reality. Any closure that is realised is at the point of realisation existentially true, whether realised from an individual mark or from a combination of marks. A name on its own without any context is not capable of realisation, for the individual has no means of determining a closure. Used in a given context, it can be realised without having a predicate attached to it. An analysis of the marks of language in terms of predication is a powerful means of explicating some patterns of reasoning, but we do not as a consequence need to enforce a strict distinction between naming and knowing.

There are two other ways in which the search for closures can realise new closures: the first, simple accretion, involves the realisation of new closures which are alternatives to current closure; and the second the provision of higher-level closures which serve to organise the closures of an individual's space. Unlike the in-filling of current texture, the provision of alternative closures results in the provision of distinct hierarchies. The beauty of a landscape for example is not found in the nested set of closures which differentiate physical space, instead it offers an alternative way of holding the landscape which in turn generates texture and with it the possibility of further closure.[2] The structure of the closures of cultural space does not consist therefore of a single hierarchy but of separate hierarchies, which may be in conflict but which are usually related and complementary. The provision of higher-level linguistic closures or stories in contrast serve to contain diversity, but they are at the same time new closures and in due course the proliferation of such closures and stories itself requires further levels of closure to contain the diversity which they have themselves generated, thus resulting in a pyramidal structure. Such an outcome would theoretically be brought to a close by the realisation of a single story that contained all others.

The different modes in which closure can be realised each have their impact on the character of knowledge, with the result that knowledge is pyramidal, containing separate hierarchies each without limit. This structure, which can be graphically witnessed in our encyclopedias and libraries, can be seen therefore to be a character of closure and not a consequence of the character of the world.

The influence of closure on the character of knowledge can be seen not only in the description of physical space but in the description of time and the events deemed to have taken place in time. The mechanism by which, in the context of personal space, present experience is placed in a wider set of closures that provide a story of prior events and likely future events has already been described. This can be regarded as a story that provides the individual with an account of what is happening. In the context of society as a whole and the closures of cultural space, news coverage in the short term, and history in the longer term, provide a similar function: the means by which society tells to itself what is happening. The character of closure has the consequence that a description of the events in a day, like a description of the objects in a realm, can have no limit. It is not simply that there are countless alternative events that might make up the news on any particular day, but that for each 'event' there are in addition an unbounded number of alternative closures. The closures which individuals realise to provide themselves with a continuous story of events, and the closures that are realised in the public domain which serve to provide a society with an account of current events, do not identify actual discrete events in openness any more than closures of objects identify discrete material things.

The stories, or sets of closures describing events, with which we operate are a means therefore of containing the diversity of possible closure. History appears as a pattern of discrete events: the Second World War, the Cuba Crisis, the fall of the Berlin Wall. These events are closures that seek to hold openness as a singularity. We may imagine the fall of the Berlin Wall as a specific event, remembering perhaps the images of the night when the Wall was wrapped in people. There are other equally effective closures for that night. It was for idealistic party members of East Germany the end of a dream; for some Turkish families who were shortly to be forced to leave it was the rise of a new German nationalism; for American hawks it was the defeat of communism. Nor are these alternatives to an underlying physical reality, as if there is an irreducible substance to the event. The Wall did not fall, and alternative descriptions such as: the abandonment of the border by the East German guards, the euphoric meeting of East and West Germans on the border in Berlin, the mass chipping away at the concrete wall, are no closer to some supposed real event that lies beyond each and every description. Each closure can be seen to have its own internal requirements that will under examination require defence. Each closure provides material and thereby requires that the event be something in particular. Yet however tightly and precisely we seek to describe the event, it will evade the description both because it will fail to encompass the event as a whole, and because the individual closure will collapse under scrutiny.

Lying behind the attempted descriptions of the present there is sometimes the implicit assumption that the present could in principle be captured through the unlimited adoption of currently available closures. As if the present could be accounted for by an infinite news wire service. The profusion of closure that resulted would however be no closer to openness and its very diversity would threaten the capacity to intervene effectively. As a result the search for closure demands higher-level closure to provide security from this infinity. Instead of countless possible closures a distilled and agreed closure or set of closures is sought. News provides such a service on a daily basis, and history seeks to provide a means of containing further still this already limited diversity. History thus offers a series of stories which seek to bring stability and order to the infinity of closures that can be realised to describe the happenings of a time. For example, the description, 'The English Civil War', seeks to encapsulate a disparate and potentially limitless number of closures which could be realised in the context of the lives of a few millions of people living in England in the mid-seventeenth century. In place of the welter of possible closures there is offered a single unifying idea through which all other closures of the time can be understood. Descriptions of the civil war in terms of the rising power of a trading, if still aristocratic, class seek to unify further the plenitude of apparently unconnected closures. History can be seen therefore as the forgetting of the complexity of possible closure, so that a single version or perhaps competing versions emerge. It is one of the familiar paradoxes of history that the present, about which we surely know the most – with the exception perhaps of a handful of secret documents – is not susceptible to the historian. It is not however that history becomes clear over time but that the competing closures of the present can be forgotten, or rewritten in a new guise, in order to take their place within an overall closure. A similar principle can be discerned in every area of knowledge. Higher-order closures and stories serve to organise the marks of cultural space and to contain its diversity. Such unifications are only possible by ignoring closures that do not accord with such overall stories, or by framing lower-level closure in the light of the story, in an attempt to move towards a greater order of closure within the system as a whole.

The closures of history provide an example of the pyramidal character of knowledge but they also illustrate the interdependence of knowledge. The way the search for closure pushes forward cultural space has the consequence that low-level closures in the form of facts, cannot be separated from theory, the stories that link and organise them. High-level closures or stories contain the diversity of lower-level closure, but as with all closures they realise texture as well as material, and the texture cannot be exhausted by the closures it seeks to contain. Like low-level linguistic closure, theories fail because they are not adequate to the diversity which they seek to enclose, nor could they in principle be adequate. The search for closure has the consequence that we modify low-level closure in order to safeguard the higher-level story. The story however only has content through the closures it contains. In addition to theories being dependent on facts, the facts are dependent on the theory. Once a period has been characterised as, say, 'the English civil war' all of the events within the period will be interpreted in this light. A period of peace across the nation, or a region of the country that was calm throughout, will require further closure to ensure that the higher-level closure is maintained. Similarly the proposal of a higher-level theory of history as some have sought to do, in the form for example of a class analysis,[3] will require further shifts in lower-level closures if the theory is to be retained. In either case what is understood to be taking place at a particular location on a particular day will be dependent on these higher-level closures.

Knowledge is not therefore the result of a myriad of individual closures offering individual items of data, but is the outcome of a system of closure in which each closure has an impact on the other closures in the system. Embedded in the everyday closures of language are stories, in the form of linked sets of closures, of which we are rarely aware for they appear self-evident. They are only self-evident however because the system of closures has been developed over time so that each supports the other. As an illustration of this interdependence, it can be seen that the divisions used to categorise the physical world are not only hierarchical but serve a range of other functions. Fields and hedges, houses and streets, towns and countries, are identified because these closures enable us to draw attention to boundaries, and thus to ownership and personal rights. Of course the higher-level stories of ownership and rights are only possible in the light of these lower-level descriptions of physical space, but as closures of cultural space they will have been realised together over time. The one influences the other, so that the simple everyday closures that we acquire as a child carry with them the higher-level stories of which they are at the same time a part. The closures of cultural space are therefore constantly moving forward and developing in the light of individual closure, but they do so interdependently.

A related but further consequence of the way the search for closure operates is that knowledge often appears to be on the verge of completion, because cultural space as a whole is constructed in such a manner that it is necessarily relatively secure. As it can now be seen the extension of cultural space is usually carried out in such a manner that it also provides a defence of prior closures. In the process some closures will be abandoned, but the search for closure on the part of individuals aims to incorporate the gradual accretion of closure by attempting to both squeeze out openness held within texture and to provide a single overall story within which the diversity of closure can be contained. The result is a system of closure which appears to have countered most of the threats to its stability, and to have explored most of the possibilities offered by its closures. There will be areas of cultural space where the rapid accretion of closure makes it apparent that further closure is either possible or necessary, but taken as a whole cultural space must have ensured that its closures are self-reinforcing, for otherwise the system of linguistic closure as a whole would be unstable providing a threat to closure in general.

Such an outcome enables us to account for the fact that individuals, despite the evidence to the contrary, largely have the impression that the closures of their particular cultural space are both convincing and in many respects close to completion. The outlook for example typical of the world of the medieval church, the cultural space of the South Sea Islanders in the 1930s,[4] or the Hopi Indians, or the contemporary scientific perspective, are radically different but they share the assumption that their particular perspective is in large part both persuasive and accurate. Threats to the closures of cultural space appear to be at the edges, and to be minor in character. In the case of the contemporary scientific outlook, we may not know the origin of the Big Bang, but it is widely believed that we largely understand the workings of the world, just as those living in the medieval period believed for the most part that the Bible in conjunction with classical texts provided them with a framework of knowledge which although incomplete was as extensive as it was possible for mortals to possess. The basic framework of the closures of cultural space are self-sustaining for they have been realised over time and the search for closure has already sought to exclude openness wherever possible. Each society therefore has the impression that it has largely understood the world – even if its understanding is in the form of a religious faith. The search for closure thus often appears to be nearing completion. Such an impression is evidently an illusion, both on the empirical grounds that each culture cannot be right in this respect, but also more fundamentally that every closure from the most elementary to the most general can be shown to fail in the face of openness.

There is a further characteristic of knowledge which stems from the nature of closure: namely the unpredictability of the future development of the closures of cultural space. It is the interdependent character of closure that has the consequence that the future direction of cultural space is unpredictable – even though some of its closures are realised with a specific purpose in mind. For example, the initial closures which were realised to divide physical space, enabled land to be enclosed, bought and sold, and fought over. As it has been indicated the provision of these higher-level stories in turn influenced low-level closure. In a cultural space that has enclosed all land and left no residue, and has identified each person uniquely, a whole apparatus of social order and control can be exercised. Such an outcome is not the necessary consequence of identifying people and locations nor could we suppose that such an aim was implicit from the outset, but it is an outcome that has become possible through the evolution of cultural space.

Similarly the Victorian desire to 'label' the natural world, and identify and categorise the flora and fauna can be regarded as being, for the most part, driven by the blind accretion of closure itself rather than in the pursuit of some specific goal. The labelling of the natural world in turn required a taxonomy to contain the profusion of low-level closure. In its turn the taxonomy produced new closures which themselves served as the basis for further accretion.[5] The search for closure thereby resulted in a familiar pattern – the accretion of closure, the attempt at completion through the in-filling of texture and the provision of higher-level closure, and as a consequence the further accretion of closure in defence of closures already realised. Each category of flora and fauna served to unify disparate individual low-level closures. Yet at the same time it forced further closures in order to secure closure. Further higher-level closures were also sought in order to provide categories of the categories and thereby attempt the completion of closure. The resulting orderliness whereby each plant and animal appeared to have its place was not a reflection of the character of openness, or even less the sign of a master builder at work, but the outcome of the search for closure. As a result of the cataloguing of fauna a yet higher-order closure was made possible in the form of Darwin's theory. Yet those who were engaged in the initial task of identification could hardly have known that their closures would result in a story that would widely be thought to undermine the uniqueness of humanity. Seemingly insignificant low-level closures thus led to a significant reorganisation of cultural space. Although therefore the closures of cultural space appear persuasive and relatively complete, any new closure is potentially threatening since the system of closure is interdependent and any one closure can impinge on any other. Like the weather therefore the system of closure can be regarded as being chaotic, and as such is unpredictable. In addition therefore to being pyramidal the closures of cultural space are seemingly successful, interdependent, and unpredictable.

As with the individual, the closures of cultural space can be driven by accretion alone. We can imagine a prisoner in a cell, naming the bricks or patterns in the walls, providing differentiation for its own sake. The activity is not driven by a purpose other than the desire for closure. There would be no limit to this process as the provision of new closure would itself offer the possibility of further closure, and from the closures realised a whole array of new combinations of closure could be generated. In the same way much of cultural space is initially realised without purpose, as in the case of the naming of the natural world, other than the realisation of closure. Once realised however these closures carry further possibilities for closure and opportunities for intervention.

There are many closures of cultural space, however, that are not the consequence of the desire for closure alone, but are directed to particular ends — although as it has been argued these ends are in turn the consequence of closure. Moreover, as it will be argued later in the book, control over the realisation of closure is the means by which individuals and organisations exert power.[6] In the context of personal space we can simply refuse to realise new closures. There is no equivalent barrier to the closures of cultural space, but they are nevertheless capable of being manipulated and in part controlled by institutions and individuals. In personal space we use closures to justify, explain, and defend behaviour and opinion. A similar mechanism applies to cultural space, and politics is in this sense the battle for closure. 'The divine right of kings' was a closure employed by monarchists to support the crown; 'the rights of the individual' is a closure employed by liberals to attack government; 'the Cold War' served to defend the arms race; 'wealth creating' is a closure that enables the defence of large and powerful corporations or individuals; 'professional' can be used by individuals and groups to extend and maintain a special status with resultant privileges.

Although such closures are realised to achieve specific ends, it is still the case that the future outcome of their realisation remains unpredictable. For there lurks in every new closure potential risk. Just as the early-nineteenth-century taxonomist could not have known that the way was being paved for Darwin, so the politician or institution cannot be sure of the consequences of a closure instigated for a particular purpose. An ideal closure for those in current positions of power, which will be examined in greater detail in Part V, is to have a self-sustaining system of closure that legitimises the present social hierarchy, while appearing to be have a different function. The case will be made that all institutions rely to some degree on such systems of closure, but the failure of closure and its interdependence ensures that change is always possible. It is the unpredictable way in which closures interact that limits the capacity for an individual or institution to control cultural space, although some institutions, such as the medieval church, have been remarkably successful in their attempt to do so.

In summary therefore the search for closure can be seen to have the consequence that knowledge is hierarchical and subdivided, unlimited and pyramidal. Its closures are interdependent, and have the appearance of being largely complete; and the future structure and use of the closures of cultural space is unpredictable. These characteristics can be seen to apply to all of the closures of cultural space, from the most everyday to the most arcane: from our personal knowledge of our immediate environment to the geography of the universe, from our understanding of the things around us to sub-atomic physics, from our knowledge of our place in a community to the political structure of our culture. At every level therefore the search for closure is responsible for the structure of how we understand the world.

10

THE CLOSURES OF SCIENCE

The success of the stories of science stems from the abstract
character of the closures involved and the incorporation of
idealised mathematical relations.

Like other spheres of knowledge and belief the closures of science are
driven by the search for closure. In our culture however they have a special
place for they are held with great authority. If we are to look for an explanation for the authority of the closures of science it is probably in part to
be found in the method by which they are adopted but more tellingly
perhaps in their capacity to enable effective intervention. For the closures
of science and technology have been spectacularly successful and have
transformed our capacity to do things in the world, and as a consequence
have seemingly extended our capacity to understand its workings.

An obvious challenge to the framework of closure that has been
outlined is to question how it could be that our knowledge, and the
stories of science in particular, are capable of enabling such precise and
powerful means of intervening in the world, if they are built on the fragile
and accidental character of individual closure as it has been claimed. If
closure has nothing in common with openness, how is it that, for
example, Newtonian science is so good at predicting the paths of the
planets of the solar system? The success of science is accounted for if we
regard its laws as descriptions of an underlying reality. If science is the
uncovering of truth its success in enabling intervention is to be expected.
Yet, in the context of closure science is no more capable of having a
unique means of accessing openness than any other set of closures, since it
is, along with the rest of cultural space, the outcome of the process of
closure.

In response to this argument I have already sought to demonstrate that it
is not necessary for closure to have something in common, be it form or
content, with the world in order for it to be useful. A mark or combination

144

of marks is presented as a cue for closure, and the cue serves to say 'hold the world as if it was like this'. Since openness is not closure, the closure always fails if it is understood to be a description of openness, and its failure requires further closure. For as it has been frequently asserted, openness does not consist of tables and chairs, houses and people, air and earth, but we hold it as if it did so. Closer examination of any of these closures forces further closure to avoid the evident collapse of the closure in question. Although the process never gets any closer to achieving a complete closure, the initial closures are given a semblance of stability by the protecting closures that surround them. As a result of this temporary stability we are able to use them to intervene in openness.

While any particular closure has nothing in common with openness, the account of the relationship between closure and openness that has been offered does not have the consequence that there is a frictionless relationship between closure and openness, as if openness had no purchase on closure and the system of closure was free-spinning.[1] Closure fails but the manner in which it fails is the means by which openness finds purchase on closure and is able to act as a constraint. The resulting pattern of closures does not thereby get closer to truth, in the sense of moving closer to an accurate description of openness, for the closures are never more or less in common with openness since they are something else entirely, but the pattern of closures is capable of being used to intervene to effect, as judged by the subsequent impact on the system of closure.

In response, it is likely to be argued that if closures have nothing in common with openness they are not in a position to fail any more than they can be accurate or true. The failure of closure however is not shown by examining openness to see if it matches closure but by realising the closure and using it to intervene. We do not judge closure against openness but against other closures, for the world is the outcome of closure. If we see a face in a page of dots we realise a closure. The closure fails when we either provide alternative closures or when further closure is required to defend the closure 'face'. We are thus able to refine closure through its failure but we do not do so by having direct access to openness. Although therefore openness ensures that closure fails, it does not prescribe how the failure is to be understood. For the failure of closure is observed in the context of closure. The closures of cultural space are thus the result of the failure of previous closures and attempts to rectify these failures through further accretion and the provision of an overall story or theory. Through each such manoeuvre the manner in which cultural space as a whole is able to handle activity shifts, and as a rough rule of thumb it usually shifts to enable more effective handling of activity – although defined in the context of the current system of closure – but in so doing it does not get closer to the nature of openness.

As an analogy to this mechanism of closure consider the means by which a computer operating on the principles of closure might distinguish between objects. The computer receives data from a video camera. It arbitrarily determines the first set of data as object A, and the second set of data as object B. In doing so there is no connection between these 'objects' and the physical objects which as humans we are accustomed to identifying through the realisation of our system of closure. The computer can be understood merely to be holding the first set of data as one thing, and the second set of data as another. Let us suppose that the computer is then provided with a new set of data from the video camera. The computer compares each of the pixels, its colour and luminosity, with the sets of data provided by objects A and B. This new data will not be the same as the preceding data even if it is an image of the same object, for even if one set out to produce an identical image differences of lighting and distance would influence the outcome. As a consequence, if the computer compares the new image with either of the prior data sets no identity will be found. However the criteria for identity could be modified so that there would need to be a sufficient number of points of similarity, however defined, in order for the new data to be classified as either A or B. In the event that the new data was not deemed identical it would be stored as a new object, C. In parallel with the account of closure that has been outlined, once a new set of data was determined to be the 'same' as a prior set, the new 'object' would include the data from both sets so that further sets of data would be compared not with the original data set but the new combined data set. By such a process the computer operating on principles of closure could gradually discriminate between sets of data on the basis of notional objects that were derived from the data. The character of these notional objects would depend on the 'physiology' of the machine, in the form of the camera specifications, and the system of closure, the character of which would depend amongst other factors on the criteria for identity that were operative. Such a proposal is not hypothetical: computers operating on this principle can currently discriminate more accurately than humans between photographs of people providing they are taken under similar conditions at a set distance. They do so, not by utilising our system of sensory and linguistic closures, but by employing their own sets of similarities. Instead of making discriminations based on categories such as eye colour, hair length, size of nose, and so forth, the computer has its own categories which yield their own pattern.[2] Although the pattern is wholly different from ours, it enables the machine to intervene in a particular respect more effectively than a human.

If such a strategy were extended the pattern of similarities with which the computer distinguished objects would not as a consequence of its gradual sophistication slowly approximate to the world, as if it converged on truth. The final pattern might in form be quite different from the human pattern of closures, and it would as a consequence enable different interventions in openness – some of which might be more useful and others less so. The way for example that the video camera responds to openness is different from the way the eye responds, in the same way that a leaf responds to the wind differently from a flag. In neither case does the outcome reproduce openness, but instead responds to openness through the provision of preliminary closures. The system of closure built on the initial preliminary closures holds any particular pattern of preliminary closures as a thing. If the computer was able to intervene on the basis of its closures the closures could then be refined in the light of their failure. Suppose a robotic computer, operating according to the principles we have outlined, tried to move an object it had identified as 'A' but failed to do so, it would then be in a position to re-examine the similarity it had found between this 'object' and other objects identified as 'A'. Dissimilarities could always be found that could then be the basis for further closure. Such a system would result in a gradual refining of closure but at no point would this pattern become equivalent to openness nor would its form or structure be equivalent to openness.

Many attempts to get computers to recognise objects have tried to apply human concepts to the data available to the computer, by, for example, looking for shapes and patterns that reflect the world of physical objects. One can surmise that this approach has been largely unsuccessful because it assumes that human concepts correctly identify real objects in the world and that the criteria for identifying these objects can be reduced to a set of specific rules. The inevitable failure of any closure undermines the ability of the computer to find such closures, defined by a set of specific criteria, in the data. The everyday application of human concepts overcomes this problem by allowing the concept to act as cue as well as tag. We do not therefore seek to impose a complete closure but allow a high degree of flexibility to the concept. If a computer seeks to identify an object as, say, a chair according to a strict set of criteria, there will always be circumstances in which the object is mistakenly identified. For chairs are not found in openness and the application of a strict set of rules assumes that this differentiation is found in openness. In everyday life, instead of seeking to identify an object as a specific thing identified by a set of criteria we apply concepts by holding the object as if it was this thing with the consequence that the nature of the concept is flexible thereby allowing us to overcome the inevitable failure of closure. The resulting complexity of an object, such as a chair, makes it extremely difficult to construct any set of criteria which a computer could apply in a systematic manner to replicate the human categorisation. Instead, as I have argued, a strategy that is likely to prove more effective will be one in which the computer adopts a similar approach to ourselves whereby the object, in this case a chair, is held as if it was the concept rather than seeking to impose the concept on the input data. If the current criteria for the object are not applicable then the criteria for the object will need to be extended to allow this identification to be made possible. Adopting strategies along these lines appears already to have had some success.[3] The framework of closure would suggest that there will be few limits to the ability of a closure machine, even including the possibility of an awareness of self. Such a conclusion which would normally be associated with a thoroughgoing materialism can instead be seen to be the outcome of a systematic anti-realism.

It is widely assumed that there is something in common between language and the world because otherwise it at first appears mysterious how it is that language is able to be directed successfully towards the world. We have sought to demonstrate that there need not be any such similarity, but the following further example may help. Let us suppose that we sit in front of a black table with four wooden legs. With the aid of this closure we are able to move the table, place things on the table and so forth. Now let us suppose that instead of these closures we operate with closures referring to a visible flat surface, a vatace, an irregular surface, an irrace, and a pointed surface, a pointace. The tabletop is thus a vatace and the legs a type of pointace.[4] We can suppose that further closures are used to describe divergent types of each surface. When a cup of coffee is placed on the table, the vatace changes from a continuous vatace to a vatace surrounding an irrace, or perhaps a vatace surrounding a circular pointace containing a vatace. Both systems of closures will enable communication and intervention even though the systems operate with different objects and different relations between the objects. Thus adding milk to the cup will change what is on the table but it will not change the vatace. Placing a book on the table will leave the table intact but it will not leave the vatace unchanged. The manner in which the different closures generate further closures also varies. The table and cup encourage further differentiation into three-dimensional physical objects. The vatace, irrace and pointace encourage further differentiation as to the type of surface, its shape and position in relation to other surfaces. Both sets of closure will have their strengths for they will make certain interventions in the world available and others more obscure but in neither case do they have anything in common with openness. They do not have content in common for the 'things' to which they refer are different; nor do they have form in common since the relation between their things is also different. In both cases openness acts as a constraint on the closure through the failures that it generates: the vatace is partly an irrace, the table is also a bench. As a result of these failures further closures are required to retain the initial closure, but the system of closures does not get closer to openness. It does however become more effective at handling activity. We are encouraged to think that openness is something in particular because the failure of closure makes it look as if the failure occurred because something else was the case. Failure is instead the consequence of the attempt at closure in the face of openness. Closures fail because they have nothing in common with openness. They do not fail because they are wrong and a correct closure might be uncovered.

We can therefore explain the success of closure on an everyday basis without being committed to any congruence between closure and openness either in form or content. There remains however the seeming exactitude of science and mathematics in describing the world. For if these are merely the outcome of our particular pattern of closures how is it that they are so unerringly accurate? The gradual refining of closure through an empirical process can account for an increasing ability to handle activity, and the appearance of greater precision in our description of openness. It does not explain how it is that precise mathematical relationships apply between some of these closures. The discovery of mathematical laws that underlie the behaviour of objects and do not require correction seems to imply the uncovering of a complete closure and thus an accurate description of the world. The special place we accord to physical laws stems from the apparent completeness of closure. The laws of science have an appearance of permanence and precision which suggests that closure has indeed been completed. If this were the case the closures would have uncovered openness. The closures would not only have something in common with openness they would actually be openness. It will be argued however that the physical laws of science are closures and have the same characteristics as other everyday closures.

In order to understand how the laws of physics are capable of describing the world with such accuracy it is first necessary to understand how the closures of mathematics and logic are related to those of language, and it is to this issue that we shall now turn. The closures of mathematics and logic have an unchallengeable quality that makes it look as if they are indeed unassailable truths. If this were the case the closures would have eradicated texture and would thereby have described openness. Mathematics and logic it will be argued are not capable of describing openness any more than any other closure can describe openness. It will be proposed that they succeed not because they have in fact eradicated texture, but because they assume the possibility of closures that eradicate texture and limit themselves to addressing the question as to how such closures are related.

It has been shown that both practical and formal linguistic closure contain texture. In the case of practical closure, when what we take to be reality is held in the manner of a tag or set of tags there is seen to be a residue. Since reality is the outcome of the system of closure, involving for example preliminary and sensory closures, this residue is itself the outcome of prior and lower-level closures. Nevertheless, the residue cannot be eradicated, for the closure, in holding that which is different as the same, results in there being a difference which remains, a difference that is shown in texture. If for example the closure associated with the linguistic marks 'this is a house' is realised in circumstances where the individual concerned realises sensory closures that can be held as a house, the closure holds as one of many different sensory closures that could on the one hand have been held alternatively and which on the other hand will in any case require further closure, such as doors, windows, handles, bricks, which in turn require further closure in the pursuit of complete closure. In the case of formal closure a tag or set of tags is held as one with another tag or tags. Once again texture cannot be eradicated because every tag holds within it difference and thus one tag is never the same as another. The example previously given was that of the apparent analytic truth 'a bachelor is an unmarried man'. Yet it was possible to demonstrate that differences held within the tags 'bachelor' and 'unmarried man' mean that, even in cases where the tags are apparently defined as the same, texture cannot be entirely eradicated.

So what is taking place in the case of logic and mathematics? Logic and mathematics appear to be able to generate closures that can be realised in all circumstances and are thus ideally true. Unlike natural language, either in the form of practical or formal closure, this is achieved through the eradication of texture. Ordinary language is not capable of generating ideally true sentences because each mark has its own unique set of closures with which it is associated so that even if it is said to have the same meaning as another mark, differences in the use of the marks will necessarily occur. Logic and mathematics avoid this outcome by operating on the basis that complete closure is possible and proceeding to describe the relationship between such closures were they to exist. So long as the premise of the possibility of complete closure is accepted it then becomes possible to define the relationship between such closures in a manner which appears to make them unassailable. To illustrate this process, suppose an attempt is made to eradicate any residual texture from apparently analytic claims like 'a bachelor is an unmarried man' by making the tags involved identical. Claims of the form 'a bachelor is a bachelor' or 'a chair is a chair' might therefore be thought to be ideally true, if trivially so. However it is not difficult to demonstrate that such claims are not ideally true for in order to be so it would need to be possible in any particular instance to identify one thing as 'a bachelor', or 'a chair'. Surely if there is a thing which is a chair there can be no doubt that in all circumstances it is a chair? The problem is that we cannot identify such a thing. If we point to an object and say 'the chair' we have the impression that we have uniquely referred to this thing, but on examination we will be unable to give an account of exactly what this thing consists. A chair, like all other closures, carries with it texture, which cannot be eradicated. A particular instance of a chair may also be a table and a hanger, as well as being wood and legs and organic material and so forth. As soon as the texture held within the tag is brought to the fore the sentence 'a chair is a chair' is no longer ideally true. Indeed it is immediately false for when a chair is a table it is not, in this respect, a chair.[5]

The ideal truth[6] of 'the chair is a chair' depends on the assumption that the closure 'chair' is itself complete and thus the same in different instances. Mathematics and logic extend this principle. The possibility of complete closure is assumed. Rules are defined that govern the relationship between these proposed complete closures, and the theorems that follow describe the relationship between such idealised and impossible closures. It is for this reason that mathematics and logic deal with abstract symbols. For it is these symbols that are held as if they could be complete closures. The impossibility of such a closure is thus obscured. The most familiar mathematical symbol: 'x', is the supposition of that which is not anything in particular other than it is a complete closure. Unlike the closures of everyday language the closure itself is not actually sought but rather assumed, and as a result the failure of closure is hidden. If openness consisted of things then the conclusions of logic and mathematics would be ideally true. For they describe the way that closures would combine, given an agreed set of rules, if complete closure was possible. The shortcoming of logic and mathematics is therefore hidden in its symbolism. If it were possible for there to be an x, 'x or not x', must be true – assuming that 'or' and 'not' can be defined in the normal manner[7] – but in openness there can be no x and thus logic and mathematics do not succeed in uncovering the nature of openness.

The power of logic and mathematics stems from the fact that the assumption implicit in their symbolism is also the assumption on which closure itself depends. We live in a world of closures. That is how we are able to make sense of and intervene in openness. It is how we are able to have a world. Each and every closure fails under scrutiny, but its success depends on the temporary assumption that closure has eradicated texture. Through closure we hold openness as this thing, and it is only by holding it as this thing that we are able to intervene as a consequence. Further closure will show this thing to be inadequate. The success of closure though depends on postponing the appearance of its failure. Cultural space divides openness into things which are for the time being treated as having eradicated texture. All of our closures can be challenged but at the moment that they are realised, at the moment of thought they must be held as if they were complete. Mathematics and logic seek to describe how openness would be if openness was divided into things, and while openness is not so divided, the world, as we understand it – our reality – is. For reality is the outcome of cultural space and our system of closures.

It is because the closures of mathematics and logic seek to describe the relationship between complete closures, and because our world is a reality formed out of closures which at least from moment to moment are assumed to be complete, that mathematics and logic can be applied so successfully to our reality. This point can be illustrated by reference to perhaps the most widespread example of the application of mathematical reasoning: arithmetic. In the context of the account that has been given it will be seen that the number, 1, does not in a particular instance identify any actual thing, but the possibility of a thing. On such occasions '1' is thus a symbol of that which follows the realisation of a closure. A closure is in principle one and the number 1 refers to the possibility of a closure. Not a particular closure, which would not be one thing for the closure will fail and is many things, but for the abstract idea of a complete closure. The point being made here is not the one that Russell sought to make in his analysis of number, namely that numbers refer to classes, or classes of classes, and not things themselves, but rather that the things which make up those classes are the hypothesis of a complete closure. In turn arithmetic enables us to accurately describe how things combine in the world, because the world is constructed from closures. '1 + 1 = 2' defines the relationship that would apply between classes of complete closures if they were possible. So long as the assumption of the possibility of complete closure is applicable the outcome is not in doubt, presuming for the moment that we can give precise definitions of 'add' and 'equals'. Arithmetic can be applied to our reality not because openness is arithmetical but because reality is the product of our closures. We assume our closures to be complete and arithmetic then describes how complete closures combine.

Since openness is not made up of things, and since our closures are not complete, the conclusions of arithmetic are not always applicable. One and one is two, but one monitor and one keyboard might also be held as one computer. Thus one and one is one. Arithmetic only works so long as it applies to closures that are deemed to have eradicated texture and thus are complete. When applied to openness it has no purchase for any part of openness taken with any other part of openness could amount to any number of 'things'. Thus one top and four legs are one table, but five pieces of wood. Arithmetic works so long as the things to which it is applied, themselves the outcome of closure, are taken as if the closures were complete and thus regarded as this thing and nothing else. If we add ten things to ten other things we get twenty things, providing we assume the completeness of closure; however since every thing in the world is also not this thing but countless other things and combinations of things, the successful application of arithmetic requires that we do not look too carefully at the world.

The same principle applies to logic. The success of logic depends on the assumption that the things to which it applies are complete closures. Since we can give no example of a complete closure, any logical deduction can be shown to fail when it is applied to actual things or properties. If A→B, and B→C, then we can be assured that A→C, so long as there is no residual texture held in the closures A, B, C and the implies function, →, is assumed to have been defined without introducing texture. As soon as closures are realised in the context of openness, texture cannot be eradicated and therefore the formal logical relations that hold between complete closures are no longer secure.

For example, we are unlikely to have difficulty in realising the closure 'when it is raining, it is cloudy', for without clouds there cannot be rain. Similarly we are unlikely to have trouble realising the closure 'when it is cloudy it is not sunny'. If, A, B, and C, stand for 'it is raining', 'it is cloudy', and 'it is not sunny', we have therefore A→B, and B→C, yet the conclusion A→C, 'when it is raining it is not sunny' is more likely to pose difficulties for there are clearly occasions when the sun is shining at the same time that it is raining – a rainbow being the occasional outcome of such a circumstance. The failure of the deduction is due to texture held within the closures involved. We assume 'cloudy' is a single property, but texture held within the closure associated with the mark results in failure for the deduction. We could of course seek to eradicate this texture by defining 'cloudy' more precisely. In doing so however we will undermine the usual closure associated with the mark, and closures that we would usually readily realise such as 'when it is raining it is cloudy' may no longer be realisable. In the limit the attempt to define the terms precisely will make them impossible to apply, for openness will always be other.

Consider another example of a logical deduction familiar from the textbooks. 'If Socrates is a man, and all men are mortal, then Socrates is mortal.' Given the account offered so far it must follow that this deduction works providing we assume each of the closures is complete, while in practice none of them can be. Since a precise definition of 'Socrates', 'man', and 'mortal' is not possible the deduction cannot be foolproof. This can be demonstrated by exchanging the name Socrates for another name. If the name 'Socrates' is replaced in the deduction with the name 'Jesus' the deduction is at once seen to fail. 'Jesus is a man', and 'All men are mortal' does not enable us to conclude that 'Jesus is mortal.' The problem with the deduction is not with the initial conditions: according to the tenets of Christianity, Jesus was indeed a man, and it would be false to deny the claim. Nor do we wish to dispute the statement 'All men are mortal'. The problem stems from the failure of the closure of the tags 'man', 'mortal', and 'Jesus'. As with all closures, none of these marks is associated with a complete closure. No definition, however carefully and specifically designed will be capable of eradicating texture and leaving merely the material generated by the closure thereby allowing the deduction to be secure. Does 'man', for example, include the physical body only, or also the mind, or, for those with a religious set of closures, the soul? Is mortality purely a lack of physical function, or should it have some non-physical meaning? Does 'Jesus' refer to a textual figure or an historical one? It might be argued that such questions can be defined away, but they will only result in further conflicts with other closures, and textural residue will remain however closely defined the terms become.

As a result the difficulties with the deduction are not limited to the replacement of the term 'Socrates' with 'Jesus', but can equally be demonstrated in its original form. We do not have difficulty in imagining a lecture on Socrates that begins: 'Socrates' teaching has echoed through Western civilisation. He is alive and well today.' In such a context Socrates would indeed still be a man, and men might still be regarded as mortal, but the deduction would fail, for Socrates would not be mortal. It fails because the closure associated with Socrates carries with it texture and is not complete. In addition to being a man, the mark 'Socrates' can be associated with the ideas of Socrates, and these ideas are not mortal. We can thus find elements of the closure which allow Socrates to be both a man, and to be immortal. In order to secure the deduction in this respect the initial claim 'Socrates is a man' would have to be restricted to the physical body of Socrates, but even then there might be difficulties in excluding the ideas of Socrates if in the context of some theories of the mind they were deemed to be part of a particular brain state. For the deduction to be secure the closure must be regarded as being complete, while in practice, such completion cannot be achieved. The same argument applies to all deductions and as a result there can be no deductions of which we can be certain, aside from those which assume the completeness of closure and which, using logical symbols, are therefore concerned solely with the hypothetical relations between such impossible 'things'.

Describing the relations between hypothetically complete closures under a defined set of rules has the result that the conclusions of logic are in this respect secure, as are the results of mathematics and arithmetic. Given therefore the possibility of an agreed set of definitions, and rules, the conclusions of logic and mathematics can be taken to be ideally true. So long as the system does not introduce texture, no future closure can bring about the abandonment of its deductions. As it has been argued such certainty is not possible when applied to the closures of ordinary language, for such closure can never be complete; but, if we are able to treat such closures as if they were complete then the certainty of the logical deduction can be applied to language and thus to reality. In the limit this will fail but its failure can be postponed or evaded and by this means the precision of mathematics and logic seemingly found in reality. It is because in the limit there can be no secure linguistic closure, that logic and mathematics provide the archetype of closure to which most areas of knowledge aspire.

It is widely supposed that it was the Pythagoreans who were the first to use logical reasoning to deduce a theorem that was not at once apparent from the initial conditions. Having used a logical proof to derive, amongst others, the famed theorem whereby the square of a hypotenuse triangle is equivalent to the sum of the squares on the other two sides, the Pythagoreans are said to have believed that they had thereby uncovered the magical secret of the universe.[8] For here was a set of closures, a realm of knowledge, quite unlike any other. A closure that worked in all circumstances, and was capable of withstanding any attack. A closure that was not only safe from future closures, but whose safety was known in advance. Little wonder that they sought to extend this bridgehead of certainty, and little wonder that they believed that in the process they had discovered the essential character of the world. This strategy and belief has echoed through cultural space ever since. The builders of great rational systems, like Spinoza or Hegel, have of course employed this strategy, believing that from a secure rational starting point a whole edifice of knowledge can be constructed; but it is also the strategy of science which seeks to uncover the mathematical structure of the world and thereby to derive unchallengeable results. It is thus in a continuation of the Pythagorean dream that it is possible for contemporary astrophysicists to suppose that it may be possible to describe the initial conditions of the universe in mathematical form which can then be held to determine the structure of all that follows.

The role of mathematics as an archetype of closure extends beyond grand philosophical systems or science and pervades cultural space as a whole. There are many disciplines, such as the social sciences, that aspire to mathematical certainty, and throughout post-Enlightenment culture the security of mathematical and logical closure has provided an endpoint of comparison with other forms of closure. As a consequence, in many academic disciplines there has been a desirability attached to the formalising of the material under consideration. Nothing is quite so impressive as a few mathematical formulae. Philosophy has not been exempt. The very phrase 'analytic philosophy' seeks to put the subject on a more solid, logical, foundation.[9] The security that attaches to the closures of logic and mathematics gives them a machismo not found elsewhere, and which has been sought by many who have aspired to knowledge.

So what is taking place in the discovery of mathematical laws that apply to the physical world? In the same way that arithmetic can be applied to the world because we divide openness into things through the realisation of closure, so mathematical relationships between things in the physical world can be discovered because we have already imported those relationships into the world through closure. In the context of geometry therefore, triangles and squares, spheres and ellipses are not found in openness prior to closure. Through closure we impose these regular shapes on the world. As we do equally with such notions as 'object', 'line', 'force'. It should not be surprising therefore that we subsequently find mathematical relations between elements of reality since those elements make the assumption of the completeness of closure and mathematics is the working through of the necessary consequences of combining complete closures. It might be said therefore that through physical laws we discover the structure of the closures of cultural space rather than the structure of openness.

The Pythagoreans in discovering a relationship between the lengths of the sides of a triangle did not therefore discover something about the nature of openness. Instead they discovered something about the relationship between the closures point, line, angle, triangle and so forth. In the same way that the relationship between numbers can be applied to the closures that make up the world, so geometrical relationships can be found once closures incorporating these relationships have been realised and thus imposed on openness. Once the closure 'right-angled triangle' has been realised we will inevitably find the Pythagorean relationship applies to it. The accurate prediction of the height of a pyramid based on knowledge of the lengths of its other sides is not therefore evidence of the mathematical character of openness but evidence of geometrical closures that are related in a precise mathematical manner. We can calculate the height of a pyramid because the idea of a pyramid is a closure that already incorporates a whole series of other geometrical closures, including that of a right-angled triangle. If we measured the height of the actual Egyptian pyramids and found them not to correspond to the predicted height from the length of their bases, we would conclude that these were not in fact geometrical pyramids. The mystery is not therefore our ability to determine the height of the pyramid but the means by which through our realisation of the closure associated with 'pyramid' our intervention in openness is achieved.

The application of Pythagorean geometry to the world is not however the example that most clearly illustrates the precision of science. The capacity for Newtonian physics to predict with remarkable accuracy the behaviour of physical objects is a more telling case. It will be argued however that the success of Newtonian physics is to be explained in a similar manner. The laws of motion are not discoveries about the nature of openness but definitions of the relationship between forces and bodies. The mathematical consequences of these definitions are described in the first volume of Newton's *Principia*. In so far as these closures can be applied to the world they will produce results in line with the mathematical results. This does not mean that any laws of motion will prove as effective as Newton's. It does mean that if we define a system of abstract closures, that assume the possibility of closure, and deduce results from the definitions and axioms of the system, and then find a means of holding the world in the manner of these closures we will also be able to make deductions according to the system devised and 'find' those results in reality.

If, as we have argued, Newton's laws are not true because they accurately describe openness but because they offer a formal system that can be used to intervene, it may still be objected that there remains the question as to what it is that makes a system of abstract closure such as Newton's an effective means of intervention. We can offer at least a preliminary response to such a question. At the heart of the Newtonian system there is a definition of force in terms of mass, distance, and time.[10] It is Newton's notion of force from which all else follows. Newton's central idea, although expressed differently in the 'laws of motion', is to use force as the means to account for change, in contrast to Aristotle who uses force to account for movement. In an everyday sense, and certainly in the context of classical Greece, the Aristotelian closures are more easily realised. If you push a cart it moves, if you stop pushing it, it stops moving. The Newtonian closure by contrast is only realised with difficulty, namely with the aid of further closures. Since the cart stops moving when it is no longer pushed, in order to realise the Newtonian story, we have to realise a further closure, for which we have no sensory evidence, that an unseen force stops the cart with as much force as we used pushing it.

We can realise the Aristotelian and the Newtonian notion of force and the systems of closure in which they are embedded. As Paul Feyerabend argued in *Against Method*,[11] both the Aristotelian and Newtonian stories can account for motion. Each account requires further closures in order to defend the story in the face of openness. The underlying weakness of the Aristotelian system is that since a force on its own is not sufficient to account for change, a separate notion of a tendency to rest must also be added. The consequence of this is that it is no longer possible to determine precise mathematical relationships because change is a product of the force and the tendency to rest, the latter having no precise means of being quantified in all circumstances. It can be seen therefore that the success of the Newtonian system does not stem from it having uncovered the essential workings of the universe, but from having defined precise mathematical relationships between its central concepts all of which have the important characteristic of being similar to the logical variable 'x'. For while the notion of force appears to be a closure like any other, it is however as close to the abstract variable 'x' as one can get while still having content.

It has been argued that the abstract variable 'x' assumes the abstract possibility of closure and allows us to operate with that possibility. It is possible to derive theorems that relate such variables because there is no texture held in the symbol, the general possibility of closure being assumed. Similarly in the case of the concept 'force' we can find no thing in the world to which it applies.[12] We are not therefore in a position to discover the failure of closure or offer an alternative. Like 'x', the mark 'force' assumes that complete closure is realisable. The Newtonian notion of force is akin to an abstract variable because it has no sensory content.

We do not witness a force, nor is there any sensory evidence for a force – other than the outcome of the force. No matter how closely we examine a force, and no matter how finely the world is differentiated we still come no closer to the identification of a something which can be held as a force. The same principle applies to space, or time. We can provide ever finer differentiation of space and time, resulting in ever smaller distances and periods of time, but we come no closer to the identification of a something that is space or time nor can we imagine what it would be to do so. These closures are almost entirely abstract – so that when the closures are realised there is no residual texture which might offer a threat to the theory. If for example we ask what force or mass, space or time, are made of, we can give little response. For these are closures that have largely succeeded in eradicating texture. As soon as it begins to look as if texture might be provided the closures are at once under threat. If an attempt is made to give an account of force in terms of an interchange of particles, or the passage of something to something else, the closure is at once open to challenge. What makes the particle move from one to the other if not another force? If scientists discovered a graviton, a sub-graviton would be required to explain the movement of the graviton. The strength of the notion is in its very abstraction and avoidance of texture. The same case can be argued for each of the core closures of the Newtonian system.

Moreover the system of definitions that Newton sets up allows for the maintenance of the claimed relations between the closures irrespective of the empirical findings. If the observed movement of a physical object does not accord with that predicted by the theory given the observed force that has been applied, the theory allows for the presupposition of another force, so far unidentified, that is responsible for the outcome. By this means Newton's laws are always capable of being defended with relative ease. Suppose, for example, in the case of the cart, that we push on the cart and it doesn't move. Such a result is a challenge to the predicted outcome that the rate of change of movement is proportional to the force. However, Newton's system at once allows the further closure that another force of precisely the same degree – for which once again we have no evidence – is acting in the opposite direction to our own force. Even though we can find no grounds for supposing such a force to be acting the pattern of Newtonian closures is self-sustaining. The Newtonian system thus ensures that, in the event of an outcome contrary to the one at first predicted by the theory, a further force can always be supposed, which like the other forces is not independently detectable and thus secure, through which the theory can be maintained. As a result the

Newtonian system can consistently account for any outcome irrespective of the actual behaviour of matter. The only downside of the system being the potential proliferation of separate and different forces, none of which can be explained or observed directly.

The empty content of the Newtonian closure 'force' and the other core closures of the system, and the self-sustaining character of the system of definitions, not surprisingly led some early critics of Newton to regard his system as absurd.[13] Understandably they argued that if to any counter-example to the theory it was proposed that another force was operating which precisely accounted for the error, little was being advanced. It was one thing to argue that an apple falls to the ground because of gravity – a force that cannot be observed in any way other than to observe the outcome – it is another to argue that until it falls it is held in place by a precisely equal and opposite force that also cannot be observed. What Newton's critics did not realise however was that the strength of the Newtonian system is precisely due to the ease with which the system could account for any outcome due to the ephemeral character of the notion of force which is almost empty and without texture. For it is this circularity which allows the core closures of the system to be held in nearly all circumstances and thus to appear to be ideally true. As a consequence the system of closures gradually built through empirical observation on this secure foundation is able to modify and refine its account of the observed forces of the natural world so as to enable ever more effective intervention without bringing into question the starting point.

As a result we can conclude that the accuracy of the Newtonian system in certain specific circumstances need not lead us to imagine that it has thereby uncovered the essential character of the world. The accuracy can be seen as a product of the application of the initial framework of closure which then produces results in accordance with the logical system of which they are a part. If the correct result is not forthcoming the framework has a ready-made explanation: another force is acting, and accordingly a new closure will be made in defence of the overall account. For some purposes this system of closures is very powerful, for others it is hopelessly complex. Newtonian mechanics is very good at determining the movements of the planets or the speed an apple will fall to the ground – in ideal circumstances, which are those that approximate as far as possible to complete closure. It is quite useless at predicting when the apple will fall to the ground. A farmer who grows apples is likely to have a set of stories that provides a more accurate prediction of when the apples are liable to fall than Newtonian physics. If the success of Newton's theory was dependent on convincing apple-growers of its capacity to enable effective intervention it would have long been forgotten. As with all closure therefore Newton's theory enables certain types of intervention and not others. What makes the system of closure unusual is that the core closures are sufficiently abstract for them to have strict mathematical relations, with the consequence that where the closures can be realised the mathematics can also be applied. We should therefore be no more surprised by the ability of Newtonian physics to precisely determine the behaviour of, say, rockets in space, than we are by the ability of arithmetic to precisely determine the amount in a bank account after a deposit. In neither case should this be put down to having uncovered the character of openness, but rather to a system of abstract closures whose logical relations are determined in advance and which are then applicable when the closures are realised.

If Newton's physics is self-sustaining in this manner, the question arises as to why it has been replaced with Einsteinian physics. For it would appear that Newton's theory cannot have been disproved by the evidence because it could always have been modified to take account of the evidence. In order to account for its abandonment we need therefore to look for constraints on the theory. These are not to be found in the relationships between the core closures for such relations can always be defended, but in the requirement that the closures are capable of being realised. The reason that Newtonian physics has been superseded by Einsteinian physics is not empirically based but is due to the abstract relationship between the theoretically complete closures of Newton's system and the consequences this has for their realisation. Newton's theory assumed that the core closures, force, mass, distance, and time, could be independently measured and would have the same quantitative value from any standpoint. If, to use Einstein's example, the distance a train covered in a given period of time varied depending on how the distance and time were measured, the Newtonian system would rapidly collapse for the mathematical relationships between its central terms would no longer hold. This was what took place when the assumption that the speed of light was infinite was called into question. Einstein retained the basic structure of Newtonian physics, but adopted the assumption that the speed of light was not infinite and therefore that the determination of the absolute level of force, distance, mass or time was relative.[14] Although the measurement of the speed of light and the Michelson Morley experiment no doubt played a part in encouraging Einstein to develop his theory, the shift from Newtonian to Einsteinian physics is not evidence of the response of the theory to observed outcomes, as if the failure of Newtonian physics was clearly apparent in the light of the observed behaviour of matter, but instead can be seen to be due to a modification of the closures to allow for their realisation. The modification was able to take place because Newton's closures were not in fact entirely empty and lacking in texture. His closures of 'distance' and 'time' incorporated a notion of measurement which allowed an absolute determination of their magnitude. Einstein's modification was possible because of this implicit texture. The strength of the Newtonian system is however due to its nearly having succeeded in eradicating texture and thus enabling a framework that was self-sustaining and capable of continuous development.

Can we imagine therefore that the laws of motion could have been defined differently? The logic of the relationship between force, mass, distance, and time is already constrained by the way these closures are used in everyday speech, the possible logical relations that might hold between these elements is thus limited. Other formulations of the laws of motion within this framework are possible – as previously mentioned Feyerabend has indicated how we could operate with the Aristotelian framework[15] – but whether these formulations might prove to be more effective in handling activity cannot be decided in advance. Given the characteristics of closure it could be predicted that no other system of closure will directly reflect the strengths of the Newtonian system and its capacity to enable intervention,[16] but in different cultural spaces there is every reason to believe that systems of closure are possible which rely on logical relations and which would enable similar and possibly more powerful types of intervention. We can however predict that any other set of closures would need to have the strengths of the Newtonian system. The initial core closures will need to be abstract and as devoid of texture as possible; the system of closure will need to be self-sustaining so that any outcomes contrary to those predicted are easily accounted for in a manner that is not capable of being undermined; and the mathematical relations should be simple so that outcomes are determinable.

The ability of science in general to describe the world can be understood in this light. Science can be seen therefore to progress by employing the same technique that operates in the case of everyday linguistic closure. In the face of failure, new closure is adopted that provides a defence by enabling the maintenance of the overall theory. If mathematical relationships between a set of abstract closures are defined, so long as the abstract closures can themselves be realised it will also be possible to realise the outcome predicted by the theory, providing that it is possible to respond to any threat to such a realisation in a manner which sustains the overall story or theory. In practice, the realisation of each of these closures results in the provision of residual texture which potentially undermines both the specific closures and the theory in general. A continuous defence of the closures is therefore required to sustain the theory without damage. An established theory however can be seen to operate on the basis that such defences are possible and therefore threats to the theory are often ignored. It is for this reason that paradigm shifts, in the sense identified by Thomas Kuhn,[17] can take place. Within a certain paradigm it is assumed that the completion of closure is possible. At some point an alternative story is offered which calls into question the assumption of the completion of closure and draws attention to 'facts', which are now in the light of the new story identified as being important. These 'facts' may run counter to the initial paradigm and were previously ignored on the grounds that a defence could be found if it was pursued, or were perhaps not even identifiable in the context of the prior story. The theories of science are not therefore tested against openness, but against a reality which is itself the product of the closures of cultural space that incorporate the current theories and beliefs of science. Science is not therefore a search for the true theory or even a more accurate theory than the one currently available, since any theory, or closure can be defended if sufficient additions are made by way of new closures or alterations to the current structure of space. It is rather the search for a set of closures that enables us to intervene with success.

In passing it can be noted that the Darwinist scientist, Richard Dawkins, has proposed that concepts can be treated as memes that compete in the same manner as genes.[18] In this respect, closures function like Dawkins' memes. For the success of a meme is the consequence of survival, the effectiveness of the strategy with regard to the maintenance of the system, and not truth. However, although the success of memes is driven by their capacity to survive and not their truth, Dawkins appears to wish to retain the notion that the pursuit of science enables the provision of a true, or more modestly, a 'more true' theory. If Dawkins' theory sought also to take account of self-reference it would be necessary to abandon the implicit realism which underlies its defence of science as the pursuit of truth, and in doing so would approach a similar starting point to the theory of closure.

In this chapter therefore I have sought to show that although closure has nothing in common with openness and although the process of closure is at first a random process it is nevertheless capable of enabling the handling of activity and thereby proving useful. The precision with which this occurs in the application of mathematics in science should not lead us to assume that our closures have thereby captured some real aspect of openness, be it in content or form. Instead this precision can itself be seen to be the outcome of the closures themselves whose logical relations are embedded in their initial realisation.

11

WHAT IS THE WORLD MADE OF?

> Science no more uncovers the true nature of the world than
> day-to-day description, and its closures are subject to the
> same pressures as less precise and defined terminology.

I have argued that the structure of the closures of cultural space is determined by the mechanisms operational within the process of closure and
are not the outcome of the character of openness. The theories of science
are part of that system of closure and as a result are extended, maintained,
and defended in a manner that can in part be predicted independently of
the outcome of experiments which are supposedly designed to test the
theories concerned.

In order to illustrate the way in which the process of closure influences
the character of the stories of science I shall take as an example a closure
central to contemporary science and to which passing reference has
already been made. The history of this closure and its application spans
two millennia, but it will be argued that the pattern of its use and development, and the theories with which it has been associated, are implicit
from the outset. The pattern that I shall seek to bring to the fore follows a
similar structure to that which has been identified in science generally. The
closure concerned is associated with the linguistic mark 'atom', and once
served to provide an answer to the question: 'What is the world made of?'
It will be argued that the closure was able to serve this function so long as
it was able to remain abstract and lacking in texture, but, once its realisation provided material and texture a point came when the texture
exhibited by the closure undermined the closure itself.

It was in the mid-to-late fifth century BC that Leucippus and Democritus
are believed to have provided the initial use of the mark 'atom' in a manner
that would in some respects be familiar to us. Since no works by these
figures remain, we are dependent on references to their writings elsewhere, the primary source being Aristotle and his student Theophrastus.[1]

Since Aristotle and Theophrastus had their own and conflicting philosophical perspective any conjectures regarding the initial employment of the mark needs to be treated with caution. Leucippus is a more shadowy figure than Democritus, and some have even suggested he did not exist. The basic tenets however of Democritus' atomic theory are widely agreed.

Democritus proposed that everything is made of atoms, and that these atoms were indivisible and the same all the way through. Atoms were all made of the same 'stuff' and come in different shapes and sizes. Between the atoms was a void. The closure Democritus associated with 'atom' is thus remarkably similar to the contemporary closure. The similarity is not however to be explained as a chance guess by Democritus turning out to be empirically valid. Democritus' atomic story was a largely abstract closure that functioned to contain diversity and move towards complete closure. Its formulation was not accidental but designed to achieve a goal. In so far as the contemporary closure is concerned to achieve the same goal it should not be surprising that it has a similar character.

Since Democritus had no immediate evidence for his atoms, one can reasonably presume that they provided an organisational function and a defence of other closures. The accepted explanation is that atoms provided a means of solving the paradox of Democritus' contemporaries in ancient Greece, which held the necessity of the One but which was contradicted by the Many of the senses.[2] In order to solve this paradox and thus provide a stable closure, the atoms had on the one hand to be all the same thus providing the One and yet account for the variety encountered in reality, thereby explaining the Many. Hence the proposal that atoms are all made of the same stuff but come in different shapes and sizes. Each atom must itself be indivisible and the same all the way through, it must in Democritus' description be 'full'. Such a closure, while helping to overcome the paradox of the One and the Many, at once raises problems when it comes to explaining movement. For if atoms are the same all the way through and thus solid and contiguous, motion would involve the movement of one atom into the position of another: everything must therefore either be static or moving together in a linear or circular direction. As a result Democritus proposed the addition of a void between the atoms into which they can move. He also maintained that the atoms are initially moving for without this starting point motion would never get going.

Democritus' atom and the accompanying theory thereby aimed to solve the paradox of the One and the Many, explain the possibility of movement, and give an account of what the world is made. As a higher-level formal closure the atomic theory of Democritus has the advantage that it moves in the direction of completing closure by enabling the diversity of things to be the outcome of a single thing. Its attraction in the context of closure is not difficult therefore to appreciate. In the light of the characteristics of closure that have been described however it could be predicted that each of the elements of Democritus' atomic story will fail under scrutiny and that each of these failures will require further realisation of closure if the story is to be retained. With the benefit of hindsight the specific detail can be provided that serves to illustrate these inevitable structural consequences.

Historically speaking perhaps the problem that was thought to be most immediately apparent was the notion of the void, a something which is at once nothing. The paradox of the void was traditionally regarded as a logical paradox that proposed that what is not, must also be. In the context of Newtonian science the solution to the puzzle consisted in making a distinction between space and matter, so that the void consisted of space but without containing any matter. The problem with this solution being that the character of space is no more comprehensible than the character of the void. Newton's absolute space has existence, but it remains unclear in what this existence consists or how it is to be conceived. In the context of post-Newtonian physics, Newton's absolute space has been abandoned but the character of space remains unattainable. If the void proposes a circumstance in which no closure can be realised, such a proposal is indeed paradoxical, for the void is itself a closure and as with all closures contains texture that is available to further closure. Closures can always be found — even in the void — if we look closely enough: as contemporary physics has demonstrated with the profusion of effects and particles that penetrate the emptiest vacuum or deepest space, an outcome which appears contingent but which, as it has been argued earlier,[3] is merely the product of the character of closure. In this sense Democritus' void is indeed problematic.[4] Further elaboration whereby the character of the void is defined as the absence of material objects, rather than the absence of realisable closure in general, might in a preliminary manner overcome some of these difficulties, although it could be predicted that they will re-emerge on closer examination of the notion of the closure associated with the tag 'material objects'.

The failure of Democritus' closure is not however limited to problems with the notion of the void, but is quickly found in the atom itself. All practical closure is capable of further differentiation but if the atom can be further differentiated it no longer functions to achieve the unifying role for which it is designed. As a consequence atoms are forced to have the character of an abstract closure which thereby avoids texture. If

Democritus' atoms were not purely abstract they would like all other things be divisible and themselves consist of further things. For such is the character of closure. Such an outcome would at once undermine the function of this particular closure. Democritus therefore places atoms beyond any possible practical closure by making them invisible and defines them as indivisible.[5] These moves can be seen therefore to be necessitated by the requirement to devise a theory which is not capable of being at once undermined and which allows for a degree of security. They have the consequence of turning the atom into a hypothetical object which cannot in principle be accessed. Nor, more importantly, can we provide an explanation of what is meant by the definition.

Democritus defines the atom as 'full and solid', and wishes thereby to give the impression of a thing which is itself all the way through, and which could not as a result be made of something else. There can be no imperfection in the atom. It cannot contain any part of the void for example, for in doing so it would no longer be itself throughout and as such it could no longer be seen to make up the world since it would itself be a composition. Democritus' closure looks realisable, for we have the impression than we can comprehend the notion of a full, solid, and indivisible atom, by comparison with everyday objects. The comparison on closer inspection however can be seen to be misleading with the result that Democritus' closure cannot be realised.

A stone may be considered full, solid and indivisible, and no doubt Democritus had such an analogy in mind when describing his atoms. The indivisibility of a stone however rests on the practical problems we have in dividing it, not on its essential character. Its fullness and solidity rests on the difficulty of separating the stone into different components, but it is not a theoretical constraint. In principle the stone can be divided into any number of pieces. With the atom Democritus is proposing its character is not dependent on our capacity but is inherent to its being. Little content can however be given to this description. It is not sufficient to imagine that the indivisibility of the atom consists in the impracticality of dividing it, for we require a theoretical constraint to ensure that the world only consists of atoms. We cannot comprehend such a theoretical constraint for as soon as we imagine a thing we can imagine it divided. The same is true for the other characteristics Democritus used to describe the atom. He tells us that it is solid, but a perfectly solid thing must not be capable of being distorted and must therefore be impassable. Once again we can imagine this by analogy with everyday objects, but we cannot imagine an object which by definition cannot be distorted, for once a thing is imagined it can be imagined as distorted. Solidity and distortion are relative notions.

Something is solid because it cannot be cut by something else. By extension it looks as if we can propose an object which cannot be cut or distorted by anything else, but we cannot give an account of what characteristic the object would have to have for this to be in principle impossible.

The atom of Democritus sets itself up therefore as a necessarily abstract formal closure being 'so small as to elude our senses'.[6] In doing so it superficially evades the failure of closure but at the cost that it cannot undergo practical closure. The formal closure with which the mark is associated is achieved by sleight of hand. It is supposed that we can imagine such an object, or that we understand in what such an object would consist, even though the notion is incoherent. The atomic story proposed by Democritus had the advantage that it moved in the direction of the completion of closure, while obscuring its own failure. The strength of the story was in its capacity to contain the diversity of things within a single thing and thus seemingly to move in the direction of the completion of closure. Its weakness is that in order to make the closure sustainable it needed to eradicate texture with the consequence that the closure cannot be held as one with any part of sensory closure: it cannot be applied. One could predict therefore that as soon as practical closure was attempted further closures would be required to maintain and support such a closure, and that in due course these would eventually undermine the initial notion. The strength and weaknesses of the story have remained unchanged. What has happened is that the weaknesses have been exposed by the attempt to realise practical closure and in the process the strength of the story is now in danger of being lost.

Both Galileo and Newton toyed with the atomic hypothesis but in neither case was it used to provide a unifying closure. As a result the descriptions they gave of the atom owe much to Democritus but the failure of closure is not really tested. Galileo in *Dialogues Concerning the Two New Sciences*, describes atoms as a mathematical abstraction of points, lacking any dimension, clearly indivisible and uncuttable.[7] In this respect he rectifies the failure of closure implicit in the Democritus atom by abandoning dimension and therefore making the notion of indivisibility more comprehensible. Having solved one threat to closure he succeeds however in only introducing another. For if Galileo's atoms are to account for the variety of the world, and thus to answer the question of what the world is made, it would require that objects with dimension can be constituted from something that lacks dimension. This threat to closure would appear to be insurmountable, and it has resurfaced in a different form in contemporary physics.

Newton at one stage proposed an atomic hypothesis that at least addresses the problem of the relative character of solidity and divisibility: 'It seems probable to me that God in the beginning formed matter in solid massy hard impenetrable moveable particles ... and these primitive particles being solids ... so very hard as never to wear out or break into pieces, no ordinary power being able to divide what God himself made one in the first creation.'[8] It does not however provide a solution, for these atoms are no longer an abstract closure but are an empirical hypothesis. The proposal that they are practically indivisible does not solve the theoretical question of what they are made, or what it would be to make something the same all the way through. In any case Newton later abandoned the atomic hypothesis on the grounds that it was not compatible with force at a distance – the centrepiece of his account of the world.

It is not until Dalton therefore in the early part of the nineteenth century that the atom once again found its way into the centre of a higher-level closure. In 1808 Dalton writes: 'all bodies of sensible magnitude, whether liquid or solid, are constituted of a vast number of extremely small particles, or atoms of matter bound together by a force of attraction, which is more or less powerful according to circumstances. ... No new creation or destruction of matter is within the reach of chemical agency.'[9] Dalton was led to this proposal on the basis of his own work which had noted that chemicals combined in the same ratios and the weight of the product was equivalent to the weight of the initial elements, along with the behaviour of gases that for some time had been most easily explained on the basis of their being constituted of separate particles.

Dalton's atom has much in common with the atoms of Democritus but there is an essential difference. Instead of being the outcome of a desire to describe the ultimate constituents of the world, Dalton's atom functions to provide an explanation of specific empirical results. Dalton sought to provide a higher-level closure which would limit the profusion of closure in general, but he was less concerned to propose an endpoint. As a consequence, from Dalton onwards the mark 'atom' is not exclusively an abstract formal closure – an abstract closure being one that cannot in principle, either directly or indirectly, be applied to reality. Democritus' atom functions like the symbol 'x'. It serves as a notional endpoint which cannot be found in practice. Dalton's atom also serves as a high-level formal closure that functions to contain the diversity of lower-level closures, but it is no longer purely symbolic but is put forward as having content which is to be applied to reality. The empirical content of the Daltonian atom necessarily undermined the intent of Democritus in providing an endpoint to the question of the nature of the world, but in practice the use of the term still continued to entertain elements of this goal. Frequently therefore the notion of 'atom' has, in its post-Daltonian mode, still been used to provide the function that Democritus sought: namely to describe the ultimate elements of the world – a goal that in principle is beyond closure. In this respect the modern tag can be seen to be caught between its empirical function and its formal function, and there have been points in the development of the use of the tag when these two functions have been confused. At one stage, around the turn of the twentieth century, it was widely thought that the atom could not be split. Such a view can in retrospect be seen to have been merely a leftover from Democritus' theory: a requirement of an ultimate particle. If an ultimate particle can be split it is not an ultimate particle since the resulting pieces are more elementary. With the splitting of the Daltonian atom, the atom of Democritus was renamed first as protons, neutrons, and electrons, and in the contemporary Standard Model[10] as quarks and leptons. The old Daltonian atom is now made up of separate fundamental particles – the nucleus consisting of quarks while the electrons are a type of lepton. Dalton's atom in this context therefore does not even consist of smaller similar units, but of fundamentally different constituents. Yet it can be predicted that these new atoms are no safer than the Daltonian atom of chemistry.

The Standard Model of current sub-atomic physics has this in common with Dalton: it proposes a closure on the basis of empirical observations as a means to explain those observations. Its success will be determined by its ability to handle activity. Some of the supporters of the model give the appearance at times of also seeking to give an ultimate explanation of matter and presumably for the same reasons that Democritus himself put forward the initial theory, namely that a higher-level closure is required if we are to defend the profusion of closure in general. The closure cannot however have it both ways. If it is empirical it will have the character of all closure in being capable of further differentiation. If it is purely formal it is not a description of the world but a postulate around which to organise our other closures.

As a response to this historical summary of the application of the closure 'atom' it may be argued that it is only with hindsight that the weakness in the notion becomes apparent. However a similar argument can be made with reference to contemporary closures that have yet to be undermined, for the pattern of development is driven by the characteristics of closure. Dalton put forward his version of atomic theory in order to account for his own results along with previous experiments with gases. Although his theory provides an explanation of these results it incorporates closures using the tag 'atom' which, like all closures, can never be made safe. Whatever the results of consequent experiments it must always have been the case that Dalton's atoms would prove to be themselves composed of other things or to be themselves part of something else. For the closure can never rest, since within the circle of material texture is found which is the basis for new material. Furthermore it must have always been the case that other constituents of the world would be found. These need not have turned out to be the electrons, photons, and neutrinos of contemporary physics, but further constituents were inevitable, since Dalton's atoms were not contiguous. As a result the space between them, identified by Democritus as the void, must itself have been capable of further closure. The material generated through Dalton's atom thus creates texture both within the skin of the physical entity, within the circle of material, and outside of that entity. As a result both the inside and outside of the atom are capable of further closure.

The notion that Dalton's atom could have provided an endpoint was therefore from the outset a misunderstanding of the character of closure. Just as today the notion that physicists might uncover the ultimate constituents of matter, or describe the laws of the universe, let alone provide a complete story from the Big Bang to the present, will necessarily be seen to fail since it is at odds with the characteristics of closure. Complete closure, however narrowly targeted, has never been achieved, nor could it be achieved, nor can we imagine what would be involved in achieving it. There have always been those who in their enthusiasm for their own system of closures have suggested that the completion of closure is in some respect nearby. Such a notion is, as we have sought to demonstrate in many different contexts, an illusion. For not only is the system of closure as a whole not nearing completion, but under scrutiny every element within the system of closure fails and requires further closure in its defence. At the same time this chance set of closures is but one in a universe of countless alternatives. To declare that completion is near is like a child who having built a house from Lego bricks insists that this is the only construction possible, and the bricks are the only possible building material. If those who promote the Standard Model believe, as some of them appear to do,[11] that we are on the verge of the discovery of the essential constituents of matter, once and for all, they are surely deluding themselves as Dalton and Democritus did before them.

In the context of the characteristics of closure there are two further aspects that are illustrated from the history of the use of the mark 'atom'.

These concern the ephemeral character of material. For the circle of material realised by closure is precariously balanced between the openness of the texture found within material and the openness of space surrounding it. The process has been described whereby the search for closure results in a filling out of the texture held within the initial material. This process is in part a defence of current closure, but with the provision of additional material the contribution of the original closure becomes less clear. For the circle of material provided by a closure has in itself no substance, it being texture that provides material with content. So it is that with the provision of new material from texture the initial material is in part undermined. There is a further threat to the circle of material which comes not from the provision of new closures within texture, but from shifts in the surrounding space. Both of these consequences are illustrated in the manner in which the notion of the atom developed in the wake of Dalton's initial closure.

Once the atom was taken to be an actual thing, as opposed to an hypothesis, it could be divided and in due course would be found to consist of other things. However, with the realisation of protons, neutrons and electrons it becomes less clear in what the atom actually consists. If the atom is completely accounted for by a combination of these elements, an elementary form of Occam's razor applies, and the entity implied by the tag 'atom' abandoned altogether. Even as a term for a collection of more elementary things the closure is under threat. What collection of things is to be defined as an atom? It is not as easy as defining a particular atom as a precise combination of protons, neutrons, and electrons, since most atoms are found in combination with others in which case their electron fields are no longer the same. In practice, the tag 'atom' has the characteristics of other everyday tags in that it incorporates a variety of alternative closures. 'Atom' is used sometimes to refer to the nucleus alone, sometimes to the nucleus and the electrons, sometimes to something independent, sometimes to a combination.[12] Given the mechanism of linguistic closure it can be predicted that it is not possible to make the tag precise. If an attempt is made to do so it is likely to result in damage to the way the tag is currently applied, and in consequence result in a negative impact on the ability to handle activity. Scientists get by without needing to fix the closure, in the same way that we all get by with everyday tags without ever knowing quite what we mean.

The alternative threat to the material realised by Dalton's closure comes not from the texture contained within the circle of material but from the character of the surrounding space. For example, the widespread acceptance of Einstein's theoretical framework changed the character of cultural space within which the closure 'atom' was to be found. Einstein's theories allowed for the replacement of a universe of matter with a universe of energy or events. For if mass and energy are interchangeable, energy is as much a thing as a particle. Consequently the notion that everything is made of Daltonian atoms, or contemporary sub-atomic particles, begins to look rather anachronistic. As with all closure therefore the Daltonian atom can be undermined not only from the texture held within the material realised by the closure, but by shifts in the character of the surrounding space resulting in the possibility of the provision of an alternative closure altogether.

It can be seen therefore that the contemporary use of the tag 'atom' owes in part to the empiricism of Dalton, and the rationalism of Democritus. While it is no longer the case that physicists hold atoms to be the fundamental building blocks of matter, and look instead to quarks and leptons, the intention of atomic theory to account for the ultimate constituents of matter is still to some extent retained in the new terms. Empiricism encouraged a recognition of the failure of the Daltonian closure by bringing to the fore the possibility of differentiating the texture held within the circle of material. In this respect it merely highlighted the weaknesses in the initial closure. The original intention however of Democritus to provide a higher-level closure and thereby contain the proliferation of things in the world is nevertheless maintained in the continued search for the ultimate constituents of matter. Once again therefore the new closures are employed in pursuit of an impossible combination of goals.

Like Dalton's atom, quarks and leptons find themselves providing a dual role. On the one hand they are the response to the application of the current closures of science and seek to provide a means to limit the profusion of outcomes. In this respect they serve the function of an organising empirical closure. On the other hand they seek to provide a final answer to the question of the constituents of the world. Given the characteristics of closure one can predict that the two functions of the new closures will face the same threats as those that brought down the Daltonian atom. While the nature of the texture associated with these closures is different from that enclosed by the material realised from the tag 'atom', it will nevertheless be capable of further closure. This need not echo the division of the atom into smaller particles it may come in some other form, but it cannot be avoided. The bundle of energy which constitutes the contemporary particle need have no more unity than matter. What is inside this bundle? Where, or when, does it begin and end? The texture held within the closures, 'quark' and 'lepton' is extensive. The smallness and strange character of these particles, some with no mass, some with no dimension, for example, does not mean that the underlying problems of the Daltonian atom can be evaded. Our capacity to explore these closures is limited and as a result the texture held within the material is temporarily obscured. In this respect the current closures are in part protected by an abstract formal character. However, although they function in part at an abstract formal level they are also put forward as an empirical theory. In this respect means are likely to be found to explore these closures, and the more they are explored the more new material will be generated and the closures themselves be threatened. Meanwhile the framework of space in which the closures operate will continue to evolve and have its own impact.

A summary history of the tag 'atom' has been used in order to illustrate the manner in which the character of closure influences the development of science but equally it would have been possible to have chosen other scientific terms. There is of course nothing particularly special about the closure associated with the tag 'atom'. In a similar manner it could be demonstrated that physical space, for example, is not limitless because the universe is just made like that, as if reality simply happened to be infinite, as if empirical observation might uncover a boundary. Instead, its character in this respect can be seen to flow from the nature of closure. 'Physical space' seeks to realise a closure for the world as a whole. The closure associated with the mark 'physical space' seeks to realise a framework within which everything takes place. In so doing it faces the same constraints as all closure, but in this case because the closure seeks to realise a framework for the world as a whole its inability to provide a complete closure is made more immediately apparent. As a closure, 'physical space' is held as this one thing, yet any thing has limits for otherwise it is not one, nor is it a thing. Physical space must have a boundary if it is to realised as a closure, yet it cannot have boundary if it is to encompass openness. The paradox of space, to which Kant drew attention in his *Antimonies*,[13] can be seen therefore as a consequence of the inevitable failure of closure which can be observed in all closures but which in the case of attempts to describe the world as a whole is brought to the fore. Some have argued that we can escape the failure of the closure 'physical space' through the notion of infinity, but as it can also be seen in the context of Zeno's paradoxes,[14] the importation of infinity serves to hide the failure of closure in a mathematical term which incorporates that failure. As a result the failure of closure is not evaded, although it does allow us to imagine that the matter has been dealt with so that it can be put to rest.

It can be concluded therefore that as with the remainder of personal and cultural space the stories of science are driven by the search for closure. Since the marks of science and its closures are subject to the same pressures as less precise and defined terminology, they also exhibit the characteristics of closure. As a result there is no reason to believe that science uncovers the true nature of the world any more than day-to-day description.

12

STRATEGIES FOR CLOSURE

> There can be no strategy for the provision of ideal closures, but we can discern strategies that increase our potential to intervene.

There is no reason to suppose that there are, in principle, limits to the capacity for closure to enable intervention, and thus no limits to what can be achieved through intervention. The question that will now be addressed is whether there are strategies that can be employed for realising closures which enable effective intervention and a desired outcome. In a realist framework the success of a theory is a function of the world as an independent reality and the accuracy with which the theory describes that world. In the context of closure, the success of a theory is no longer driven by whether it accurately reflects an independent reality. Might it be possible therefore to frame guidelines for closure that would enable appropriate intervention? Can we for example learn from the success of the Newtonian system, or the history of the attempt to describe the ultimate constituents of matter, as to how closures should be employed in order to achieve a particular end?

Some limits can at once be drawn to this question. Since any cultural space will realise its own closures, there can be no account of success, or effectiveness, that is independent of the particular cultural space in question. It is not possible therefore to frame rules of closure which will result in successful closure in any given circumstances. Success is a function of closure, and what is successful is the outcome of a particular space. Even closures that enable interventions that potentially bring to an end the process of closure itself, such as the physics required to build a weapon, or a handbook on euthanasia, can be regarded as successful closures within an appropriate space. Moreover, any rules that might be offered that seek to describe the operation of closure must themselves be expressed through closure and cannot therefore be understood to be applicable in all circumstances. To suppose otherwise would be to suppose that openness could be described. It is only with these constraints, and thus in the context of the account of closure that has been offered, and in the context of our current cultural space, that any possible guidelines for application of closure might be proposed.

If, given these limitations, there are to be any general principles that might apply to closure it can be presumed that they will need to follow from the character of closure itself, for only in this instance will they apply in all relevant contexts. Taking this approach it is possible to distinguish a number of such principles. The first principle can be seen to stem from the ability of any closure to enable intervention. While we cannot know in advance whether a closure will result in an intervention which we desire, or in an intervention which will prove useful or successful, each closure does however increase the possibilities to intervene. It can be concluded therefore that given any cultural space, additional closure will increase the potential for intervention. The increased potential for intervention does not mean that in the context of a particular cultural space the realisation of additional closure will result in a greater capacity to achieve a desired end. The profusion of closure might obscure the realisation of an appropriate closure to the task in hand. Nevertheless the potential for intervention is increased by the realisation of closure.

Closure realises material, and the provision of material can be regarded as increasing the density of what can be referred to as 'the geography' of space. This is most easily identified in the context of physical space where the greater the number of closures in a given physical space, the more material and the more detailed its texture. If we look at an unknown landscape we see perhaps hills and valleys. With increasing closure further details are provided, such as fields and farmhouses, which differentiate the space into further divisions and new patterns. Each additional closure enables a whole array of related interventions. For each thing that is realised, can be found and referred to, and used to aid the location of other things. The increase in geography that follows the provision of closure can be identified in a similar manner throughout cultural space. An individual who has particular experience of a task, someone who might be described therefore as an expert, or specialist, calls upon additional closures, which realise additional material and thus a more detailed geography than is commonplace. For example, while the closure associated with the sentence 'the car has broken down' may be useful to the driver in communicating the situation, this closure is of limited use to the mechanic who will call upon further closures that divide the initial closure 'car' into smaller and smaller pieces. Similarly, for everyday purposes we may refer simply to the roof of a house, but a builder is more likely to talk in terms of trusses and purlins, tie beams and posts. Such closures increase differentiation and improve geography. The same principles apply to closures that operate outside the framework of physical and material things. The geography of space used to describe human behaviour, or emotion, is similarly made more dense by the realisation of additional closure and in so doing enables new intervention. Suppose for example new closures were realised that distinguished between types of anger, these closures would provide new material which could then be the basis for intervention which was not previously possible. A distinction between defensive and aggressive anger might, for example, lead to an alteration in our response to those expressing such types of anger.

While the provision of a new closure increases the density of the local geography of space and thereby allows for intervention not previously available, there is no guarantee that such intervention will result in a desired outcome. There are occasions when the reverse may be the case. For example, if a landscape was highly differentiated so that each leaf and twig were realised as separate linguistic closures, it might make describing the route from one location to another more difficult. Someone who knows an area well can be distracted by the extent and detail of their closures so that directions offered to a newcomer are confused and of little use. The provision of new closure is not therefore a guaranteed basis for achieving a particular end. However, if a particular end is desired and we are unable to see how this can be achieved, the provision of further closure will at least increase the potential for intervention and from these new possibilities it may become possible to determine a course of action that will achieve the appropriate outcome. If we do not know how to proceed in a given circumstance therefore a first strategy that might be employed is to look for new closure so as to make the geography more detailed, filling out the texture of prior closures with additional material. While it cannot be predicted what will be the outcome of realising any one of these additional closures, each closure will at least increase the possibilities for intervention.

The provision of additional material as a means of enabling intervention does not only apply in the adoption of new low-level closure but throughout space in the provision of new stories. Our capacity to use new material to intervene successfully depends on our ability to choose from the closures available ones appropriate to the relevant task. This in turn requires the realisation of organising closures or stories which link together sets of lower-level closure. Linguistic closures that serve to catalogue and describe our reality are already formed into nested hierarchies simply as a consequence of the mechanism by which they have been realised. These hierarchies ensure that each closure is found in the context of a local space and is thereby linked to other closures. In order to intervene successfully we require organising stories which function as high-level closures relating diverse lower-level closure. It is through these stories, typically of the form 'in the event that closure A is realisable then closure B is also realisable', that the profusion of closure is contained and charted. The provision of new closure is thus tempered by the requirement to be able to identify organising closures that contain the diversity. Thus while increased closure improves geography its complexity is itself a possible impediment to intervention unless that complexity is itself contained within a wider closure, or story. If we imagine a world that consisted of an infinite number of low-level linguistic closures, but an absence of stories, all interventions would in principle be possible, but we would be lost in the infinity of the task – as if in a version of a Borges' story we were lost in the corridors of an infinite library seeking the solution to a problem. The discovery of a catalogue will not in itself solve the problem if the solution requires elements from a variety of the books. Instead we require a book that contains the story of how the contents of each of the books in the library are related.

The capacity to intervene successfully depends therefore not only on the density of space but on the organisation of these closures through the provision of higher-level closure in the form of stories. As with low-level closure the greater the number of stories the greater is the potential for intervention, but equally we cannot determine in advance whether a new story will enable a desired outcome, merely that it will increase the potential for intervention. The provision of a new story is no guarantee of achieving the desired end, but as with low-level closure, in a circumstance where we are unable to determine how to intervene to achieve an end, the realisation of new high-level closure increases the possibility that we might identify such a means of intervening. We can therefore identify the principle of the realisation of new closure, either in the form of increasing the density of the local geography of space, or in the provision of new organising closures, as being a strategy for increasing the potential to intervene. This first rule of closure has much in common with empiricism, for it encourages new closure and new stories which are tried, and refined or abandoned as a consequence. It differs in so far as empiricism makes it look as if the possible closures are limited and given in advance, while they are unlimited and a function of our current space.

There is a second principle which can be discerned from the structure of openness and closure that has an impact on our capacity to intervene. It concerns the stability of closure. We have seen that in the context of logic and mathematics abstract closures are realisable that are secure through the elimination of texture. When such closures are applied to that which we take to be reality, texture is generated and the predicted outcome is no longer necessary. Closures fail because the world is open. In so far as closure is able to eradicate texture it is more secure for it has removed openness. The second rule of closure can be formulated from this outcome: the more abstract a closure the more secure it will be.

At the point of realisation all closure is stable, for the realisation consists in the holding of openness as 'this thing'. It is because closure is at the point of realisation 'safe for the time being' that there is the impression that closure is complete, while in due course closure fails. The stability of closure is called into question over time and the importance of such stability is that it provides a basis for further closure and thus for the refinement of closure. When we look at a page of dots and find a face, the capacity of the closure to organise the remainder of the page depends on its relative stability. If we are able to realise further closures, such as eyes and ears, these will function to defend and maintain the closure we associate with the mark 'face'. A challenge to any lower-level closure potentially threatens the higher closure, but the interim closures can be refined accordingly in order to maintain the higher-level closure. If we cannot realise the closure associated with 'nose', we can still retain the closure 'face', by seeing it as a face without a nose. At some point the failure of lower-level closures forces the abandonment of higher-level closure, but it is more secure than the closures lower in the nested hierarchy of closure. One outcome of this second principle of closure is therefore that the greater the level of abstraction the greater the potential stability of the closure, and furthermore that the provision of additional levels of closure increases the stability of space.

Closures or sets of closure offer a way of holding openness. They provide a description of the world. Stories are a higher level of linguistic closure for they seek to provide descriptions of the relationship between closures or sets of closure. Stories offer an explanation of the world rather than a description, and by linking lower-level closures stories enable more sophisticated intervention. Stories do not therefore hold one linguistic closure as another in the form of a new linguistic closure but realise a closure from the relationship between closures. A simple story often proposes a temporal relationship between separate linguistic closures. While 'the sky is blue' proposes a closure which holds the sky as one with blue, the story 'when it is raining, it is cloudy', does not ask us to hold 'it is raining' as one with 'it is cloudy', but instead proposes a closure which consists of realising a relation between the closure 'it is raining' and the closure 'it is cloudy'.

We do not fail to realise a story because we fail to find the relation claimed between the closures involved, but because we are unable to realise the closures in the appropriate circumstances. The relation proposed by a story is not found in practical closure, for it is a relation between closures that are assumed to be complete, and it is for this reason that logic and mathematics can in so far as they deal with the supposition of complete closure provide nearly ideal truths. The failure of stories is due to our inability to realise closure, and is not due to the inaccuracy of the relationship proposed. Thus we do not realise the story 'when there is a full moon the sky is red' because we cannot realise, or find it difficult to realise, the closure 'the sky is red' in the event that we realise the closure 'there is a full moon'. Our failure to realise the story is not due to our inability to find the connection between the closures involved, for we have nowhere to look to find such a connection. The connection or relationship is a relation between closures that are taken to be complete and not a thing found in the reality realised through intersensory closure.

It follows therefore that our ability to realise a story is increased if the closures involved are themselves more secure. If the closures involved can always be realised the story will itself be capable of realisation whatever it proposes. The less texture and more abstract the closures involved the more likely it is that we will be capable of realising the story. Newton proposed the story that whenever we find a force we find a proportional change. As it has been argued we can however almost always realise a force because it is so abstract a closure that it has virtually no texture. As a result the story can also be realised. Another example of a closure which has virtually no texture is the ancient notion of 'evil spirits', and in a similar way a story involving such a closure will in almost all cases be capable of being realised. So that the story 'when disaster strikes evil spirits are present' can always be realised for there are almost no circumstances in which the closure 'evil spirits' cannot be realised, it being an almost entirely abstract closure. In a scientific age, we are inclined to imagine the evil spirits story is not capable of enabling effective intervention. While of course it does not enable the same interventions as a contemporary explanation, the culture of medieval Europe and many non-Western cultures provide evidence of the capacity for such a closure to influence the manner in which intervention is attempted, which in the context of the cultural space in question can be seen to be effective.

In the same way that the stability of closure is increased as a result of the abstraction of closure from openness, so the stability of a story is increased if it incorporates abstract closure. It can be seen therefore that a consequence of this second rule of closure is that our explanations of the world are more secure the more abstract the closures involved. The more closely we approximate to the eradication of texture, the more likely it is that the closures, whether descriptions or explanations, will be realisable. 'Force is proportional to change' and 'God made the universe' are just two examples of stories that incorporate such abstract closures that they can be maintained in almost any circumstances and are thus relatively secure.

The stability of a story does not mean that it will enable effective intervention, any more than additional closure ensures a desired outcome. It does however mean that the story can be used to organise other stories and closures and provide a fixed framework for space. Within a fixed framework the stories and closures of space can be refined and modified so that the system of closure can be improved in its capacity to intervene. It is only once the whole structure of space is ordered according to the story in question that it will be possible to ascertain how effective the closure is in enabling the achievement of desires which are themselves the outcome of the same space. Even then, since the space will have been developed in order to be self-sustaining, it will be possible for the system of closure to appear successful in its own terms.

Returning therefore to the question with which we began, namely whether it is possible to derive a strategy for the realisation of successful closure, it can be seen that the provision of new closure in order to increase the potential for intervention, and the realisation of abstract closures in order to enable a relatively secure framework, are strategies that can be employed in an attempt to find closures that yield desired outcomes, but they do not in themselves assure a desirable outcome. Nor could they, for the desirability of the outcome is a function of personal or cultural space and cannot therefore precede the realisation of closure.

Over the last few hundred years, in the context of the attempted provision of an accurate description of an independent reality, in the form of science, two strategies have been advocated in the pursuit of truth: empiricism and rationalism. In the light of the principles or rules governing closure that have been identified it becomes possible to provide an account of the extent to which the strategies of empiricism and rationalism have on the one hand been successful, and on the other have failed.

By proposing that we should examine the world and modify our theories accordingly, empiricism encourages the realisation of new closure, for an examination of the world through observation is liable to bring to the fore the failure of current closure and throw up the possibility of new closure. In this respect therefore the success of empiricism can be traced to the first rule of closure: that the provision of new closure increases the potential to intervene. Furthermore by encouraging accretion empiricism is also potentially a revolutionary strategy. For new closures may threaten the current structure of cultural space. As a result empiricism by encouraging the accretion of closure can be a means of halting the stultification of cultural space and of opposing the dominance of a framework of abstract closure that cannot otherwise be dislodged. It need not be supposed therefore that the success of empiricism is due to its uncovering of the truth through a painstaking observation of reality, but rather that its success stems from its encouragement of new closure that both increases the potential to intervene and has the potential to unsettle long established and possibly limiting higher-order closures.

While it can be concluded therefore that the success of empiricism is understandable in the light of the character of closure, it equally follows that it cannot provide the ideal truth that some of its advocates have proposed. Empiricism cannot enable us to uncover the facts let alone provide a theory to account for these facts. For this requires the realisation of closures without texture. The only closures that have this quality are abstract closures that cannot be applied in the context of openness. Experiments do not therefore yield unquestionable results, for the outcome can undergo an unlimited number of alternative closures. Experiments fail to dictate a particular closure or set of closures, and it is always possible for an experiment to be made consistent with any previous closure, either by a modification of the closures involved or the reorganisation of space. Much of the work in the history of science over the last thirty years has served to illustrate these conclusions which can be seen to flow from the characteristics of closure.[1]

An illustration of the inability of empiricism to dictate an outcome is demonstrable in the capacity of the Vatican to avoid the adoption of Galileo's closures for four centuries. There is no reason to suppose that it could not still do so today and at the same time retain a workable framework of closure. For, in the context of closure, there can be no facts that are not disputable and that cannot in principle be reformulated to comply with the higher-level closures realised by members of the Christian church. The abandonment of Aristotle in favour of Copernicus is not therefore the result of the presentation of undeniable facts uncovered as the result of experimentation. Rather it is the result of the accretion of new closures, some of which are adopted blindly, that offer the possibility of new higher-level closure, or stories. These new stories and explanations of the world are not unambiguously more effective, but they do make possible some interventions that were not previously available. Historically, some people found the new closures powerful, others found them unattractive. The gradual adoption of the Copernican system was a result therefore of the influence of those individuals for whom the closure was, on balance, preferable.[2] This circumstance now encompasses almost everyone.

Turning to rationalism, it can be seen that, as with empiricism, its success stems from the character of closure and from the second principle governing the application of closure. Rationalism in the strong sense – that the true nature of the world can be uncovered through the application of logic alone – is not a currently fashionable view. In the more modest sense that any successful account of the world must be consistent it is however widely held. In either form rationalism can be regarded as an attempt to emulate the success of logic and mathematics by seeking to apply similar principles to the closures of ordinary language. The success of rationalism follows therefore from the second rule of closure for it is the move to abstraction that allows a system of closure to approach stability and in so doing enables the system to be refined and improved over time – mathematics and logic being a special limiting case in which security is largely achieved.

While the success of rationalism stems from the eradication of texture, the failure of the great rationalist projects to uncover the nature of the world from logic alone stems from the reintroduction of texture when such abstract closures are applied to openness. While therefore it is possible to build systems of logic and mathematics that approximate to being entirely secure, we cannot derive from any such systems closures that can be realised in the context of openness without reintroducing texture. As it has been argued, while deductions within a system of logic generate incontestable results, no deduction can be applied to the world with similarly incontestable results. Even the simplest deductions will result in conclusions that do not carry the certainty and necessity we assume of the logical system itself. In the context of propositional or predicate calculus one way of explaining this outcome is to say that the initial propositions are not capable of having a determinate truth value. Propositions are both true and false, since they are at one and the same time a closure and a failure of closure, and from such a starting point any conclusion can be drawn. As a result there can in the context of linguistic closure be no certain deduction, for it will rest on the erroneous assumption of the completion of closure. If for no other reason, attempts to provide a purely rational system of knowledge can make no headway.

Like empiricism therefore, rationalism is not capable of delivering ideal truths, other than those that have eradicated texture and therefore cannot be applied to openness. Like empiricism also however, the strength of rationalism stems from its role as a strategy that can be directed towards the provision of closures that enable desired interventions. Although therefore the principles governing closure that have been identified do not have the consequence that a combination of empiricism and rationalism will bring us closer to the truth, they do have the consequence that empiricism and rationalism are strategies that can be employed in the search for means to achieve a desired intervention – at least in so far as empiricism and rationalism can be seen to equate to the principles governing closure that have been identified.

While therefore there can be no strategy for arriving at an ideal set of closures for no such ideal set of closures is possible; and while there can be no strategy for realising effective closure for effectiveness is an outcome of the particular space that is employed; it is possible to discern some general strategies that can be used when a desired intervention is sought. The realisation of new closure on the one hand, and the move to abstraction on the other, will not in themselves deliver the desired intervention, but they can be employed to increase the potential for intervention and provide for stability and thereby allow for the possibility of building a system of closure which is self-sustaining and capable of continual refinement. The outcome of such a system of closure is that it is more likely to yield closures that enable desired interventions.

13

THE CLOSURE OF 'CLOSURE'

> The process of closure is responsible for the pattern of life
> and the desire for knowledge and as such is equally respon-
> sible for the theory of closure itself outlined here.

It has been argued that the process of closure is responsible for the realisation of the things that make up the material world, and the provision of
subjectivity and thus experience. In due course it influences the character
of personal and cultural space and can be consciously pursued as a search
or desire for closure. The desire for closure for example accounts for the
excitement we find in the new, and the fear we have of the unknown. The
process of closure realises thus both the content and the structure of
reality, and the stories that we use to contain and order this reality, thereby
enabling precise intervention in openness in pursuit of specific desired
ends.

The story that has been offered of the mechanism by which the process
of closure is responsible for the form and character of knowledge must
itself be a product of that process, and must itself share the limitations of
the closures it has sought to describe. It could not be the case that the
theory of closure set out in this book itself manages to avoid the characteristics of closure that it has outlined. It does not therefore pretend to be a
description of openness, or a final and true theory of the world. It is a
story, a closure that is itself the outcome of a linked set of closures, that
offers a way of holding the world – and in the context of closure it could
not be otherwise. So it must be that the closures internal to the theory,
material, texture, and so forth, exhibit the characteristics that they themselves describe. They will, for example, themselves prove to be incomplete
and will generate their own texture which requires further closure.

The theory of closure is reflexive, like the theories examined in the
Prologue, but it is formulated so that the reflexivity of the theory of
closure is not self-destructive. The non-realist philosophies examined in

the Prologue appeared self-destructive because if the non-realist claim was taken seriously it was no longer clear how that claim could itself be made. If 'we are trapped in a language game' it is not apparent how this can be expressed from within the language game for it appears to be a claim about our underlying metaphysical circumstances and yet precisely because we are trapped there is no means of expressing these circumstances. The theory of closure aims to move on from the circles of self-referential paradox by providing an account of the means by which intervention and understanding are possible even though language, and closure in general, does not picture or map openness. The non-realism of the theory of closure is not therefore something that has to be hinted at, or mysteriously pointed towards; it is not something that has to be expressed poetically, or somehow shown in the structure of the text. Instead it openly seeks to provide a theory which enables us to account for our ability to intervene in and understand the world. From the vantage point of this theory it becomes apparent how it is that the theory – along with all other theories both scientific and non-scientific – is able to say what it does say and yet not be able to describe openness directly. The description of human experience as the outcome of a hierarchy of closures, and the account given of intersensory closure and linguistic closure in particular, are the means by which this is achieved. The non-realism of the theory of closure is not expressed therefore in opposition to realism. It does not as a consequence require an implicit realism to make sense of its non-realist stance, as the non-realist philosophies in the Prologue might be considered as doing, and it is for this reason that it escapes the self-referential paradoxes that typify the contemporary predicament. In this context the non-realism of the philosophies identified in the Prologue, rather than being simply mistaken, can be seen to require an account of how they are possible; an account which the theory of closure aims to provide.

The theory of closure can state its position because it gives an account of what it is to state a position. While no statement, no claim of any sort, is able to identify that which is not closure, nevertheless such statements and claims enable us to intervene in and understand our circumstances. The theory of closure offers itself therefore as a way to hold the world. It does not thereby claim to have uncovered the structure of the world. In doing so it is not denying the possibility of truth from a perspective of truth, for it seeks no such authority. Instead the theory itself sets out to show how a framework that does not rely on truth can even so account for our understanding of the world and our ability to intervene in it. The theory is thus an explanation of itself and how it is possible, while being at the same time an account of its limitations.

Each additional element in the theory of closure is for these reasons in part a description of its own possibility. Material and texture are for example terms introduced to explicate the character of closure, in the process they are also an explanation of themselves. For the closure 'closure' must itself generate material and texture, and that texture can be filled with new material; in this case the material provided by the closure 'material'. It can be seen therefore that the theory of closure has a hidden constraint that is throughout determining the character of the theory, which in the terminology that the theory rejects might be described as its capacity to map onto itself. The account of closure that has been proposed however denies a mapping relationship and instead proposes something that might be thought to be akin to a metaphorical relationship. The story of closure is in this sense a metaphor for a theory of metaphors. To describe it as a metaphor however is to carry with it the baggage of the everyday use of the term 'metaphor' which places it in a realist context. The notion of a metaphor is at once degenerate on the literal, suggesting imprecision and fuzziness. Yet the story of closure aims to be exact, in no different a manner than the theories and laws of science, even if it and the theories of science are unable to live up to their goal. It is to escape from these familiar paradoxes that the story of closure introduces a new terminology, a terminology that in due course aims to be able to account for itself and its capacity to be effective. In this context it can be said that the theory of closure is itself a set of closures that seeks to describe the relationship between closure and openness. While such a description is itself a closure, it is not thereby undermined by its incapacity to claim that it is an accurate description of an independent reality, for the reasons that the theory itself has sought to demonstrate.

While the theory of closure does not claim to be an accurate description of an independent reality, it is not therefore weakened by this admission. For there are no theories that can legitimately claim to be able to provide a description of openness – the theories and laws of science being no exception. Indeed, it is only by recognising the limitations of our closures that it is possible to frame them in such a manner that they are no longer self-destructive. In the Prologue it was shown that the assumption that our theories might provide a true account of the world is one that from the outset leads to paradox. The theory of closure accepts its inability to provide a true description of openness but at the same time provides an account that, despite such limitation, enables us to explicate why our closures are capable of enabling precise intervention in that which we take to be the world. It does not seek to deny its claims, or represent its claims as non-assertoric, for such strategies implicitly assume that an assertoric mode is itself possible, and that claims can describe how things are. As a result it provides an account not only of the means by which our theories of the world in the form of science, mathematics, and our systems of knowledge generally, are successful, but it also thereby demonstrates its own capacity to provide a general account of the nature of our circumstances.

While the theory of closure is not self-destructive or paradoxical, without the capacity to describe an independent reality it may appear to some that the account of closure makes out that the world is a figment of our imagination, that it is not real but is merely a construct in our heads. As a result there will no doubt be critics who regard the theory of closure as idealist. Such a charge is wholly misdirected, if it is thought that the theory of closure argues that the world is an invention of our own.

According to the account of closure that has been offered, openness is not realised by closure nor is it a product of the mind. Openness is independent of closure. Closure is merely the means by which it becomes possible to intervene in openness. Nor should it be supposed that the account of closure makes openness into an inaccessible other that lies outside of our experience, outside of language, as if we are cut off from the world in an irretrievable manner. For openness is only outside closure in the sense that it cannot be described by closure. Through texture openness is embedded within every closure of the system.[1] Every closure at once realises both material and texture and unlike material texture is open. In this respect therefore we are in every aspect of our perception and understanding up against openness. Openness is not some distant other but immediately present. Furthermore it is openness that constrains the realisation of future closure. Not only therefore does the world as openness exist independently of our closures but openness constrains the realisation of closure. It does not do so because the world is already differentiated into things, but because the character of openness makes it easier to realise some closures than others. Why and how this is the case lies beyond closure, but this does not thereby make the theory of closure idealist, if by such a charge it is meant that the world is an invention of our own.

Closure would not be possible without openness, and within every closure is found texture which is open. Closure may provide the form but it is openness that provides the content of the world. Moreover, while the pursuit of closure provides us with knowledge, science and mathematics, and enables us to intervene in openness to remarkable effect, it is the pursuit of openness that is in many respects more highly prized; and it is to the pursuit of openness rather than closure that the next part of the book is devoted.

Part IV

THE SEARCH FOR OPENNESS
Art, religion, and the unknown

Introduction: the search for openness and
the process of closure

> The search for openness is the search to escape the limita-
> tions of closure, and is itself the outcome of the process of
> closure.

Since openness is not a thing, or a combination of things, since it has as a
consequence no characteristic that we can point to or describe, the
concern up until now has been to demonstrate how through closure we
are able to have experience of a reality, and how moreover we can describe
and intervene in that reality. In addition it has been shown how the
process of closure, through the provision of linguistic closure and its
success in enabling intervention in the world, leads to a conscious search
for closure which can be witnessed in the lives of individuals and in
culture generally. This search for closure determines much of our
behaviour and is a central aspect of our lives.

In contrast to the search for closure, however, there is a seemingly
contrary desire that in some ways is equally significant in our lives and in
our culture, namely the search for openness. In the same way that the
search for closure has an impact throughout our culture, determining the
character of knowledge and providing the goals of our personal and social
lives, so it will be shown that the search for openness has a wholly
different but similarly profound impact on many aspects of experience and
culture.

The search for openness can be understood as a move away from
current closure and towards openness. In speaking of 'a move away' this
does not involve, nor could it involve, standing somewhere other than
closure, but consists rather in a facing outwards from current closure, in

search of an other. One way of understanding this turn away from closure is as the result of the failure of closure, and our awareness of that failure. The search for openness is itself however an aspect of the process of closure, for outside closure there is no thing. It is because the failure of closure is inherent in the character of closure that the move towards openness is also inherent in the process of closure itself.

It may seem paradoxical that a move away from closure could be the outcome of the process of closure, but elements of this move away from closure have already been implicit in the account of closure that has been given. Until this point however the concern has been to highlight the role and importance of the provision of new closure rather than to identify as part of the process the move away from previous closure; and it is this move away from prior closure that is the starting point for what in a more general context can be regarded as the search for openness.

When describing the mechanism of linguistic closure it was argued that new linguistic closure is not realised from material, but from texture. Unlike material, texture is open. Texture is not a particular thing and can be closed in an unlimited number of ways. When we come across a new mark, or a familiar mark in a new context, it is realised by seeking to hold the world in the manner of the mark. We can have the impression that this simply involves the naming of material already realised, but the realisation of a new mark always carries with it new material and that material has been realised from texture. As we have seen this is most apparent in circumstances where a new mark calls for a realisation that is very different from our current closures. In order to realise material from a new mark it is not possible simply to observe the material of our current reality and use the new mark to name some aspect of that reality for there would be nothing new in the outcome, there would be no realisation of some thing. When we point to a particular chair and call it a table, we do not simply swap one mark for another, but hold what we saw as a chair as a table. We find in the texture of the chair that which is not chair: that which can be held differently. The texture that is held within the material of our prior closures is itself the product of previous stages of closure. In the case of the chair we can observe its colour, its purpose, its constituents. In this sense we can return to the sensory closures from which the realisation of 'chair' was made possible, and in the return to these prior closures we can seek new closure which would hold them as something other.[1] These prior closures are also however the outcome of holding that which is different as the same, and can equally be held open. Sensory closure being an outcome of patterns realised from preliminary closure, which could have been realised differently, and preliminary closure being one of an unlimited number of ways of holding openness.

The realisation of a new closure therefore must at some point in the realisation of the closure have involved a move towards openness. For it is only through a move towards openness that is possible to escape the confines of prior closures. Only by returning to texture can we find that which is new. A requirement of the realisation of new closure is therefore a turning towards texture from material. Texture is not in itself openness, for it is already the outcome of prior layers of closure, but within these constraints it can be held open in the context of the next layer of closure. To realise new linguistic closure we examine the texture of our current closures. This texture is made up of other linguistic, intersensory and sensory closures. Each of these prior closures themselves contains texture and it may be that in order to realise a new linguistic closure we have to examine the texture of these prior closures as well. In so doing we move in the direction of openness from the constraint of closure. Thus having realised a face in a page of dots we move towards openness and away from this particular closure if new closures are to be realised.

It is possible to distinguish two distinct ways in which the move towards openness can occur. We can return to the closures of the dots from which the face was realised and hold them open in search of a new closure: we look once more at the page of dots and seek to find a new thing. We hold the page of dots open in search of new closure. Alternatively we may seek additional closure that provides detail to the closure 'face' that has already been realised – say, the shape of the nose, or the expression. In order to achieve these closures a move towards the openness held within the texture of the face will be required. We look at the face and seek new closure. To do so we examine the texture of the material realised by the mark 'face', which is to say we hold the page of dots in the manner of a face and seek new closure from the prior closures that were involved in the realisation of the face.

The move towards texture, and thus towards openness, can be seen therefore to be part of the process of closure and the realisation of mate-rial. However, we are not to be primarily concerned here with the provision of a more detailed account of this mechanism, but in its higher-level consequences. The search for openness with which we are concerned is not the move towards texture inherent in the provision of all new closures, but the higher-level desire to escape from the constraints of closure. This higher-level desire itself flows from the process of closure and the inherent move towards openness within that process, and just as the move towards openness is driven by the limitation or failure of closure, so the high-level desire for openness is driven by an awareness of that limitation or failure of closure. The search for closure and the search for openness are not so much therefore aspects of the process of closure but products of that process, which can be found both in individuals and across culture as a whole.

The search for closure has been described as a desire for knowledge: a desire for material in the face of texture. In contrast, the search for openness can be seen as a desire for the unknown: a desire to escape the constraints of material and approach the openness of texture. We have only to consider the response to the discovery of a complete and final description of the world to appreciate the importance of this desire. While possibly initially greeted with enthusiasm, the provision of a total theory would soon become irksome, and attempts would be made to demonstrate its lack of completeness. If it was possible to know everything, and possible to exhaust the world with our closures, there would be nothing left to discover. We would be consigned to the monotonous application of known categories. While complete closure is a goal pursued by individuals and society, and while it consequently determines the character of reality and the structure of knowledge, the provision of complete closure, were it possible, would nevertheless be the denial of much that we value.

In mathematics we come closest to complete closure and in its character we can see what it is that we desire in closure and what in our desire for openness we wish to escape. On the one hand mathematics provides us with certainty and knowledge. On the other hand the orderliness of mathematics makes it restrictive. Mathematics may be certain and it may be secure, but it has a 'what you see is what you get' quality, which makes it dry, and to those not capable of extending its closures, its very knownness has a lifeless and barren quality. The process of closure requires the interplay of openness and closure, texture and material. The search for closure is the search for material alone, but if successful it would bring the process of closure to a halt. Closure provides the frame, the material skeleton of the world, but it is the openness of texture that provides its content, and its potential. In the precision of mathematics, we can see how the provision of complete closure would eradicate openness, and all that would be left would be an empty shell: a frame without content.

As it has been shown, human desire can be regarded as an outcome of the complex layered system of closure that realises our experience. We desire openness because the move towards openness allows for the continuance of the process of closure. As a result, in the same way that the realisation of closure can be experienced as desirable, so also can we find desirable the absence of closure. It is, to use familiar examples, a feeling we can sense under the stars on a dark night, or facing out to sea as a storm develops. A sense of an immensity that cannot be captured through closure. An awareness of the inexplicable, of the failure of closure, and in that moment a sense that there is something deeper in the abandonment of closure than in the habitual framework that surrounds us. Some of our deepest feelings are thereby associated with the absence of closure. In silence and stillness, in the abandonment of the here and now. Not the moment of the closure but the moment in its wake.

The influence of the search for openness can be seen in many aspects of our culture. Three distinct stages in the desire for openness will be distinguished. Firstly, there is the desire to approach the edge of closure in search of the unknown. This is the desire that makes the mysterious enjoyable, and the edge of closure an exciting location. It is a desire that motivates intellectual and physical exploration, and influences much of popular entertainment culture. Secondly, there is the attempt to avoid closure through the presentation of marks whose realisation is then undermined. I shall argue that this is the identifying characteristic of painting, poetry, and literature and more generally of the contemporary notion of 'art'. Thirdly, there is the attempt to identify that which is beyond closure, to seek to describe openness directly. This third stage in the search for openness is typified by aspects of religion and philosophy. In these three aspects, the search for openness can be seen to be associated with what some have regarded as the highest forms of human endeavour. It is also linked to those very activities that it is often claimed separate ourselves from those of a machine. Computers have learned to play chess, an activity whose success depends on a rigid application of closure in the form of codified rules; they have yet to write poetry – an activity that is seemingly quintessentially human, and which relies at least in part on a search for openness.

The first stage in the search for openness is a direct outcome of the turn towards texture that is part of the process of closure. From the outset of our acquisition of linguistic closure, we are aware of the failure of linguistic closure to exhaust the prior closures of our perceptual world. For there is both more to the perceptual world than our linguistic closures describe, and alternative linguistic closures are possible. The identification of the failure of linguistic closure in any particular instance is typically followed by a move towards openness, in the form of texture, as a preliminary to the provision of new linguistic closure. With the development of personal space to the point of self-consciousness we are able to become aware of the value of closure in enabling intervention. As a stage in the process of closure we therefore also value the turn towards texture. This desire for openness is however a prelude to the realisation of new closure. In this sense we find the unknown attractive, and seek out what is unknown as a means to new closure. We do not in this respect desire the unknown in itself, but as a site of potential closure.

The second and third stages of the search for openness are a more complex outcome of the development of linguistic closure. For in these instances the search for openness is not a prelude to new closure, although it is still the outcome of the process of closure. These stages of the search for openness are instead the product of an identification of the failure of linguistic closure in general rather than its failure in a particular instance. The identification of the failure of closure in general is not obtained by comparing the closures that make up our experience with openness and finding them lacking, for we have no means of accessing openness. Rather it consists in the inability of linguistic closure to describe what we take to be reality; a reality that is itself the product of closure. The recognition of the limitation of closure can be seen therefore to be a recognition of the limitation of higher-level closure to describe prior and lower-level closure.

It may be helpful to give an example of the manner in which it is possible for us to identify the failure of linguistic closure as a whole. One of the types of circumstance in which the failure of linguistic closure can be made apparent is that in which our capacity to provide linguistic differentiation is small, and where as a result the inability of higher-level closure to encompass the variety of lower-level closure becomes identifiable. When we look up at the stars or out at the sea we can provide linguistic closures that seek to describe and encapsulate our experience, but aside from the stars and the night, the sea and the sky, the realisation of further linguistic closure is not perhaps immediately apparent, and as a consequence we can become aware of the gap between our capacity to provide linguistic closure and our experience of reality. We can as a result identify the general failure of closure, and realise the notion of that which is beyond not merely this closure, but linguistic closure in general. In doing so the provision of a new closure with which to secure the identification of failure is denied. The supposition of a beyond to closure is not of course an awareness of something, or some characteristic, but an encouragement to escape closure. This need not take place in explicit terms but consist in a sense of the inadequacy of language to describe the world. As a result we wonder at the world, rather than think something specific about it. Such a supposition of the general failure of closure might be considered an intuitive hunch, but it has its rigorous counterpart in the attempt to complete closure and the identification of the impossibility of doing so. An identification in which self-reference plays an important part.

In so far as the search for openness is an attempt to escape the constraints of closure it is concerned with the escape from linguistic and intersensory closure, because we have no parallel with which to compare sensory closure, and no alternative by which to see its particularity. While we can identify our inability to adequately describe the night sky, we are less likely to identify the limitation of our sensory experience. Although we can hypothesise that other animals or beings might be able to interface with the world in different ways from our own, we have few means of identifying from our own experience the limitation of our sensory closures, and no means of identifying the limitation of our preliminary closures. It is by analogy therefore that the search for openness can be extended from linguistic closure to closure in general. It is perhaps because the identification of the failure of closure comes at a relatively late stage in the provision of a system of closure that we are misguidedly inclined to regard the search for openness as being the characteristic which separates our experience from that of a machine.

The search for closure, as we have seen, is not able to realise complete closures. In a parallel manner, at each of the three stages of the search for openness the desire for openness cannot be fulfilled. The desire to move to the edge of closure is compromised by the provision of new closure. The desire to avoid closure is compromised by the requirement to offer closure in order for something to be expressed. The attempt to identify a beyond to closure is an attempt to name that which cannot be named, for the naming is at once a denial of that which it seeks to identify. In the same way that the search for closure is consistently undermined by the uncovering of its failure and the appearance of openness, so the search for openness is undermined by the emergence of closure. The search for openness and the search for closure are thus mirror images of each other. Both are sought and both are unattainable, and for the same reason. They are the product of the process of closure, and the process of closure involves the play of both closure and openness.

We desire openness because openness is inherent in the process of closure. The search for openness is the desire to escape closure, but the true abandonment of closure would involve the loss of everything we have including ourselves and our experience. The search for openness is desired because in openness there is the potential for closure and in the postponement of a particular closure we have a sense of the immensity of the closures that are possible. It is for this reason that we associate the abandonment of closure with a feeling of richness and depth. The depth however that we associate with openness is the potential for closure. So it is that in the search for openness we find closure, and in the search for closure we find openness. Those who set out with the specific intention of realising closure, such as scientists and philosophers, can therefore find themselves closest to openness, and those who seek openness directly, such as mystics or religious figures, can find themselves entrapped by closure.

In an examination of the search for openness, the following three chapters will now consider in more detail the three stages already outlined: the move towards the edge of closure; the avoidance of closure; and the naming of the unnameable. While at each stage the search for openness is already compromised by its involvement with closure, the pattern of that involvement is different as are the consequences for our personal and cultural space.

14

THE EDGE OF THE WORLD

The desire for openness is shown in a move to what is seen
to be the limit of closure, the edge of our world.

The first stage in the search for openness is found in the desire to move in
the direction of that for which we do not already have closure, to move in
the direction of the unknown. If we only desired closure we might be
expected to avoid the edges of cultural space, an exploration of which may
undermine current closure. Such a strategy, however, would if pursued
bring the process of closure to a halt around currently available closures. As
it has been argued, in one sense all closure involves a move in the direction of openness for it must have involved a turn towards texture. Yet
closure is not generally an outcome of the conscious desire for openness.
The process of closure involves a turn towards openness but that move is
only on occasion desired in its own right. The desire for openness is made
explicit in circumstances where there is a conscious attempt to move to
the boundaries of knowledge. Since texture is intimately bound into
closure, the boundaries of knowledge are found in the realisation of every
closure, yet the structure of closure obscures this circumstance. The desire
for the unknown is shown therefore in a move to what is seen to be the
limit of closure, the edge of our world. For it is at the edge of our world
that the unknown becomes most apparent. At the edge of space, we do not
experience openness, for we can only experience that which has undergone closure, but it is easier to access texture.

The desire to move towards the edge of closure as an expression of the
search for openness can be illustrated in our relationship to physical space.
Our experience of physical space is already the product of closure, for we
have divided it into many things, and the expression of the desire for
openness is found in a desire to explore the limits of that space. For
exploring the edge of our physical space is a means to escape the confines
of our current closures. As children we can be thrilled by the prospect of

exploring; through the gap in the garden fence: a world of untold mystery and excitement. It might be argued that the desire to explore is simply the desire to know what lies beyond our current space and as such is an expression of the search for closure. Exploration however involves a deliberate move away from the known, while the search for closure might be expected to rest with current closures and defend their realisation. To intentionally explore the edge of our physical space expresses therefore a desire to step outside of current closure, with its known and familiar forms, and enter a space without containment: a desire for openness. This aspect of exploration is more typical of childhood, because as adults the limits of our physical space are largely known, and we are no longer in a position to approach its edge. As a result adult exploration is more likely to be illustrative of the desire for closure, the desire to know aspects of our space that have so far eluded us, to find out what is between one place and another, to find the way from one location to another. By contrast, as a child the exploring of a wood, for example, can appear to offer the potential of a magical world that unlike the everyday has seemingly no limits. Such a relationship to the edge of physical space can have an echo in adult exploration, if it is carried out not to achieve some particular end but for its own sake. It should not be surprising that in a social context this form of exploration is at times regarded as irresponsible. The case is often made that, for example, the exploration of inter-planetary space should be directed to particular ends and judged accordingly. One of the attractions of such exploration however is the attraction of the edge, and the move towards the unknown. Its irresponsibility stems precisely from its abandonment of current closure. It is a move away from safety, a move that might even come to threaten current closure, and, thus for those for whom the primary concern is the defence of current closures, it is to be avoided and criticised.

In a more parochial fashion the desire to travel for its own sake expresses in some small degree a search for openness. Travel that is desired for itself attempts to escape from the known, where we can feel trapped by familiarity. As an expression of the search for openness, however, travel and more generally the exploration of the physical world has limited potential. For it is a search that at once cannot succeed. The excitement that as a child we associate with the possibility of openness is difficult to retain, for every exploration brings new closure, and with it new material that separates us from openness. Like the rainbow the edge of closure recedes from us. No matter how far we run closure maintains its grip. We can travel to be somewhere else, to be free from closure, but we can never arrive. Instead of openness we have new closure and new limitations, and the desire to throw off the limitation of our current closure remains. It is for this reason therefore that the exploration of physical space as a search for openness is a limited strategy. In the search for openness we find closure; a pattern that is repeated in more complex forms wherever the desire for openness is found, although in this case it is perhaps most immediately apparent. For as an expression of the search for openness the move to the edge cannot arrive at openness any more than the desire for closure can find a final resting place.

There is a further aspect of the exploration of physical space that illustrates the general characteristics of the desire for the unknown and the way in which this is expressed in the move towards the edge of closure. So far it has been implicitly suggested that the edge of physical space is found at a distance from ourselves at the limit of what we have previously encountered. The edge of physical space is however found at the point of the failure of any of the closures that participate in the identification of that space. The edge of physical space is found therefore at the outer limit of that physical space but it is also found throughout physical space. For every part of physical space allows for further exploration, further ventures into the unknown. We can seek to know and map the universe, but the bark of a single tree has unseen crevices and nodules whose further examination allow for new closure and which if mapped would have no less detail than the plenitude of stars in the night sky. In the same way that the edge of physical space is found throughout as well as at the limit of physical space, so also is the edge of closure in general found throughout the linguistic closures of our personal space. The desire to move to the edge in search of the unknown can be expressed in the move towards the texture held within any element of material, but it is more typically expressed in the move towards the limit of that space, for it is here that the potential to escape current closure is most readily apparent.

In an age that largely accepts the scientific story of the world, the provision of mystery is often found, in the context of popular culture, in the supposition of unexplained scientific phenomena. Instead of an unknown which is found throughout cultural space in the texture associated with each element of material, we have flying saucers, aliens, ghosts, extrasensory perception and so forth. The desire for openness, the desire to escape the constraints of current closure, can be seen to drive belief in these supposed mysteries. The widespread interest in such things is a response to a world in which everything is seemingly explained: a world of closure. There is excitement in the unexplained and unknown, for here is a part of the world beyond our current closures. A part of the world that reintroduces possibility, that makes the world seem a larger and more exciting place. The mysteries of popular culture propose a set of events that is unexplained, and in doing so seek to introduce an element of openness. These mysteries are however pale and insipid examples of the mystery of closure which surrounds us. The mysteries of popular culture are however easily accessed, for they suppose an event or combination of events which are describable in the context of currently available closure and for which there are no immediate explanations. Yet every closure is precarious, and as a result we are surrounded by the profoundly mysterious. Yet the illusion of knowledge is so convincing to many that these thin mysteries of popular culture are taken up as a desirable relief from the tyranny of closure.

More significantly, the desire for the unknown and thus the search for openness can be identified in every arena where individuals are engaged in the pursuit of knowledge. Whether in history or science, what is perceived as the edge of closure is an interesting and desirable location; nor is it desirable simply because it offers the potential for new closure. It is desirable in itself because it suggests the possibility of escape from current closure. A scientist may look towards the edge of current closure, towards the unknown, as a prelude to closure. If a scientific mystery is solved with the provision of new closure, it does not bring the desire to move to the edge to an end, but merely shifts what is understood to be the edge to a new position. The pursuit of knowledge consists in the provision and extension of closure. Yet the provision of new closure requires a move towards the edge, in its most obvious form at the limit of current space, but also in the examination of texture held within current closures, or in the provision of alternative closures to those currently held.

The first stage in the search for openness can be seen therefore in the conscious pursuit of the move towards texture or openness, itself implicit in all closure. Cultural space makes it look as if we have largely understood the world, and the desire to move to the edge of the world is a desire to recapture the possibility of new closure, and thereby maintain the process of closure that was more readily available to us as children when we were in the process of forming our personal space. It is the first and highly limited form of the search for openness because it remains firmly embedded in the framework of closure. The move to the edge looks out beyond current closure but as a prelude to further closure. Later stages in the search for openness seek to look out from closure not in search of further closure, but in search of that which is not closure.

15

ART AND THE AVOIDANCE
OF CLOSURE

What distinguishes art from knowledge is the acceptance of
the failure of closure and the avoidance of an attempt to
complete closure.

The move to the edge of closure is an expression of the search for openness, but for the most part it is a preliminary to further closure. As a search
for openness it is therefore a shortlived strategy. The second stage in the
search for openness consists in a more thoroughgoing attempt to escape
closure, for it is not intended as a step towards further closure. It consists
in the provision of marks that systematically seek to evade attempts to
realise closure. As a strategy it plays an important role in our culture, and I
shall argue that this strategy, the avoidance of closure, typifies art in its
contemporary sense, providing its motivation, and determining its character.

Art in its traditional Greek sense was engaged in an attempt to mimic or
copy aspects of the world. In this respect it can be regarded as a craft
which is reliant upon technique. There is an element of such a meaning of
'art' in our use of the word 'artful', but otherwise the meaning we attach
to 'art' has shifted so that it largely excludes that which it once identified,
and is instead associated with the pursuit of a less tangible goal. The traditional notion of art as the attempt to provide a copy of some aspect of the
world is an illustration of the desire for closure, for it seeks to achieve the
same goal as knowledge, namely an accurate description of reality. I shall
argue however that this aspect of art has largely ceased to be that which we
value in painting and literature, and although works of contemporary art
may have an element of reproducing the world, copying has for the most
part been reduced to technique and no longer has a significant role in
characterising examples of art.

The case I shall make is that what distinguishes art from knowledge,
both of which operate with visual or auditory marks, is the acceptance of

the failure of closure and the avoidance of an attempt to complete closure. Art in this sense is the pursuit of openness and the avoidance of closure. As a result the artist is one who is not engaged in an attempt to provide closure but seeks instead to point towards the residue that lies outside of closure. Thus we can say that for the artist closure cannot exhaust the world. Another way of expressing this would be to say that the artist is more interested in texture than in material. One advantage in seeing art in this way is that it explains why it is not sufficient for a literary or visual work to have been created for it to be regarded as art. So long as the work has only sought closure, however brilliant the technique or painstaking the task, we are reluctant to refer to it as art. It is with the intentional avoidance of realising closure, and the successful replication of this intention in the observer or reader, that the work becomes artistic.

There may be objections to this description of art. It is not however important to the argument that this description is accepted as an accurate portrayal of our current actual use of the mark 'art'. However there are good grounds for thinking that it is perhaps a good first approximation to its current use, and might function to clarify and make more precise that use. Its aim is not so much to give a definition of the term, but to explicate the avoidance of closure as a stage in the search for openness, by identifying a mode of human endeavour, important in our culture, which operates in this manner. Such human endeavour may or may not be coextensive with that which we call 'art', and if some wish to argue that art is either more broadly or more tightly defined, this can be easily conceded without impinging on the main thrust of the argument.

In order to illustrate what is involved in the avoidance of closure, three different examples of artistic endeavour will be considered: painting, photography, and the written text.

Painting

Painting serves many functions but the one with which we are concerned here is the attempt to avoid closure, and it is this function that enables us to identify painting as art. Painting can be decorative, it can involve fine detail and exquisite technique, but the case being made is that these have no part in themselves in the avoidance of closure, and as such are not relevant to our identification of the work as 'art'. What are we to understand in this context by 'the avoidance of closure'? Not simply the provision of marks that are with difficulty realised as material, but rather the denial or the destabilisation of material that is offered, with the consequence that the marks remain in part open. In the case of painting it is possible to discern a number of distinct ways in which the avoidance of closure has been attempted: the marks provided by the painter can be ambiguous in their provision of material; the material provided from separate parts of the painting can be combined to different effects; the material provided from the marks can itself function as a metaphor that defies precise closure; and finally the material generated from the painting can take its place in a specific web of metaphors available in cultural space that themselves evade closure.[1]

As with language, the marks provided by a painter can function as tags or cues and take place in the context of cultural space. The most easily identifiable visual marks are those that can at once be realised as persons or things. Paintings that offer only familiar combinations of such tags are unlikely to regarded as art. Pictures, for example, of horses, or boats, whose aim is simply to provide a representation or copy of a physical thing, make no attempt to avoid closure. As a result we are disinclined to refer to these pictures as art. In order to be identified as art the marks need to be seen to seek to avoid closure. There are a number of strategies that have been used to achieve this aim. The first strategy that we shall identify is for the marks to be ambiguous. This functions to achieve the goal of avoiding closure for through ambiguity the marks continue to function as cues rather than as tags to specific things. Perhaps the most obvious example of this strategy can be seen in the move from realism to abstraction that is usually regarded as having taken place in European art at some point in the mid-to-late nineteenth century. In the classic Impressionist paintings of Monet and Turner marks are provided so that they do not dictate the realisation of a particular thing, either in any instance or taken across the painting as a whole. The term 'Impressionism' itself encapsulates this notion: a painting that provides an impression but that does not dictate. As a result the painters sought to avoid marks that would be regarded as familiar tags and instead put forward marks that could be held as cues for an indeterminate scene, thereby escaping the familiar closures of physical objects in favour of colour and light.

A similar strategy can be observed in the paintings of Cézanne and the Cubists. Such paintings are commonly interpreted as an attempt to uncover underlying form in the world, and thus might be thought to be an example of the search for closure in seeking to express a deep truth about the nature of reality. Cézanne's many pictures of the Mont Sainte-Victoire cannot however be interpreted as a desire to describe the true nature of the hill, in the manner of a scientist, as if the paintings are presented as accurate and precise copies of reality. Instead they propose an alternative perspective to the familiar tags of realistic painting, and in the context of the late nineteenth century forced the viewer to treat the marks as cues. The marks deliberately fail if we attempt to realise them as portrayals of particular physical objects, not having the detail and the specificity that we require. Instead therefore the viewer is encouraged to look for something else. The artist thereby seeks to escape the constraints of familiar closure, and offers a means to use the marks to create new closure. In this exercise can be seen a search for openness expressed through a move to the edge of closure. Unlike the astronomer however who peers towards the edge of the universe the artist is not hopeful of finally uncovering reality but seeks to explore, on the basis that exploration has no limit and that any version of reality conveyed by the marks on the canvas is a misrepresentation. Cézanne shares with the scientist the desire to uncover reality but unlike the scientist Cézanne does not propose that the uncovering will deliver a final version.[2]

Whether a mark is recognised as a tag depends on the linguistic closures currently available in cultural space. In the same way that the realisation of a sentence contributes to linguistic closure either by confirming the current use of tags or by providing a new tag, so an image contributes in a similar manner to the closures of personal and cultural space. We can identify the brushstrokes of a painter as marks, similar in character to the marks of language; and, as with the marks of language, the closures associated with painterly marks can undergo change through their combination with other painterly marks. A mark used by a painter in one cultural environment can for example function as a tag but in a different environment as a cue. As cultural space has become used to the images of Impressionism their capacity to avoid closure has become more limited, for we have become used to the notion that the marks of a painting need not depict an object, and instead they have acquired a new familiarity. As a result the marks once again risk becoming tags, not of objects but of an impressionistic effect which is at once identified. With repetition, the new closure becomes a habit like the old closures of pictorial realism and the marks cease to function as cues. In the same way that linguistic marks begin as cues and solidify around a given set of associated closures, so also therefore do the marks of painting progress from cue to tag. In the context of this account it can be seen that there is a good reason that the copying of a Cézanne or cubist style or for that matter any artist or style is not itself regarded as art. For such a copying does not illustrate a search for openness on the part of the painter, but instead a search for closure. The painterly marks of cultural space have shifted since the turn of the century, and the brushstrokes that once unsettled familiar realisations now provide their own tags. The timeless baroque character of Cézanne's landscapes is no longer a venture into the unknown, a crack in the wall of closure that surrounds us, but a renowned style that can be rapidly identified.

In order to avoid closure a painter cannot simply offer a confusion of marks or avoid the provision of marks that realise material, for such a strategy would not provide anything. If the avoidance of closure was simply achieved by the provision of marks whose realisation is either difficult or impossible, any random combination of marks could be described as art. The attempt to avoid closure cannot therefore simply abandon closure altogether, but requires the suggestion of closure: the offer of closure that is then denied. If insight alone were the defining characteristic of art it would apply equally to the scientist; what distinguishes the artist is the offer of closure that cannot come to rest and so cannot be completed. The Impressionists can be seen therefore not to abandon the closures of objects without offering us something in their place. A smudge will not do. What is offered might be described as a new perspective on reality, one that seeks to escape the rigid framework of objects, by focusing on patterns of light and colour. The impressionists put forward such images as an insight into the nature of the world, as if to tell us that it is not what we thought it was, trapped as we are by the familiar closures of material objects. This new vision is not however presented in the manner of a theorist who would claim that this is how the world really is, as if stripped of our objectification the world consists of patterns of light. Rather it offers an alternative as if to draw attention to the failure of closure and the limitation of our understanding. As we look at Monet's water lilies, or Seurat's pointillist vision of a crowd on the banks of the Seine, we do not imagine that these portrayals are intended to show what water lilies or the turn-of-the-century crowd really look like, rather they serve to unsettle our habitual closures and offer an alternative not as a replacement but as a means of looking out from our current closures and escaping their grip.

The search for openness illustrated by these paintings is a consequence of the mark being held as a cue, with the consequent avoidance of closure. The power of the paintings comes from the uncertainty associated with the realisation of material from the marks. If on the one hand, the material realised from the painting was unambiguous, the marks of the painting would be tags and they would have no value as art. If on the other, we were unable to realise material at all, the painting would serve no function and have no impact upon us. It is because material is suggested and at the same time denied that, as a consequence, the paintings have the capacity to give us the impression of approaching openness and therefore of saying something deep about the nature of the world. In the transition to abstraction paintings suggest the familiar material of an object world but at the same time deny it, forcing a search for an alternative closure and offering a new closure which is itself uncertain.

Our inability to provide material in the form of physical objects which is brought to the fore in the transition to abstraction can be seen therefore to encourage the viewer to return the mark to a cue. In this respect a badly drawn painting might have the same effect, but it seems reasonable to suppose art is distinguished from poor technique by the intention of the artist and the capacity of the marks to suggest closure. In the case of Cézanne most of us nowadays come to his paintings knowing in advance the value placed upon them, if only because they are likely to be found in museums, and as a consequence assume that closure can be realised. Our experience of the failure of attempts at representational closure then functions to stimulate a search for alternatives, and echoes Cézanne's own desire.[3] Faced with an unknown painting from an unknown source we are likely to be more sceptical and more inclined therefore to account for our failure to realise material as the inadequacy of the painting and its inappropriate use of marks than our own inability to find closure.

The debate that surrounds new contemporary painting highlights the difficulty that we find in separating works that intentionally avoid closure, and those that are empty of any worthwhile content. On the one hand the distinguishing characteristic of art is that it avoids closure, on the other hand bad technique can also yield a similar consequence. Indeed since in the limit any set of marks is capable of realising material, and all closures can be seen to fail, there is a sense in which all marks can be seen to both offer closure and deny closure. It is hardly surprising therefore that there can be much debate over whether a particular work is or isn't art.

If we define art as the avoidance of closure, so long as an individual seeks to express the search for openness in a work and believes this to have been successfully achieved, it is, at least for the artist, art. Whether it is recognised as art more widely will depend on the character of cultural space. It is because all closure fails in the limit, and thus for example the most highly representational painting could be seen to demonstrate the failure of closure in its inability to complete closure despite its precision, and because at the same time any set of marks can be realised as a closure, that the intent of the individual may give us an indication as to whether we should look for a deliberate avoidance of familiar closures or not. The intentional pursuit of openness would seem therefore to be important to our identification of a work as art. The identification of a visual image as art involves therefore not only that we realise closure which is then undermined, but that this transition is deemed to have been intentional on the part of the individual who created the work.

Wholly abstract painting, a dominant feature of contemporary painting, might be thought to avoid closure altogether, since like music and unlike most texts it cannot be realised as a copy of some aspect of the world. This is however not so. For while abstraction avoids the closures of realism it offers alternative closures of shape, colour, and pattern. At the time of the first world war, Kandinsky's abstracts merely by their abstraction could be a threat to familiar closure,[4] while today an abstract image in itself is likely to offer little threat. To avoid closure a contemporary abstract painting like its figurative predecessors must avoid or undermine familiar categories that we might impose. In this context there can be seen to be a reason for the importance of tension and balance in a painting: to stop the painting becoming still and therefore one thing, this closure. As with a figurative painting, for a mark to become a cue it must be unclear how it is to be realised. Abstraction does not in itself guarantee that the mark is taken as a cue. If it did so, to engage in art would be a trivial exercise. Instead the successful artist must constantly escape closure, by eschewing familiar categories or by undermining them. It is for this reason that art is constantly on the move, for the boundary of cultural space is itself a product of the immediate past. A work that today escapes closure may if repeated tomorrow be realised as material in a routine manner.[5]

While the ambiguity of the mark as a means to avoid closure is perhaps most easily identifiable in the move to abstraction, it is also found in traditional realist painting. The most obvious, if hackneyed example, being the *Mona Lisa*. The enduring attraction of the painting can in the context of the account that has been offered be traced to the ambiguity embedded in the image with a consequent lack of closure and thus a sense of reaching out to something beyond closure. The ambiguity of the mark forces the viewer to play with alternative closures, and consequently the painting avoids closure even though the material image is at once clear.

Introducing ambiguity into the mark, or more precisely we should say drawing attention to an ambiguity in the closures with which a pictorial mark can be associated, is an important mechanism by which painters have sought to avoid closure, but it is only the first of a number of strategies that have been used to approach openness. Unlike a literary work, a painting is not linear and the order of combination is left to the viewer. As a consequence the composition of the painting and the manner in which the viewer is encouraged to hold it together and to stray from one aspect to another is a further means by which the artist can avoid closure. In the same manner that marks are inherently ambiguous, and that closure fails, no painting can force a particular order of reading and thus must at some level avoid closure, but just as we assume that marks are tags and that closure is possible, so we assume that a painting portrays something – even if that thing is intangible, such as an emotion rather than a physical object. One of the means by which an artist can draw attention to the avoidance of closure is therefore to make the order of combination of the images within the canvas unclear so that the eye cannot rest on a single subject but is forced to play with the relationships in the picture thereby teasing out that which escapes closure. An image in which the elements are conventionally aligned, in which the subject is at once apparent and is not challenged by further viewing has the impression of being static, as if the first glance tells us all that we can know. In short the closure realised is immediate and suggests its completeness. Composition that introduces doubt as to the subject of the painting, and that forces the viewer to question the relationship between its elements, moves in the direction of avoiding closure. At its most simplistic, for example, closure is usually encouraged by placing the subject of the painting at the centre of the canvas, with the consequent effect that such pictures are likely to lack life and vitality.

Any such generalisation with regard to composition will however have counter examples. There can be no general rules of composition, for, as with the ambiguity of the mark, the role of composition will be dependent on the current closures of visual and linguistic space. Placing a single, unambiguous image of a material object in the centre of a canvas might in certain circumstances lead to a questioning of the image on precisely the grounds that it offers no more than itself and thus, in the context of a museum for example, forces further readings. Warhol's 'soup can' could be cited as an example.

Composition can be seen therefore to play a role in the avoidance of closure by unsettling the manner in which we would expect habitually to read an image. As with the ambiguity of the mark a closure must be attempted for the abandonment of closure altogether provides no content. A composition that is read simply as a mess does not propose any closure that can then be undermined. The avoidance of closure requires the suggestion of closure but at the same time its undoing. Composition as a means of avoiding closure requires therefore an ambiguity of subject rather than its simple absence.

The highlighting of the ambiguity of the mark or the use of composition as a means of unsettling closure are strategies employed by abstract and realist painters alike. In addition it is possible to discern two further mechanisms that apply more particularly to realist painting. In neither case do they rely on the ambiguity of the image to unsettle closure: the realisation of object-centred material is assumed and the avoidance of closure operates at the level of the meaning that is realised from the image. The image itself can seek to step outside of its own particularity and hint at some other meaning; or, the elements of the painting can be deliberately chosen to echo a specific web of metaphors which either themselves avoid closure or whose combination within the image avoids closure.

The first of these two strategies can be illustrated by Rembrandt's late self-portraits.[6] These portraits can be seen as not merely providing a copy of the individual's appearance but as seeking to move beyond the appearance to the experience of the individual and perhaps of all individuals. We are not being asked to look at an image of an old man as if to comment on the painting's likeness to the individual's appearance, but are encouraged to ask what the painting tells us about what it is to be old. The painting suggests a lifetime of events and a response to those events. Yet we cannot complete the closure as if we could read off unambiguously what was meant or intended here. We are left with the image and the sense that the image is more than an image. We can weave many tales about this character and it is in the complexity and variety of those tales, and the assumption that the image was intended to generate them, that closure is avoided. The image although itself not ambiguous cannot rest, for it is caught in a web of metaphors of its own making. The image thus stands for something which is itself open.

An alternative strategy for avoiding closure in the case of realist painting can be identified when elements of the painting are deliberately intended to be held as specific metaphors. This occurs, for example, in the use of religious or classical images in which the individuals portrayed and the scene depicted are known to the viewer. The avoidance of closure is achieved in such cases not by the image but by the metaphor or the manner in which the metaphor is conveyed. Religious painting, for example, can be regarded as seeking to portray something other than this world, to indicate something of a world beyond, so that familiar images of people and events are given a quality over and above their immediate appearance. This task is made possible by a known set of beliefs which themselves carry an element of mystery. In a secular age these images have for many lost their force, and are sometimes read as mere parodies of a false account of the world, but in so far as the paintings seek to express that which is inexpressible by use of a specific known belief, they can be interpreted as an attempt to avoid closure.

The employment of the elements within figurative painting as metaphors is not limited to religious or classical work but can be found in twentieth-century painting. Magritte for example can be seen to develop his own metaphors which are used throughout his work. A blue sky dotted with idealised fluffy clouds; a formal frontage to a row of houses or the internal panelling of a room; a strange machined bannister-like pedestal or biblioquet; these images recur throughout his paintings and act as metaphors – each standing for an aspect of human experience. We can seek to give them literal meaning. For example, throughout Magritte's painting runs the theme of the artist in an impossible attempt to convey what is really there. The image of a perfectly blue sky with idealised white clouds can in this context be interpreted as Magritte's metaphor for the false image of reality he is able to paint while what he wishes to capture is beyond the canvas. Many of his paintings can be seen to depict this predicament. Some well-known examples consist of a scene in which we see an artist painting a canvas placed on an easel in front of the subject or landscape being painted. One of Magritte's paintings, however, consists only of the image of an unreally perfect sky, with nothing else in the frame.[7] Here we see the same familiar idealised fluffy clouds on a perfectly blue background. Magritte can be regarded as offering us a painting which consists of nothing other than his metaphor for the unreal reality he is capable of painting. It is a circular, self-reflexive work. Since the painting consists of a portrayal of the predicament of the artist, a predicament with which Magritte was concerned throughout his life, the title of the painting, *The Curse*, is less enigmatic than might at first be supposed.

There are many further examples that could be offered to demonstrate the use of this pictorial metaphor by Magritte. One of his late paintings in particular can be seen to illustrate its use. An artist is depicted painting a window through the panes of which we see the familiar idealised blue sky and clouds.[8] On closer inspection the blue sky and clouds are not beyond the window but are painted on the surface of the windowpanes for it is slightly ajar and through the space between the open window and the frame there is only black. It is as if Magritte wishes to say that the artist seeks to peer through the sky, through the images of the world that can be conveyed, in search of what is really there, but in the opening is found nothing.

Magritte's metaphors are a personal affair. Their first use can be found in his paintings of the mid-to-late 1920s, and they evolve over time. The popularity of his paintings in large part ignores the metaphorical character of the images and relies on a Surrealist shock value. The use of realistic images in an unreal manner is itself an attempt to challenge closure, but the repetition of this procedure taken at face value loses its force as we become accustomed to the effect. As a result, without their metaphorical edge Magritte's images can quickly lose their vitality, and instead of an escape from closure the viewer can assume a closure associated with the surreal style.[9] We can witness this transition in the popular culture of the 1960s and 1970s. At first, Magritte's images were fresh and surprising, challenging the familiar closures of materiality, suggesting another surreal world combining known objects in unexpected ways. The very ease however with which openness was seemingly approached undermined his paintings' capacity to do so. The prevalence of such images resulted in a supposed familiarity with their apparent theme and as a result the style for some became hackneyed in the manner of a piece of music that is played too often and as a consequence no longer has emotional force. In such a transition we can see the shifting of the boundary of cultural space.

Magritte's use of metaphors can be interpreted therefore as seeking to uncover that which is not uncoverable, to present a predicament whose representation dissolves the predicament. As such he strives to undermine our cosy familiar closures not in aid of a vision that would itself be just another closure but in aid of the avoidance of closure. Some of his paintings can be read as an attempt to express this very thought itself. In one of his paintings the walls of a room are filled with six of his most frequently repeated metaphors: those that elsewhere he will call the elements of life along with two others.[10] In the centre of the room stands a cannon. The title of the work On the Threshold of Freedom encourages the notion that the work itself expresses the desire to break through the categories that surround us and find something outside of the room, beyond closure.[11]

In seeking to give an indication of the manner in which Magritte uses his images as metaphors to avoid closure, there is at once an undermining of that attempted avoidance of closure. For by providing a specific literary account an attempt has been made to fill the cracks that Magritte's paintings seek to peer through. It would be foolhardy therefore to suggest that Magritte's visual metaphors are specific or could be described, for if so they would immediately have failed to avoid closure. The purpose of this description is not therefore to propose a definitive description of Magritte's paintings, as if the code could be cracked, but to indicate how a painting that relies on realist imagery can use its own metaphors to escape closure. To parody Magritte's painting of a pipe which includes the words 'Ceçi n'est pas une pipe', it can also be said that 'the last four paragraphs are not a description of Magritte's painting'.

The photograph

The photograph has the appearance of providing a perfect copy of the world, a complete closure. If visual art consisted of the attempt to provide a copy of reality the photograph might thus be considered the high point of art. Instead the immediate closure offered by the photo-graph, and the mechanical process involved in its production, is an impediment to it being held as art since both are antithetical to the avoidance of closure. The seemingly precise copy of reality does not encourage ambiguity and the mechanical process on which it relies undermines the notion that an observed ambiguity is deliberate. Since art requires the deliberate avoidance of closure photography has had an uphill task in being regarded as art. It would be a mistake to imagine however that the photograph does in fact provide a perfect copy of the world, for like all closures it must fail. The photograph is the visual equivalent of the textual fact. Both presume to be a true copy of the world, and closure is thus assumed to be complete, but under scrutiny the closure fails.

The strength of our belief in the accuracy of the photographic copy is so great that it is hard to uncover the failure in the closure. The Wittgensteinian point can be made that unlike a photograph the visual field of experience is not bounded.[12] It is not possible to see the edge of our visual space in the way that we can always see the edge of a photo-graph. The edge of visual space lies outside of that space. This difference between a photograph and experience – let alone the divergence between a photograph and the world – is an indication of the failure of photo-graphic closure. The bounded character of photographs is one of many differences between the photograph and experience, but these differences are obscured by our attachment to the closures of physical things. The bounded character of the photograph is only a hint of the radical discon-tinuity between the photograph and experience. Another indication can be found in the detail of the photograph. Experience has the character of a fractal image, an endless opening can take place from any point. The photograph can be similarly opened but in the process the divergence with experience becomes evident. It is for this reason that the resolution of a photograph is not increased by making the image larger. It would be mistaken to imagine that with a perfect lens and a perfectly responsive film this failure of closure could be eradicated. The fractal character of experience and the photograph cannot be the same, for they are not the same thing.

Reference has been made to 'the closure of the photograph' but correctly speaking it is not of course the photograph but the viewer that realises closure, and it is this realisation that is deemed a copy of our visual field. However, the closure realised from the photograph is learnt but it has become so habitual that its divergence from experience has largely been lost. The mechanical character of the photograph encourages this assumption since we are used to the notion that subjectivity, driven by a desire for closure, 'lies'. It can be supposed therefore that a mere mechanism must be true. While it is the case that the mechanism does not seek a closure and cannot therefore be seen to impose order on the world, the mechanism is no closer to the original, any more than a flag waving in the wind is the same as the wind itself, or the same as the experience of the wind on our face. The similarity between the photograph and visual experience is a closure of our own making, and a closure closely allied to the notion of the photograph itself. As with all closures it seeks to hold two things as one. There is the photograph and the visual experience, and the closure proposes that we hold these two things as the same and operate on that basis. With the use of this closure the photograph can serve many purposes which themselves reinforce the closure. If the closure is shown to fail we promptly make allowances as a means of retaining the closure itself: the lens was inadequate, the processing poor. Over time therefore, as with facts, we come to take the photograph as reality, so used are we to its power to handle activity that we have difficulty in even catching sight of its failure. Yet as with facts and reality, the photograph is not a representation of the visual experience. They are held as one through closure. As evidence of this argument, it can be seen that in circumstances where the closure of the equivalence between the photograph and the visual field is absent, as it is in animals, babies and in rare cases where adults have never encountered or known about photographic images, the behaviour of the animal, baby, or person, indicates that the similarity is not immediately discerned.

Such an argument at once raises a question. If the photograph is not a representation of the visual field what is it then that makes the photograph capable of being held as the same as the visual field? Could we not equally hold some other object as a copy of reality? A first response to such challenges would be to point out that the apparent similarity in form between the photograph and our visual experience can be accounted for as a consequence of applying the same closures to the photograph and to the visual field. Our ability to realise similar closures from the photograph and from the visual field enables us to realise the closure that the two are equivalent, or at least, similar. Such a response is however unlikely to satisfy the hypothetical critic. For the question remains as to why we are able to realise similar closures in each case. In the limit this question seeks an answer that must lie outside of closure. In the context of our current space and our current scientific account of reality we can offer a further explanation. The response of the film in the camera to openness can be held as similar in some respects to the response of the retina to openness because both mechanisms are responsive to photons. Such an explanation or a further elaboration cannot tell us what the nature of the similarity is, for the similarity is always the outcome of closure, but it might for example provide us with a story which enables us to intervene to produce 'better' photographs, and it has the appearance of providing an answer to the initial question.

In the photograph we find a similar paradox to that which besets the fact. The photograph typifies closure, in apparently presenting an image that is not ambiguous and is immediately capable of being realised. An artist seeking to avoid closure is seemingly constrained by the photograph, held within its grip. Yet all closure fails, and the closure offered by the photograph is not complete. The humblest snapshot has thus in a sense avoided closure, and could to this degree be regarded as art. The snapshot is not however intended to avoid closure, but is more typically a deliberate attempt to capture reality. Furthermore, for the reasons that have been outlined, we approach the photograph with the assumption of complete closure. It is therefore difficult to uncover its failure. Even if the snapshot was intended to avoid closure it would be unlikely to have this outcome for the average disinterested viewer.

Photography is however capable of being art by avoiding closure using the same techniques that apply to realist painting, or, with more difficulty, by drawing attention to the failure of closure implicit in the technique itself. In this context it no longer seems accidental that the majority of photography that aspires to being art is shot in black and white. For the self-evident difference between the black and white photograph and a copy of the world makes it easier for the photograph to escape closure through the use of composition or metaphor. The colour photograph has a tougher task in overcoming the embedded assumption that it simply replicates the world, but it is of course still possible. The cracks in the closure offered by the colour photograph have to some extent been explored by Hockney in his photographic montages whereby successive images of an object, such as the Eiffel Tower, are placed together to construct one continuous large image. The failure of closure inherent in each perspective has the consequence that the combined image is obviously distorted and different from our visual experience. As a result the work can be regarded as pointing to the gap between the photograph and the world, and as a consequence can be taken as an example of the search for openness through the avoidance of closure.

The text

The deliberate avoidance of closure serves the same function in a text as it does in a painting. The search for openness through the avoidance of closure is perhaps most evident in poetry, but throughout literature, as with the visual arts, the avoidance of closure can be regarded as the defining characteristic of literary art. The techniques employed by painters to avoid closure are echoed by the writer. As with painting, the text can seek to remain open by avoiding known tags and operating with words that seek to function as cues; a text can also seek to avoid closure by undermining its own closures, or by providing levels of meaning that cannot be fully deciphered.

Writing that largely falls within the ambit of recognised tags, like most novels, seeks to avoid closure in the same way that a realist painting avoids closure. Closure is avoided not at the level of the mark but at the level of the meaning associated with the text as a whole. The text is held open because we are unable to provide an account of it that brings the description to a halt. The text functions as a metaphor for something that is beyond a simple description. The characters have depth because they are not fully understood, the plot is open because its meaning is multi-faceted and in debate. A text that seeks to be literary art can have no single interpretation, but it requires nevertheless to offer possible closures – a random combination will not suffice. To avoid closure something must be presented but at the same time be seen to fail, so that in the half light between one closure and the next there is a sense of glimpsing openness. A great novel must offer a story, a set of closures, but at the same time these closures have in some way to be held open. The story may itself be a metaphor, or the closures within the story may interact in such a manner that the text as a whole cannot rest.

The importance of the avoidance of closure in our assessment of a literary text is also apparent in those texts where little attempt is made to avoid closure. What we might call 'bad art' is typified by the immediacy and obviousness of closure and the lack of a strategy on the part of the author to undermine such a closure. We find unsatisfying, for example, characters portrayed in a novel as simply 'good' or 'bad', because the world is not capable of closure and if it is portrayed as if closure was complete it feels false. Commonly we might say of such characterisation that it is one-dimensional and lacks complexity, this however obscures the underlying reason for our discontent. It is not a lack of detail that makes the description limited, but the presentation of a closure that is held to be complete. The same criterion operates in the case of the plot, which is an attempt to impose a closure on a series of events. The pot-boiler novel with transparent characters and a clear plot provides the satisfactions of closure. Like a crossword puzzle we can enjoy the hunting down of the closure and the elegance with which the solution is arrived at. Furthermore the readability of such books is based on our desire for closure, on our need to know what happened. As a result, if effective, they can be compulsive and popular. The reason we do not hail them as art however can in the context of this account be seen not as function of literary prejudice but as a consequence of the attempt by such works to seek closure and not openness.

The avoidance of closure in most novels that are regarded as literature tends to operate at the level of the text as a whole rather than at the level of individual words or sentences. In poetry however the avoidance of closure is often found both at the level of the ambiguity of the mark and at the level of the text as a whole. There are of course exceptions to this principle, such as Virginia Woolf's *The Waves*, or James Joyce's *Ulysses* or even more evidently *Finnegan's Wake*, but for the most part the generalisation stands. In order therefore to illustrate the means by which a text can seek to avoid closure an interpretation will be offered of a poem, T.S. Eliot's *Four Quartets*,[13] for it is a poem that can be seen to employ many different strategies to avoid closure.

At the outset it is important to begin with a caveat. As with the previous account of Magritte's painting, to seek to give a final account of a poem that is itself attempting to avoid closure is at once contradictory. The descriptions offered are therefore necessarily preliminary. It will be argued, for example, that *Four Quartets*, offers a description of our experience of time, but such a claim is made on the understanding that it makes no claim to being the only interpretation. The criticism of the poem and the extracts taken are offered as a demonstration that the poem avoids closure rather than as an attempt to say what the poem itself so carefully avoids saying.

Throughout the *Quartets*, individual marks are offered which cannot be realised as known tags. At its most straightforward Eliot generates new words in an attempt to escape the emptiness of the known tag. For example the phrase 'the moment in the draughty church at smokefall'[14] forces the reader to search for a new closure and thus to encounter that which is not closure. Since there is no English word 'smokefall' and from the phrase we have no means of determining its precise meaning, we are never able to provide a final closure. The meaning of the phrase remains open. We have a sense of what might be meant, but we cannot rest with any particular closure.

More commonly closure is avoided either because a word is used metaphorically or because the closure we would normally associate with a mark is undermined by the context in which it is offered. For example, a central theme of the first quartet 'Burnt Norton' is the character of the momentary present. Eliot offers a description of this momentary and fleeting present:

> At the still point of the turning world. Neither flesh nor fleshless;
> Neither from nor towards; at the still point, there the dance is,
> But neither arrest nor movement. And do not call it fixity
> Where past and future are gathered. Neither movement from
>   nor towards,
> Neither ascent nor decline.

These descriptions of the present are at once metaphorical for the words are not used in their familiar sense, they are offered therefore as cues rather than tags, encouraging the reader to search for a new closure. The closure however sought is at once denied. The present is the still point, but it is not still in any sense that we can grasp.

In these lines, Eliot seeks to express the character of experience: the momentary slice of time that is the present is all that we have. Trapped as we are in its grip, and separated from past and future, the present is that which constitutes our lives. Eliot offers as a description of this present, a present that he will later describe as 'timeless' for it is not in time, a contradiction that logicians might be tempted to describe as meaningless: 'Neither flesh nor fleshless'. Here, we are offered two closures: first, the closure that the present is not flesh, and then the closure that the present is not fleshless. The first of these closures can be realised in numerous ways: one attempt at closure might be indicated by saying that the present is not a physical thing, experience is paper-thin and has no substance, a translucent, ephemeral cinema screen. The present is not flesh. In contrast to this perspective, we are offered a second closure: the present is not fleshless. Again an attempt at closure might be realised in the following manner: the present is every-thing without which there would for us be nothing. The present is not therefore empty of life, it is not fleshless. Eliot elaborates this notion of the present as not fleshless, by noting that it is all that we have – it is every-thing:

> Except for the point, the still point,
> There would be no dance, and there is only the dance.

So held in the description 'Neither flesh nor fleshless' are two very different closures of the present, one that holds the present as wafer-thin and one that holds the present as everything, but at the same time through the contradiction we have a denial of either closure, and through this denial an attempt to escape the confines of linguistic closure in general to suggest thereby the unfathomable character of the present. The avoidance of closure is thus not the outcome of vagueness or an imprecise and loose use of language but is the result of the precise use of marks to offer closures which are unsettled from within. The pursuit of openness through the avoidance of closure is not to be seen therefore as an abandonment of closure or a lack of rigour, rather it requires closure to be taken to its limit and at the same time to be denied or the completeness of the closure offered to be undermined.

A further technique Eliot can be regarded as employing to avoid closure is to develop his own language of metaphors which themselves interweave to hold open the text. Words such as 'the sea', 'the rose-garden', 'fire', are introduced in one context encouraging a closure and then used again partly as a tag to this closure and at the same time in an attempt to move the closure forward. Thus 'the sea' can be realised as a metaphor for the past and future. The sea of time thus stretches either side of the present. Having introduced the sea in this manner Eliot is then able to use the tag 'sea' to allude to this metaphorical meaning and at the same time to develop such a meaning. So in the context of the initial metaphor the present can be seen as being surrounded by the sea:

the sea is all about us[15]

but Eliot does not leave the metaphor here but expands and extends it. We are cut off from the past for we are always in the present and we find only its remains:

The sea is the land's edge also, the granite
Into which it reaches, the beaches where it tosses
Its hints of earlier and other creation:[16]

Eliot plays with the necessity of the past and future in the context of this metaphor:

We cannot think of a time that is oceanless[17]

As a result in the closing passage of the final poem of Four Quartets, where Eliot can be seen to attempt to describe his vision of perfection, he is able to use his own metaphor of the sea to offer a closure that is at the same time not realisable.

> ... heard, half-heard, in the stillness
> Between two waves of the sea.

In such a manner Eliot can be seen to play with a metaphor of his own making in an attempt to avoid the text becoming static, and the closures becoming fixed. In the context of this account he can be seen to do so in search of openness which is almost glimpsed in the offer of closure and its denial.

The techniques Eliot uses to avoid closure are numerous and it is not necessary to illustrate them all, but there is one other which should be identified for it has extensive parallels elsewhere. It is used as a means to undermine closure but it does not usually function at the level of individual words but at the level of the text as a whole: it is self-reference. Through self-reference a literary work is able to move or unsettle the closures it suggests. As Magritte's paintings of an easel in a landscape draw attention to the limitation of painting in general and thus of this painting in particular and anything it seeks to 'say' about the world, so Eliot's reference to the inadequacy of words can be seen to undermine the closures that he offers and point to something beyond the text:

> So here I am, in the middle way, having had twenty years –
> Twenty years largely wasted, the years of l'entre deux guerres,
> Trying to use words, and every attempt
> Is a wholly new start, and a different kind of failure
> Because one has only learnt to get the better of words
> For the thing one no longer has to say, or the way in which
> One is no longer disposed to say it.[18]

Not only therefore does Eliot tell us that he cannot say what he wishes to say, that he is trapped by closure, but he tells us this in the poem. If he has only learnt to get the better of words for the thing he no longer wishes to say that applies equally to these few lines themselves. The self-reference succeeds in unsettling the immediate closure offered, since the notion that 'one has only learnt to get the better of words for the thing one no longer has to say' is itself something that he no longer has to say or is no longer the way he would choose to say it. He thus seeks to escape the grip of closure which limits what he can say to that which he no longer desires to say by a self-reference that does not allow the text to say anything in particular.

The move of self-reference can be identified in many literary and visual works, whether in Shakespeare's frequent reference to plays, and plays within plays, the short stories of Borges, or the almost ubiquitous self-referential gesture in French new wave cinema by directors such as Goddard, Resnais, and Truffaut. It is even to be found in Hollywood films.[19] Self-reference is not necessarily an avoidance of closure. It might for example be accounted for on the grounds of a self-obsession of those involved. It might be thought that since authors, painters, or directors spend much of their time writing, or painting, or directing, it is perhaps not so surprising that there are many references to such activity in their work. Certainly, the use of their own activity as a subject is not in itself evidence of a desire to avoid closure, but it has the capacity to unsettle all that is being said, or shown, by pointing to the limitation of the process itself, and frequently, as in the example from Eliot cited above, serves to achieve this end.

Self-reference as a means of avoiding closure has equally been a strategy of philosophers. Derrida is perhaps the most immediate and obvious contemporary example since his text is constantly self-referential with the evident and deliberate policy of undermining closure. In this respect he echoes Nietzsche who employs self-reference to the same end. In the following passage for example Nietzsche's purpose has similarities with the example from Eliot.

> We immortalise what cannot live and fly much longer, things only which are exhausted and mellow! And it is only for your after-noon, you, my written and painted thoughts, for which alone I have colours, many colours perhaps, many variegated softenings, and fifty yellows and browns and greens and reds; – but nobody will divine thereby how ye looked in your morning.[20]

In such cases self-reference serves to avoid closure and thus to emulate the pursuit of openness that typifies visual and literary art. It is not altogether unexpected therefore that these figures are sometimes criticised or ignored on precisely the grounds that they are engaged in literature and not in the more 'sound' practice of philosophy. This is a matter to which I shall return in the next chapter.

Eliot's text can therefore be seen to seek the avoidance of closure at many levels, from the ambiguity and openness of individual words to the metaphorical meaning of the text as a whole. Yet it does so not by failing to offer closure, but by teasing us with closures that we cannot fulfil. As a result we have a sense that we catch sight of that which is beyond closure, and it is this sense that gives the text its depth. It is the search for openness through an avoidance of closure which pervades the *Four Quartets*, and which inclines us to regard the work as an example of great art.

In conclusion therefore, it has been argued that the intentional avoidance of closure is the defining characteristic of artistic endeavour. In support of this case a number of works have been identified that illustrate the avoidance of closure. In response it may correctly be said that it has not been shown that all those things which are referred to as art do in fact demonstrate the avoidance of closure. Nor has it been shown that the avoidance of closure is exclusive to artistic endeavour. The primary purpose however has not been to define 'art' but to illustrate how the avoidance of closure has been attempted and therefore to indicate in what the avoidance of closure might consist. In this sense it has been incidental to the case that such a procedure is typified by what is currently referred to as 'art'. However it does seem to me that there are good grounds for at least considering such a definition.

16

NAMING THE UNNAMEABLE

It is through a combination of the exoteric and the esoteric
that attempts have been made to achieve the evidently
impossible task involved in naming the unnameable.

The final stage in the search for openness can be seen in the attempt to
identify directly that which is beyond closure. It will at once be apparent
that such an identification cannot consist in a description of openness, for
such a description would necessarily be couched in the framework of
closure. The desire to know the character of openness, to know what is on
the other side of the veil of closure, paradoxical though such knowledge
would be, is an expression of the search for closure. Instead, the final stage
in the search for openness is expressed through the attempt to mark or
identify that which is beyond closure, that which is beyond our reality.
Such an identification is not sought as a means of offering a description –
for if it is so it is at once a closure – but as a means of drawing attention to
the limitation of our understanding and pointing to that which is beyond
or other. The search for openness expressed in this manner will be referred
to as the attempt to name the unnameable.

As previously indicated the desire to draw attention to a world beyond
closure can be understood as the outcome of a recognition of the limitation of closure. This recognition can result from a general awareness of an
irreducible gap between linguistic and sensory closure, or as the result of a
determined attempt to complete linguistic closure. Examples that have been
mentioned in the relation to the former case have been experiences such as
standing beneath the stars, or facing a grand landscape. In these cases our
capacity to describe our experience appears a pale shadow of the experience itself. In such circumstances the failure of closure is not brought to
the fore by extensive and detailed attempts to achieve closure, but by the
seeming impossibility of such an exercise and the consequent assumption
that it is not possible. It is a recognition therefore of the limitation of

closure that is driven by a wonder at the world and our inability to comprehend it. Alternatively, the limitation of closure can be recognised by the deliberate attempt to complete closure, and the resulting failure to do so. In this case the awareness of the failure of closure is not the outcome of a general sense of an inability to describe experience, but can instead be seen to be the result of a determined attempt to provide such a description and the consequential identification of failure in this attempt. These two mechanisms by which we become aware of the failure of closure can both lead to a search for openness. The former is often associated with what might be identified as a religious approach and the latter with what might loosely be thought of as a philosophical stance.

The attempt to mark that which is other can therefore be the outcome of a religious or a philosophical motivation. In either case however, although the attempt to point towards a beyond to closure is not immediately paradoxical in the manner that would apply to an attempt to describe openness, it is still the case that the attempt is compromised by the provision of a mark that requires realisation to have content, but if realised is at once a closure. In the attempt to name the unnameable, the mark seeks on the one hand to identify an other to closure, but at the same time in identifying such an other is at risk of reducing it to a closure and thereby to have failed in its goal. As with the other two stages in the search for openness that have already been described, the attempt to name the unnameable cannot in the end be fulfilled. For in the limit the attempt to point to that which is beyond closure either carries with it elements of closure or it is empty.

Since the failure to describe openness is embedded in the attempt to do so it has understandably been argued that the desire to name the unnameable should be curtailed. Amongst others Wittgenstein and the positivists are perhaps the most obvious examples of such a philosophical stance.[1] The strategy is not however so easily dismissed. For the problem with the Wittgensteinian doctrine, that what should be said should be said clearly and what cannot be said should be passed over in silence, is that in the limit nothing can be said clearly, and there is nothing about which something cannot be said. Although the attempt to name the unnameable is paradoxical, it is in this respect not dissimilar from closure generally. Closure seeks to reduce openness to some thing, when it is not some thing but open. Although this procedure fails it does not have the consequence that it is pointless, for it is through failure and the response to that failure that it becomes possible for us to build the whole framework of personal and cultural space. The naming of the unnameable cannot describe openness any more than closure can be made complete, but that does not make it pointless.

What distinguishes the naming of the unnameable from closure in general is that the name is specifically offered as an indicator of that which cannot be named. Unlike closure therefore the mark is not provided in order that it should be realised as material, but is offered in order to indicate that closure does not exhaust the world. The offer of a name which is at once not a name has the consequence that the attempt to name the unnameable is a deliberately mystical activity. It is however found in many aspects of culture. There are for example aspects of literature and art that combine the avoidance of closure with the desire to identify a beyond. In order to provide a framework for the examination of the mechanism involved however the variety of means that are employed in that attempt to name the unnameable will be reduced to the two distinct approaches which have already been indicated, namely the religious and the philosophical.

Religion

Religion as a system of belief, backed by an institution, provides an explanation and a description of the world, and in this respect functions as a closure or linked set of closures. As with closure generally, religious beliefs enable certain forms of intervention, and make possible certain modes of behaviour that would not otherwise be possible. As a closure most would nowadays accept that the stories of religion are, in the context of our everyday reality, currently not as practically effective as the stories of science. As it has been argued the success of science in this respect is not due to its providing a more accurate, or true, picture of the world; nor is it the result of a fundamental difference in the character of its closures. Instead the case has been argued that the success of science stems from its encouragement of the identification of the failure of closure and the realisation of new closure in response. While science evolves therefore through empirical enquiry, religious closures are more typically bound by the authority of a primary text or texts and are less able therefore to realise new closures and thereby increase their capacity to intervene.

There is however another side to religion that has not generally been present in science and which can be regarded as central to the religious approach. From this perspective religion is concerned not only with closure but with a search for openness. What impels the religious attitude is not only the provision of a system of belief and thus a system of closure, but the desire for openness which is found in those moments when the world is experienced as a mysterious and wondrous place. In this context science can be seen to seek to give an account of reality, to explain our world, while religion in this respect can be regarded as attempting to point to that which is beyond our understanding. In this mode, religion points to the unexplained as an expression of the desire for openness, a desire to marvel at the world rather than a desire to understand it. Instead of improving on the everyday closures that enable us to intervene in our reality, religion, in this sense, seeks to draw our attention to the limitation of our everyday understanding and to point to that which is not understood, to that which lies beyond closure.

What has been identified as a religious desire to point towards that which is not closure and to thereby name the unnameable, requires a means of expression. As soon as it is expressed however it can be realised as a closure rather than as a gesture towards openness. In seeking to understand the religious response to this predicament the religious approach to the world can be understood to have both an exoteric and an esoteric side. The exoteric side consists of the religion as a set of fixed beliefs that seemingly offer an account of the world, its origin, and our place in it. The esoteric side in contrast is concerned to point to that which we cannot know, to that which lies beyond our reality, beyond closure. Inevitably, the esoteric aspect of religion cannot consist in any particular account of our circumstances, but is instead engaged in a gesture to that which is other, the aim of which is to draw attention to the limitation of our closures and our current reality.

Marx's oft-quoted dictum that 'religion is the opium of the masses' can, in the context of this distinction, be seen to have obtained its force from the exoteric characteristics of religion. Religion in this respect can be regarded as offering an unchanging system of beliefs, a set of closures, that claims to be the one true account of the world. As a result it can be argued that it encourages an acceptance of the current circumstance and thereby undermines a desire for new closure and thus a new means of intervention. Such an account of religion, as the provision of a relatively fixed framework of closure, is appropriate to its dominant institutional character, but it ignores the esoteric attempt to point to that which is beyond closure. When, under communist regimes, religion has been outlawed, the weakness of the dictum has come to the fore. The scientism that backs the Marxist perspective suggests that closure can be complete – at least in its post-Leninist phase.[2] Such a vision is however at odds with our experience of closure and has no place for the desire for openness, whether expressed through art or religion. It is perhaps for this reason that the cultural space that results from such a system can be seen to have

elements that can be associated with attempts at pure closure: a dry precision which some might describe as soulless in character. The resilience of
the desire for religion in Eastern Europe and Russia, during the communist
period, can in this context be interpreted therefore not so much as a desire
for an alternative closure but a desire for openness in a cultural space that
asserted only closure. This desire may also have played a part in maintaining the equally widespread belief in the supernatural that is a marked
facet of contemporary Russian society.

The esoteric and the exoteric aspects of religion are not independent of
each other. For the exoteric account is at the same time the means by
which the esoteric notion is conveyed. The esoteric motivation that can be
thought to be at the inception of the religious attitude requires a means by
which it can be expressed. The exoteric story is the vehicle for this esoteric
expression. As a result the relationship between the esoteric and the
exoteric is an important one for religion, functioning at many levels
within the institution. In the Judaeo-Christian tradition, for example, the
esoteric and exoteric are combined at many different levels from fundamental beliefs to the practice of specific rituals. At the level of the
underlying framework of belief, the esoteric and the exoteric are both
incorporated in the name 'God'. 'God' is on the one hand an attempt to
point to that which is beyond closure, an attempt to name the unnameable. On the other hand, 'God' is also part of an account of the world and
its origin. This exoteric aspect of the notion of God is itself a continuum,
from an unobservable ultimate cause, to the notion of a being who
controls the world and both judges and intervenes in the events of
humankind, to an image of God as a grandfatherly figure with a white
beard.

The esoteric and exoteric sides of religious belief can be seen therefore
to be bound together for each requires the other. The esoteric notion alone
without an exoteric framework would have no means of expression. On
the other hand the exoteric account proposes a description of that which
lies beyond closure, but such a description is at once paradoxical for the
other of closure cannot be understood in the context of closure. The
exoteric account requires therefore that it be understood esoterically. It is
because the esoteric and exoteric are inherently linked that each religion
has its own particular character even though each can be regarded as a
response to the desire for openness and a recognition of the limitation of
closure.

In order to examine in more detail the pattern of the relationship
between the esoteric and exoteric it is helpful to consider the manner in
which the notion of 'God' is given meaning. A first attempt to describe the

function of this term might be to argue that the esoteric notion of an other is given exoteric meaning through the mark 'God'. This exoteric meaning is not however a description of something in the world, for it seeks to describe that which is beyond closure. Furthermore it is this characteristic that enables God to be an explanation of the world. We can of course give an account of the cause or origin of some happening in the world by reference to some other thing or event, but it is not possible to give an account of the cause or origin of the world taken as a whole by reference to some thing or some event within the world. Science can give an account of the causes of events in the world by reference to other aspects of the world, but it cannot give an account of what caused the world to exist, for the cause would have to be outside of the world. Contemporary science flirts with the notion that it can provide a so-called bootstrap theory, but this does not look like a possible task. Even if such a bootstrap theory could account for the self-origination of the universe in the context of a certain set of laws and mathematics, it could give no account of what originated this framework nor why at a certain juncture these should be instituted. While the exoteric notion of God can be regarded therefore as setting out to provide a description of that which is beyond closure and thus allows for the possibility of an explanation of the world, in practice the attempt to offer a description at once embeds the description in the context of closure. If, for example, God, rather than being an esoteric pointer to that which is beyond, is understood in the specific sense of an individual, or being, responsible for the creation of the world, the concreteness of the description undermines the function it sets out to achieve. For the description makes God part of our closures, a thing like all other things, however omnipotent. If God is understood and realised as a closure, we can ask of this realisation what we would ask of any other closure. We can question its origin, its cause, its function. If we suppose that God made the world, we are still left with the equivalent question: who made God? As soon as the notion becomes exoteric therefore, it at once undermines that which the notion set out to achieve. For instead of pointing to that which is beyond closure it turns the beyond into part of closure, and the initial motivation, the recognition of the limitation of closure, is thereby undermined.

Every further detail that is added to describe the characteristics of God can in this context be seen to serve to undermine the role of God as a beyond to closure, and at the same time brings to the fore the failure of the exoteric closure. If, for example, we look at the elaboration of the idea of God in the Christian notion of 'God the father', such marks when realised make the challenge to the esoteric notion of God more evident.

The description 'God the father' when realised in the context of our current closures proposes God as a paternal individual. Such a notion is immediately at odds with the esoteric notion of God, for instead of being a beyond to closure it offers an account of the other that relies heavily on the particularity of our own worldy experience. In recognition of this it may be argued that 'father' is to be interpreted metaphorically to mean someone who looks over humankind in a protective manner, but there remains a parallel with human individuality. In order to maintain its initial function the characterisation of God the father needs to be understood entirely esoterically. Yet if the description 'father' has nothing in common with the everyday notion of family, or with the notion of individuality, little sense can be made of Jesus as 'God the son', or of the claim that we are made in God's image, and more importantly the use of the term no longer has a role other than to have attempted to describe that which cannot be described.

At the exoteric level the Christian church has sometimes portrayed God as precisely a wise, fatherly figure with a white beard, and for those seeking an easily realisable closure God can be understood in this manner. Such a realisation is however clearly incompatible with the esoteric notion of God, and therefore with the religious motivation that it is to be supposed initiated the notion. Furthermore as an exoteric account of the world and its workings such a closure is of little value when placed alongside the alternative and powerful closures of science. The exoteric story has value therefore not as a description of the beyond, but as a metaphor for the limitation of our understanding, and it is in the abandonment of the literal to the point where there is nothing left of the specific closure, that the exoteric story functions as a pointer to the unnameable.

While it is therefore the esoteric notion that gives value to the exoteric attempt to provide a detailed closure, it is the exoteric story that enables the religion to prosper. In the context of Christianity the esoteric and exoteric can be regarded as having been institutionalised, into a distinction between the private and the public, between the church and the laity, between the sacred and the mundane. The exoteric aspect of the religion apparently offers a description of that which is beyond, in the form for example of heaven and hell, angels and devils. It is a description that can have no basis in the experience of each individual for it is a description of that which is necessarily outside of reality. Since it cannot appeal to experience, nor can it point to any thing or aspect of reality as evidence of its account, in order for a religion to promulgate a particular account successfully its account must carry authority from some other source. Since there is no means of accessing such knowledge from our reality and our closures,[3] its authority can only come from the world beyond. The stories of Christianity are therefore given authority not by any person or thing in our reality but by a supposed direct link to the world beyond: through God or God's agents. This can be seen to operate both at the level of the textual basis for the system of beliefs and at the level of the organisation of the institution. Thus the authority of the Bible stems from those who have a special link to God, either in the form of prophets, or in the form of Jesus, the direct voice of God in human form. In turn the authority of the institution of the church can be seen to stem from its members also having a special link to the world beyond, which provides them with an authority not available to the lay public.

In its public face therefore the institution of the church advocates the adoption of an exoteric story. These closures offer a seemingly graspable version of reality, its origin and our place in it. There must remain however an esoteric meaning that is available to the initiated, to those on the inside. By incorporating a public and private face religion is then able to present the belief system within quotes, as if it is not to be taken seriously. It is understood that only the initiated have an understanding of the unsayable and essential character of the religion. By this means religion is able to offer a set of closures that can be taken, by those who wish to take it that way, as a straightforward closure: a description and explanation of the world and our place in it. At the same time the religion is able to use this story as a ladder to the esoteric identification of that which is beyond closure. From the perspective of the lay observer there is a mystery to the story conveyed to them which is known only to those who have access to the esoteric secrets of the religion. To the initiates, the public story is the framework through which individuals are led towards the esoteric. The network of beliefs we associate with institutional religion, and which are the means by which it is identifiable, can thus be seen to be a necessary and in a sense 'secular' requirement of the religious belief itself, which is in contrast not capable of description.

In the same way that a system of religious belief can be regarded as functioning at an esoteric and exoteric level, so also can we interpret the religious rituals, and the physical space and architecture adopted by a religion in the context of this distinction. Rituals in this context can be seen to be, on the one hand, a learnt and repeated set of actions, and as such have an exoteric character. On the other hand, rituals are an attempt to escape the limitations of everyday activity and the constraints of closure, and aim to carry with them an esoteric significance. Ritualised activity in this sense is not undertaken in order to intervene in the world but to bring attention to the other of closure, to that which is beyond our reality. The drinking of wine and the breaking of bread, for example, in the Christian communion service is a worldly metaphor for that which is not of this world. It is intended to indicate the esoteric, but it does so through an exoteric ritual with associated exoteric beliefs as to the nature and significance of the ritual. In this respect rituals can be regarded as echoing the structure of religious belief. It is perhaps for this reason that those who devote themselves to a religious life often embed themselves in a rigorous set of rules and rituals. The medieval monastic discipline which allocated to every hour of the day a task, and which insisted upon ritual prayer many times a day, provided a framework of behaviour that was deliberately not of this world, and which thereby sought to constantly reaffirm the existence of that which was other. As with the system of beliefs that make up the exoteric framework of faith, there will be practitioners who see in the rituals no more than the exoteric surface. In so doing however the esoteric aspect of the ritual will have been lost. Since the esoteric is that which distinguishes religion from other general attempts at closure, the loss of the esoteric is also the loss of that which identifies the activity as being specifically religious. As with the religious text the ritual has value as a pointer to the esoteric. If however this element is ignored or abandoned the activity is reduced to a known activity that has the character of farce. Seen from the outside, religious ritual, be it a Druidic ceremony, an ancient Greek sacrifice, an Aztec incantation, or a Christian service, is an institutional dance of learnt behaviour without practical function. As such it can appear to be merely laughable. Its value however can be regarded as coming through its association with the esoteric, for the ritual is not seeking to achieve something in the world, or express something about the world, but to gesture towards that which is not of this world. It can be seen as a reminder of a beyond to closure.

Similarly the architecture utilised by religion is on the one hand a repeated style of building, or physical space, but it is at the same time an attempt to mark out a holy place, a location which is not of this earth, a point of passage from this world to another. In this manner most religions seek to imbue their place of worship with a mystical otherness. At many levels therefore, from the most general of its beliefs to the most specific of its practical characteristics, institutional religion offers a framework that is necessarily exoteric – for otherwise it would have no content – but whose underlying motivation is esoteric. The exoteric aspect of this framework is the vehicle by which it becomes possible to encourage and make more credible an escape from closure.

In the context of this account, the extraordinary success of institutional religion can be explained in the light of its combination of the exoteric and esoteric. As an exoteric story that seeks to provide a means for inter-vening in the world, the closures of religion have been weak. Unlike science, religion operates with a largely fixed set of closures and over time its shortcomings in handling activity come to the fore. If religion had to rely on its capacity to enable successful practical interventions in the world alone, it would therefore have been unlikely to have found widespread support. Battles are not won by prayer but by superior arms and tactics. The resilience of religion, in the face of challenges to its capacity to enable successful intervention, can instead be seen to stem from its esoteric other-worldliness which enables the realisation of any specific closure to be challenged. In the West, for something approaching two millennia, each side on the eve of most conflicts has prayed for God's support in their cause. Despite the evident failure of this strategy – since it is not possible for both sides to have won – the ritual of prayer has remained. One means of explaining this resilience is to recognise that although prayer has an exoteric character, it also has an esoteric function in encouraging the indi-vidual to look beyond their current closures. A failure in the realm of the exoteric can thus be accounted for by reference to its esoteric character. Similarly, the breakdown of closure that follows a literal realisation of the Christian Bible can be offset by the text itself being treated as holy, which has the consequence that any realisation of its marks can be dismissed as a limited human attempt to comprehend something beyond us. The church can disseminate specific closures but at the same time deny their speci-ficity when challenged. For those who are satisfied with the closure therefore closure is offered; and for those, whom we might describe as the truly religious, who seek openness, the closures provided can be regarded as a vehicle to aid the escape from closure. In either case the outcome may prove useful to the individual concerned, and may in the case of a battle result in a greater determination to succeed.

Given this outlook a case can be made that in circumstances where a religion ceases to operate at the esoteric level it is liable to rapidly succumb to the failure of the closures that it offers. The relative weakness of Christianity in an age of belief in science might therefore be seen to stem not from the lack of evidence for its claims, which has always been the case, but from an attack on the possibility of the esoteric. Science has largely successfully contributed to the notion that all that can be said can be said clearly, with the consequence that the marks of Christianity are treated as tags to closures whose inadequacy in the practical sphere are apparent. The case that has been presented however suggests that the antagonism between science and religion does not centre on their providing alternative closures, but on the characteristic of science that it is often taken to deny the possibility of anything other than closure.[4]

It is one of the problems faced by those who wish to convey a religious sentiment, that while the naming of the unnameable is an esoteric pursuit, the spread of a religion requires an exoteric face. The framework of belief, the rituals, the material space and objects, which are routes towards the esoteric are however at risk of becoming everyday closures which, instead of functioning as a ladder to the esoteric, function inversely to exclude the possibility of openness. As a result, religion, whose underlying impetus might be thought to be driven by a wonder at the world and a recognition of a beyond to closure, has often been transformed into the precise opposite: an extreme form of closure. In this sense the realisation of the closures of a religion are at odds with the religious sentiment. The contemporary theologian, Don Cupitt, adopts a stance of this type, in *Taking Leave of God* where he argues that it is in the abandonment of belief in God that it is possible to come closest to God.[5] In abandoning the exoteric, Cupitt seeks to recapture the esoteric. In explicitly advocating such an outlook Cupitt risks alienating those members of the church and of the laity who wish to retain the familiar closures of the religion. Yet, in the abandonment of the closure 'God', and the advocacy of the pursuit of something other than closure, openness sought through the cue 'God' which has no realisation but instead a history of abandoned closure, Cupitt seeks to express what might well be regarded as the essence of the religious sentiment.

It can be seen therefore that although religious belief has often consisted in the adamant assertion of the specific closures of a particular text and the practices of a particular institution, the underlying motivation for religion is found in the attempt to escape closure and is not in competition to provide the correct answer. Each religion utilises its own story, its own ladder, by which it seeks to name the unnameable. This story or ladder must be discarded or remain unrealised if the religion is to retain its initial esoteric character. While each religion therefore has its own way by which it seeks to name the unnameable, in its desire to pursue openness through the naming of the unnameable it is at one with all religions. In this context there is more in common between devoted followers of divergent religions than between religious and non-religious individuals within a culture. Christian and Buddhist monks both seek to escape the closures of the world and in a similar manner wish to avoid earthly desires and goals, for they are engaged in the same esoteric pursuit. This pursuit separates them from the everyday world of closure typical of society as a whole, so that their lifestyle can be seen to have more in common with each other than it does with the rest of their culture.

Religion therefore combines an exoteric and esoteric aspect. The exoteric offers a set of closures as an explanation of the world, but it is the esoteric that identifies this attempt at closure as religious. When individuals or societies abide by the literal claims of religious belief they seek to impose closures which, on account of their inflexibility, can have undesirable and damaging consequences, as the history of religious conflict is testimony. It is however in the attempt to indicate something other than closure, a pointer, a path to another world, a way into wonder, by combining the exoteric with the esoteric, that religious belief distinguishes itself from the more familiar pursuit of closure and thereby gains its force. As such, for some, it can result in a deepening of experience with consequential effects for how individuals view themselves and the purpose of life. It is not unique in this respect, for there are non-religious means by which attempts can be made to name the unnameable, but it can be seen as one of a range of strategies, if possibly the most influential, which have been developed to draw attention to the limitation of closure and point to that which is beyond.

Philosophy

Philosophy is an activity which perhaps more than any other can be considered to be engaged in the single-minded pursuit of closure. For not only are its practitioners seeking to provide an account of reality or, an account of our description of reality, but they are seemingly also engaged in an attempt to weed out any contradictions or loose ends that might undermine such an account. It is a discipline that has sought therefore to apply a relentless rationality in a determination to uncover a stance that is unassailable. It would appear contradictory therefore that philosophy might also be engaged in the mystical activity of naming the unnameable. However, as it has been argued the single-minded pursuit of closure leads to the identification of its failure, and the recognition of that failure leads to an attempt to identify that which is beyond. This process can be seen to have been present in philosophy from its inception, with the result that there have been many different accounts given of openness, each the outcome of a particular system of closure. On the grounds of brevity, I shall not attempt to give an exhaustive history of this process, but merely to identify in a cursory manner some strategies that have been employed.

What distinguishes the philosophical attempt to name the unnameable from religion is that it flows from a specific identification of the failure of closure rather than a general and intuitive assumption of that failure. It shares with religion however the introduction of an esoteric element into an exoteric account. The mechanism of this process can be seen from the outset of philosophy, most easily perhaps in the writings of Plato. A similar mechanism has been played out in different ways throughout the history of philosophy. At its most schematic there is first an attempt to achieve closure; this is followed by a recognition of the failure of the proposed system of closure; which in turn is followed by the supposition of an other which is beyond closure. Moreover it is because the failure of closure only becomes apparent when it is pursued, that it is precisely those philosophers who have most rigorously pursued closure who have been forced to identify that which lies outside or beyond closure. Since this identification could not itself consist of another closure, the indication of a beyond to closure has necessarily been esoteric.

We can discern three primary strategies that have been employed in the philosophical attempt to name the unnameable. The beyond to closure has been identified as a goal or ideal towards which we strive but which is not itself attainable; as an indescribable other which is defined negatively in the attempt to seek limits to closure; and as a realm which is described but in a paradoxical fashion. In each case the postulation of an other to closure follows an elaboration of a framework of closure which is then seen to fail and require an other, a space outside of the closures offered.

The first strategy is illustrated by Plato, who can be regarded as having used the failure of closure, in the form of our inability to give an account of terms whose meaning we believe to be straightforward,[6] to argue for an ideal world which lies beyond the observable world and towards which we strive. In the famous metaphor of the cave Plato describes our experience as that of a chained individual watching shadows on the wall of an underground cavern cast by the light of a fire, and imagining that these are reality. Instead they are the palest of imitations of what is to be found outside the cave in the bright light of the sun. In this metaphor Plato can be taken to propose therefore that we inhabit a secondary limited world, while we should strive to escape from the cave and inhabit the real world outside. Plato proposes this framework as the result of the failure of closure, but the framework itself can only be described if Plato himself has access to the real world beyond. As inhabitants of the cave, we can identify limitations to our system of closure and thus can point to something beyond our circumstance but we are not in a position to identify the nature of our circumstance. Plato spends considerable time in seeking to demonstrate the failure of closure, but he attempts to go further by offering some form of description of that which is other. Plato's solution to this dilemma is the solution adopted by the Christian church – itself influenced by Platonic ideas – that esoteric knowledge is possible to a few special individuals. Plato therefore argues that it is philosophers who are capable of accessing the ideal world and are thus in a position to attempt to convey its truths, although such knowledge cannot be expressed directly. Plato's identification of the world of forms can be seen therefore to be esoteric, for what it seeks to express cannot be expressed in the context in which it is presented. Knowledge of the ideal, of the world of forms, is not accessible from a world of shadows. While Plato does not attempt to provide a description of the ideal he provides a description of the relationship between the ideal and our everyday experience, and that description is not itself comprehensible from our current perspective – it is therefore esoteric. Plato offers us a metaphor, a story, by which to understand our circumstance, but in order to fully understand the metaphor we must already have escaped our circumstance.

For a long period, in the aftermath of Greek philosophy, Western philosophy was largely dominated by the elaboration of a system of closure compatible with Christian belief. As a strategy for naming the unnameable it is largely subsumed therefore within the religious attempt to name the unnameable, which has already been described, through the elaboration of a framework incorporating the story of God. It is therefore only in the last three to four hundred years that Western philosophers have begun once again to offer alternative ways of seeking to name that which is beyond closure.

Amongst post-Cartesian philosophers, Kant and the early Wittgenstein offer good examples of the second strategy identified in the attempt to name the unnameable. This strategy identifies an other to closure through the limitation of closure and as such refuses to offer a description, or even sometimes a name for this other. A system of closure is offered, which in Kant's case involves the provision of a description of experience and understanding, and for Wittgenstein consists in a description of language, but in both cases the system of closure relies upon an other which lies outside of the framework provided. Kant's world of appearances, his description of experience and knowledge, cannot contain that which causes experience or that which has the experience. Reality and the self in this sense are therefore for Kant beyond the world of appearance and thus transcendent. Kant identifies these elements that lie outside of experience or knowledge as noumena.[7] In this description and with it the postulation of a noumenal world can be seen the attempt to name the unnameable. For 'the noumenal world' seeks to identify something beyond closure, something which cannot be identified for it cannot be experienced or known.

Kant's noumenal world has the familiar characteristics of attempts to name the unnameable. It appears to identify something outside of closure which in this case consists in the underlying reality which results in our experience, and the thing which has the experience. However, if the noumenal world was intended to identify these things it need not be referred to as the noumenal world but instead as 'reality' and 'the self'. Kant does not use these terms because they would at once be part of our knowledge and thus part of the world of appearance. That which lies beyond the world of appearance cannot however be understood in the context of the terms which Kant uses elsewhere to describe experience and understanding. He therefore seeks to point outwards from his system of closure to a world that cannot be identified as any particular thing but which is necessarily other. As with all attempts to name the unnameable however, by naming the other in the context of a specific system of closure it would appear that Kant has already introduced an element of closure into that which is necessarily beyond closure. Thus although Kant frequently asserts that nothing can be known about the noumenal world, this world is offered as the site of the transcendental object and the transcendental subject,[8] in which case we already have an understanding of that which is supposedly beyond understanding.

Philosophical attempts to name the unnameable can be seen therefore to be driven by the enforced recognition of the limitation of the system of closure provided. As a result the provision of the name, even though it is intended to be empty of descriptive content, carries with it the character of the system of closure from which it originated. Kant's noumenal world is offered in the context of his world of appearance and the remainder of his system of closure, but it is as a consequence no longer empty of descriptive content. As soon as we imagine the noumenal world as the underlying reality that lies behind the artifice of experience generated by the combination of intuition and concept, we have introduced into 'the noumenal world' an element of closure – even though there is no detailed description of the nature of this underlying reality. For the constraint of the notion of 'an underlying reality' is already a closure, a way of holding that which is not capable of closure. Although therefore Kant presents the noumenal world as an empty notion it has necessarily an exoteric element. By identifying a noumenal world Kant draws attention to the limitation of closure, which is at once undermined to the extent that our notion of the noumenal world carries with it elements of the system of closure from which it originated.[9] In order to understand Kant's notion of the noumenal world we must therefore not understand it, and it is in this sense esoteric.

As with Kant, Wittgentein's *Tractatus* can be regarded as offering a system of closure, in this case a description of the workings of language, which necessitated the supposition of an other which lay outside of the system. Unlike Kant, Wittgenstein avoided providing a specific name for this other, and instead referred to the limits of language. This did not however mean that he was not also engaged in the attempt to name the unnameable, in the sense that has been described: namely that there is an attempt to point towards that which is other. Wittgenstein's avoidance of an actual name for that which is beyond his system of closure, was driven by his explicit recognition that a beyond cannot be identified, but the notion of a beyond that cannot be described or referred to is found throughout the *Tractatus*.

No doubt precisely because the attempt to name the unnameable is esoteric, it has not been uncommon in a scientistic age for philosophers to seek to 'tidy up' philosophical systems like those of Kant or Wittgenstein in order to rid them of what are seen as unnecessarily mystical elements.[10] The thought was that if only one could excise these forays into mysticism there might be viable theories of knowledge or language that could be rescued and incorporated into the main body of our understanding of the world. Such attempts can be seen to be motivated by the desire for closure, which finds in the esoteric an uncomfortable lack of completeness. However it is precisely the rigour of philosophers such as Kant and Wittgenstein that resulted in exposing the limitation of their own systems of closure. Only through a lack of awareness, or an avoidance, of the incompleteness of closure does it become possible to imagine that the esoteric can be excluded.

The framework of science has been so successful in providing means to intervene that it is a seemingly short step to imagine that it might be extended to provide a complete account of the world. In this manner, some philosophers have imagined that areas like epistemology, or the philosophy of language, might be turned into a science. In one sense this is true. There is for example no reason that aspects of our current system of closure which are presently disputed might come to be widely regarded as unchallengeable. As a result it is possible that an epistemological theory or a theory of language might be proposed that comes to be realised by almost everyone. In another sense it is an impossible dream. For these theories cannot escape the characteristics of closure and will therefore necessarily fail in certain respects. If these failures are examined carefully they will in turn be seen to reintroduce the esoteric. As it has been argued, science is no more capable of providing a complete closure than any other system. It only appears possible because the success of the framework is capable of obscuring its limitation; an oversight which is encouraged because science claims to be on the way towards a complete closure with the result that it can be supposed that some future theory will solve what for the present can be regarded as a temporary difficulty. Yet a future theory will itself generate further failures. While science usually supposes therefore that the esoteric can be overcome, the esoteric is unavoidable because closure fails.

Although many philosophers have in some form sought to point to that which is other than closure, and although this attempt has been an important element in their philosophy, it has usually been the case that the bulk of their work has been devoted to the realisation of closures or a system of closure rather than an attempt to point to that which is other. Heidegger is the exception, for his work can be regarded as being almost entirely devoted to the attempt to name the unnameable. It is not surprising therefore that his writing has been of interest to theologians, or that those, who have thought philosophy could aspire to being a science and thereby eradicate the esoteric, have in general regarded his work as worthless.[11] While Kant and Wittgenstein point to a beyond as the result of drawing limits to closure, Heidegger can be taken to illustrate the third philosophical strategy in the attempt to name the unnameable by appearing to offer a description of that which is other only for this to be presented in such a manner that it is not capable of closure.

One of the consequences of Heidegger's concentration on the attempt to name the unnameable is that the descriptions that one might be tempted to offer of his philosophy are likely to be misleading. What his work tries to say can be seen to be esoteric. The forms by which he seeks to express this are necessarily exoteric but they are at the same time discarded. As a result it is possible to begin by saying that 'Being' is Heidegger's name for that which is beyond closure. His work can therefore be considered to be devoted to the attempt to uncover the nature of Being. Heidegger is however aware that by naming that which is other there is immediately a risk of providing a closure rather than offering a pointer towards openness. He is more inclined therefore to say that he is engaged in the attempt to find a way to ask the question of Being, rather than to claim to be offering a description. His early work, *Being and Time* can be taken as an illustration of the tension between these two accounts. In the work, Heidegger elaborates a highly complex framework and a new vocabulary which at times might appear to suggest an account of that which is beyond. It is a framework built in opposition to the closures of our current cultural space. It is not however a framework that can be regarded as a description of openness, for such a description must be elusive. Instead it can be seen as an attempt to show how our current closures are not complete and thus to indicate how it might be possible to approach the question of Being.[12]

Heidegger's early work is sometimes interpreted as an account of Being, and thus a description of that which is beyond. Such an account is at once paradoxical and as a result it would seem more credible to understand *Being and Time* as functioning as a ladder to the esoteric. In the same way that the exoteric stories of religion function to point to that which is other. In his later work, Heidegger abandons the detail of the extended exoteric story outlined in *Being and Time* in favour of a search for a means of calling upon the esoteric more immediately. Instead of a monolithic system, we have a variety of tactical manoeuvres to approach the naming of the unnameable. For Heidegger is at once aware that the name, however undermined, however preliminary, is at the same time an exoteric intrusion into the esoteric attempt to say what cannot be said. As a result his writing becomes increasingly mystical. He offers us a variety of ways to characterise his approaches to that which is other. For example, that he is engaged in an attempt to try to move towards the possibility of asking the question of Being, or later the question 'why?', or later again the question of language. Such formulations would however be largely without content without the framework of his earlier work which, although abandoned, still functions as an indicator to what might be afoot. It can be argued therefore that it is through offering a name for the unnameable, in the form of Being, and furthermore a whole system of related closures, outlined in *Being and Time*, that Heidegger is able to show what it is that he is wanting to achieve. It is in such a context that he is then in a position to point out that this name is not in fact satisfactory because in the provision of the name the unnameable has been obscured. As a result, for a period Heidegger continued to refer to Being but crossed it through in order to show that as soon as it was regarded as a thing that might be named it was at once redundant. It is not possible however to avoid the residual closure left in the attempt to name the unnameable, even if it is denied or crossed out, for there remains the pointing towards which already indicates a characteristic for it points outwards from somewhere. Recognising this Heidegger can be seen to take one further step by proposing that the task is not to name the unnameable but to move towards a space from which it is possible to hint at a name. In the context of the account that has been offered in terms of the exoteric and the esoteric however, such manoeu-vring can be seen as an attempt to escape the exoteric and leave only the esoteric. However no such escape could be possible while still leaving any content. In order to be able to understand the possibility of the esoteric we require a context, and that context must be exoteric.

Heidegger is not the only twentieth-century philosopher who can be regarded as having offered names for the unnameable that are then abandoned or undermined in order to preserve the esoteric nature of the enterprise. Derrida can be interpreted in a similar manner. We are offered a whole series of names for the unnameable: absence, writing, 'differance', supplement, trace, track, pathway. Each at the point at which it is introduced serves to point outwards from the current arrangement of closure. Each is abandoned because the naming itself has made that which is other part of closure, a seemingly known and identified space. Thus 'writing' is at first a means to point to that which is beyond language, beyond the logos, but in offering a name for this absence Derrida has seemingly made it present and given it decidable meaning. He therefore chooses another means of seeking to point outwards with the term 'differance', which, through its own spelling, differs from itself thereby postponing the point at which it becomes a known and understood closure.

In order to maintain the esoteric intention of the name of the unnameable, which is at once threatened by its adoption as a name in the manner of a closure, Derrida employs a series of textual manoeuvres. As with Heidegger, the names are crossed out or erased; they are abandoned in favour of new names; the shifting character of the text caused by the successive abandonment of the names is itself regarded as a means of showing what it cannot say directly, it is a pathway, a pointer to the esoteric; the pathway is itself abandoned as too overt a notion in favour of a glimpse or a hint which is not spoken of directly but inferred. It will however now be apparent that there can be no end to this progressive denial of the process of naming. The esoteric slips from our grasp as soon as we seek to identify it, no matter how carefully it is approached. The unnameable cannot be named however softly the name is called, however frequently the name is changed, however implicit and unsaid the terms. One can suppose that it was for this reason that Wittgenstein in his later work adopted the strategy of avoiding all reference to the esoteric. Yet silence was merely another strategy, another name for the unnameable which was not annunciated, but on which the remainder of his text relied.[13]

Large parts of analytic and pragmatist philosophy have deliberately sought to avoid and excise the esoteric on the grounds that it is either meaningless or unfounded. In one sense this strategy is justified, for it is only through closure that we are able to have knowledge. The provision of a name or description of that which is outside of closure will either fail or it will be empty of content. There can be no religious story or philosophical account of the beyond to closure that provides knowledge of openness. Where the denial of the esoteric is misguided is where it also implies a belief in the possibility of the completion of closure. Although openness cannot be described, the attempt to name the unnameable can be an encouragement to the recognition of the limitation of closure through the identification of the esoteric. In a more single-minded manner than religion, philosophers seek to complete closure, and as a consequence the provision of an esoteric name is not sufficient. One consequence is the provision of a series of closures that cannot achieve what they set out to achieve and which in their expression undermine the identification of the esoteric.

To what purpose then is this cyclical manoeuvring in search of the esoteric? Despite the esoteric aspect of the attempt to name the unnameable, the attempt nevertheless has an impact on the capacity to intervene. Although the approach must fail, it may for some draw attention to the limitation of closure and point to a beyond even if the beyond is not identified nor can it any longer be characterised as a beyond. The naming of the unnameable cannot be purely esoteric and its exoteric content necessarily has consequences for intervention. The role that the naming of the unnameable plays can in turn be seen to depend on the particular system of closure within which it takes place. For the esoteric cannot of course be taken to be a single place that remains the same in all circumstances. 'Writing', 'Being', 'God', are all names of the unnameable, and play a similar role in the space that they occupy, but they do not name the same thing. The esoteric other to which they seek to point cannot be imagined as a thing which they are all somehow seeking to name, while the exoteric content in each case is necessarily different and the consequences for intervention are correspondingly diverse. The esoteric is necessarily defined by reference to the exoteric and therefore each system of closure can be seen to have its own form of the esoteric. In some systems of closure the attempted identification of the esoteric may encourage a sense of humility, a sense that what we know is not so very much; in other cases it can encourage arrogance, through the notion that there is contact with a greater power. The exoteric content of the name has like all closure an impact on the whole system, the space of closures in which it finds itself. If it was possible to name the unnameable purely esoterically, it would as a consequence have no role in the world of closure and would correspondingly have no impact on the capacity to intervene in a reality that is itself the product of closure. Any naming, however tentative and however circumscribed must nevertheless take place in a context and have an impact on that context, with resultant consequences for possible means of intervention.

One way of describing the attempt to name the unnameable is as an attempt to provide a description of the location within which personal and cultural space operates, an attempt to observe the system of closure itself and place it in context. The naming of the unnameable is therefore in this sense at the apex of the process of closure, and is for this reason significant for the character of the space as a whole. Yet a system of closure cannot catch sight of itself, for to do so it would have to stand outside of itself. The name of the unnameable is thus both the framework within which everything else sits and is nothing. The circularity of the exercise does not mean that we can simply decide to abandon it, for it is at the same time a continuance of the process of closure. The answer therefore to the question of what purpose there is in seeking to name the unnameable is the same as that which applies to closure: there are as many purposes as there are names. That the names will have to be abandoned and will not achieve their goal, does not mean that they have no function any more than the failure of closure is reason to abandon the attempt to provide closure. There is a strain in our culture, perhaps itself a Platonic legacy, that has seen in the attempt to name the unnameable a high and elevated motive. The contrary belief prevalent in twentieth-century Anglo-American philosophy has seen in the attempt to name the unnameable a meaningless and purposeless exercise. It is neither. The attempt to name the unnameable is a continuation of the process of closure that can have far-reaching consequences for space as a whole. It is not possible to deny the esoteric for it is bound into the process of closure from the outset. On the other hand a pursuit of the esoteric does not necessarily involve a higher motive or a beneficial outcome. History is strewn with examples of those who in the name of the esoteric have caused harm either to others or themselves. The esoteric is pursued in an exoteric context, and that context should be judged in the same way that we judge any other closure. The search for openness is no more predictable in its resulting capacity to handle activity than the search for closure, and the names that are offered should be treated with the same care.

Openness and self-reference

There is a final point that should be made by way of a postscript to this chapter. One way in which the pursuit of closure leads to an uncovering of the esoteric is the point at which the system of closure identifies itself as such, the point of self-reference. At the point of self-reference the limitation of closure is made evident because the system seeks to identify itself, in doing so it must stand outside of itself, but to be outside is no longer to be part of the system. A system of closure at the point of self-reference therefore uncovers its general failure, and in doing so points to the esoteric.

The term 'openness' within the theory of closure presented in this book is itself an attempt to name the unnameable, to provide an outside to closure. No description has been given of its content but nevertheless it suggests a closure whose meaning is understood by reference to the rest of the system. Such a meaning is misleading, for the intent of the term is inexpressible and esoteric. No other device, however, such as erasure, or the replacement of 'openness' with some other mark in an attempt to avoid the exoteric has been employed. This has not been done because the attempt to avoid the exoteric is both unavoidable and unnecessary. The framework of openness and closure brings the esoteric into the exoteric. The simplest and most elementary of closures carries within it an openness that denies the possibility of the closure. We are caught in the play of openness and closure, of the esoteric and the exoteric. It is not that we inhabit a reality that is known and straightforward, and that a mystical other lies beyond, but that each of the elements that make up our reality has, if examined closely, an esoteric aspect. The term 'openness' is esoteric, but so also are the terms 'table' or 'chair'. We do not have to conceive of a radical divide between science and factual disciplines on the one hand, and art and religion on the other. Instead these activities span a continuum in which the exoteric and the esoteric are always to some degree present. Those closures that find themselves at the centre of our personal and cultural space have a defensive wall securing them from openness in the web of closures with which they are linked, but under examination these closures will still fail. The failure of the closures at the edge is more easily identified. There is no need therefore to cross out 'openness' in order to show that it has an esoteric aspect.

The story of closure is both an exoteric theory that asks to be judged in the same manner as any other exoteric account, and is at the same time esoteric. As an exoteric account it defends its own closures in familiar ways and seeks to demonstrate their value by indicating how they are able to encapsulate previously diverse aspects of cultural space, with the implied consequence that it will enable more effective intervention. Throughout however it also involves the esoteric, for the account embeds openness within closure, embeds the esoteric within the exoteric. It is not necessary for the story of closure to be written with a line through all of it, as if to identify its esoteric character, for the story itself expresses that esoteric character from the outset in the involvement of openness with closure. The story of closure aims to provide itself as an example of the theory it espouses. The story of closure thus self-refers in a manner which seeks to express and reinforce each element of that theory. The claim that 'we find ourselves in the play of openness and closure' is therefore itself at play in openness and closure, itself an exoteric claim and an esoteric gesture.

Part V

THE POLITICS OF CLOSURE

Introduction: society and power

Up until this point my concern has largely been with the structure of closure and its role in the context of the individual. The purpose has been to give an account of how closure enables us to make sense of and intervene in the world, despite our inability to access openness. This has involved a description of how experience as the product of sensation, thought, and language, is made possible through closure.[1] As a consequence it has been necessary to trace the role of closure in determining the character of each individual's space and more generally of cultural space as a whole; the manner by which on the one hand the desire for closure and on the other the desire for openness, themselves the product of the process of closure, function to structure that space; and the limitations of closure and the consequences of this limitation. In this concluding part, I shall turn from the role of closure in determining the character of individual experience and understanding to the role of closure in determining the relationship between individuals and the structure of society.

To help uncover the impact of closure on the organisation of society a distinction will be made between the closures of personal space, the closures of institutional or group space, and the closures of cultural space. As we have already identified,[2] closures are only realised by individuals not groups or cultures so the closures of institutional space and cultural space are not closures realised by institutions or culture but are those commonly and typically realised by those individuals that make up at a certain point the institution or culture. The introduction of the notion of institutional or group space will aid our understanding of the relationship between individuals and will unpack the, previously largely unexamined, notion of cultural space.

The case will be made that an individual's personal space is not formed in a vacuum but in the context of institutional and cultural space.

249

Furthermore, since linguistic closure is from the outset social, many aspects of the relationship between an individual's personal space and that of institutional and cultural space are not accidents of the particular circumstances but are the outcome of the underlying structure of closure. As a result, it will be argued that many aspects of the relationships between individuals flow from the structure of closure, power relationships in particular. As a consequence the structure of closure not only plays an important role in determining how individuals and institutions interact, but in determining how societies are organised, how they develop, and what they are capable of achieving.

The part begins with two chapters that lay the ground for the analysis of power and society that follows. The first of these examines the relationship between the closures of an individual and the closures of others. In this chapter the inherently social and political character of personal space will be demonstrated. The second chapter in the part then turns to the question of the role of closure in determining the desires of individuals. For it is only with an understanding of the relationship between closure and desire that it becomes possible to understand the role of closure in forming power relations between individuals and institutions, and as a result in determining the organisation of society. The next two chapters show how power is the outcome of the imposition of closure. The first of these chapters looks at power in the context of individuals and the second in the context of society more generally. The closing chapters of the part then turn to the implications for the organisation of society and the possibility of progress. As a consequence it will become apparent that closure not only determines the character of knowledge and our capacity to intervene in the world, but determines also the nature of human relations and the hierarchy and power relations of society as a whole.

17

THE SOCIAL AND POLITICAL
CHARACTER OF PERSONAL SPACE

Through cultural space we inherit the history of closure and
with it closures of authority which give individuals and
institutions power.

The system of closure that provides our sensations and thoughts, that gives
us reality and the lifetime dance that is experience, is subjective in the
sense that it is realised by an individual alone. It is also at the same time
dependent on the closures of others. For the intersensory closures of
personal space, and linguistic closures in particular, are realised with
encouragement and direction from other individuals who are acting in
accordance with their own closures. Moreover, in the case of linguistic
closure a history of prior linguistic closure on the part of others is
embedded in the marks available in cultural space. For the patterns of
closure offered by the marks of language are themselves the outcome of
countless individual closures that individuals in the culture have realised
over time. Through the marks of language therefore, a history of prior
linguistic closure is carried in every individual's space and in turn every
individual's space contributes to cultural space: what we are is dependent
on the closures that others have realised and what others will be is dependent on the closures we ourselves realise. The irretrievably social aspect of
the closures of personal space is thus played out even in seemingly purely
private moments of experience. There is no hiding place from the closures
of language, and as a result no hiding place from the closures of others. We
do not therefore have our own personal data of the world, a hot line to
openness which provides us with independent information. How reality
is, is a function of how we hold it through our system of closure. This
system incorporates at its highest level linguistic closures that have been
realised from the marks of cultural space, and thus from a cultural history
of the closures of others.[1]

Earlier, when considering the initial stages of the search for closure, it was argued that childhood is largely characterised by the accretion of new closure. At the time the concern was to identify the character of the search for closure in the context of the individual. In the context of society however, the accretion of such closures is at the same time the adoption of agreed cultural closures through the realisation of linguistic marks, and the exclusion of those linguistic closures which lie outside these accepted norms. A baby seeks closures, and these closures enable effective intervention, provide fixity in the chaos that is openness; in short, give content to the world. In the empty fullness of the open any closure will do to hold fast that which is forever flux. The social environment however does not allow merely any closure but encourages, and on occasion insists on, those of cultural space. This is most easily identified in the context of linguistic closure. Wild grasps at linguistic closure are gradually modified into the accepted closures of contemporary society. A gesture or phrase uttered by a child that does not comply with the norm, which in other words belies an alternative closure, is corrected and brought into line. Adults do not in general seek to realise the marks offered by children unless they conform to the socially accepted combination of marks, or closures, of cultural space.

The adoption of the marks of cultural space, in the form of the words of language, and the realisation of agreed combinations of these marks, in the form of the facts or truths of the particular cultural space, is not a matter of choice for the child, for without the adoption of this framework of marks and the closures with which they are associated it will not be possible to communicate with parents or guardians in order to satisfy needs and desires. Sticking to closures that are unrecognised, and therefore the utilisation of new marks, or unrecognised combinations of marks, is unlikely to prove a successful strategy. Moreover, since the closures of any proficient speaker are the outcome of a cultural history of experimentation with closure, they are likely to be more powerful in the handling of activity than those closures adopted randomly by the child. In principle a child could generate closures that might prove to be superior to the currently available social closures in enabling certain specific interventions, but it is unlikely that there will be an opportunity to develop such closures that lie outside of cultural space. There are occasions perhaps when a child may temporarily employ elements of a private language, especially perhaps with other children who have not yet fully acquired the framework of cultural space and are thus more likely to be open to alternatives.[2] These minimal excursions outside of social closure however are in general steadily eradicated by the requirement to communicate with others on whom the child's desires depend, in addition to the likelihood that social closures, given their cultural history, will prove more effective in enabling the handling of activity.

Over time therefore the child first adopts the same marks of linguistic closure as parents and guardians, and then applies these marks as tags in a manner that converges on the use of tags in the immediate social environment. In short the child adopts the closures of cultural space: the child learns language, and realises those combinations of marks that carry authority within the community. As a consequence there is gradual convergence of the personal space of the child and of the adults or more proficient speakers around them. Since all users of a language undergo the same process there is a large degree of similarity between the personal space of speakers of a language. The mechanism for the development and imposition of cultural space on the speakers of a language through the adoption of closures of authority and the formation of institutions will be examined in later chapters. In passing it is worth noting that it is to a considerable extent the convergence of personal space that encourages the notion that there is a single reality and that tags are labels attaching to parts of that reality. The capacity of language to generate convergence, and to enable communication, has therefore beguiled us into imagining that with sufficient care divergence could be eradicated and tags applied with perfect precision to a pre-existent reality that is already parcelled into discrete elements merely awaiting a label.

If we are to understand the social impact of the process of closure it is important to be clear about the relationship between personal and cultural space. It will be apparent from the preceding argument that cultural space plays a major part in determining the character of personal linguistic closures. Cultural space guides and directs the formation of personal space but it does not determine personal space exhaustively. For there remain aspects of personal linguistic closure that evade the dominance of cultural space. Communication and effective intervention requires the adoption of the marks of language, and the closures of cultural space, but this does not have the consequence that the linguistic closures with which each individual operates are identical. There are countless closures – streets visited, people encountered, conversations entertained – that contribute to one person's space that are not available to another. These closures in turn have an impact, through moulding, on the remaining closures of the individual's space. As a consequence, although language encourages and requires a convergence of the personal space of the individuals within a community, there remains a residual divergence that cannot be eradicated. The residual divergence is not restricted to variations in sensory closure, as if the framework of understanding was agreed and was merely applied to each individual's actual circumstances, but extends in a manner that has previously been described to all closures of personal space. Each individual has their own particular history of closure as the result of their circumstances and their particular make-up, and as a result, despite the convergence imposed by the requirement of communication, each individual in a society will realise the terms of language slightly differently: each word will have for each individual a slightly different meaning and over time as the personal space of an individual changes so that meaning will also change. This divergence is however covered up by our need to maintain a stable framework of marks.

As evidence for such an account it is not difficult to provide examples of linguistic marks that are realised differently by different individuals, or by the same individual in different circumstances, even though they are deemed to have the same meaning. 'Car' for a child does not generate the same material, we could say does not mean the same thing, as it will do when the child has grown up and seen many cars and interacted with them. For a child, 'car' may simply indicate a particular object, or perhaps something with wheels, or a given shape. As an adult, the term 'car' has been realised in many diverse circumstances with the result that its meaning encompasses many objects and is incorporated into many other associated tags. It may in response be argued that there is a core meaning to the term which is realised across all proficient speakers, or at least a meaning that is commonly and typically associated with the term. However there remain divergences between individuals. 'Car' for a car salesman may be associated with a set of closures involving income and success. For a taxi driver it may be associated with a contrary set of closures. For each individual the realisation that follows from a set of tags that incorporates the word 'car' will reflect their personal space and the particular local geography surrounding the tag. As a result the realised closure is not the same in each case, nor could it ever be precisely the same. Although this divergence in the closure associated with the tag is always present, it is largely hidden since the assumption of a common set of tags is a requirement for communication.

The divergence in closure can be made explicit by finding circumstances in which this divergence enables realisation for one individual and not for another. We can suppose therefore that a car dealer may realise the closure 'car' when identifying a taxi because in the relevant respects, namely that it is a saleable vehicle, it can be held as a car, while a taxi driver is unlikely to make such a closure; conversely there may be circumstances in which the taxi driver might use the tag 'car' to indicate a closure, when the salesman would not realise the same tag: in order to describe a vehicle that was so rusted and damaged that it had no value for example. These applications of the tag are not merely additions to a core meaning, as if there were a stable centre to the closure which was always retained. As if the taxi driver or salesman could be said to have failed to have understood the meaning of the tag: that despite being a taxi it was still a car, or despite its condition the vehicle was still a car. The meaning of a tag is the combination of the circumstances in which the tag is realised. As a result there is no core centre to the meaning of the word 'car' that cannot be abandoned. At any given point an attempt can be made to provide a dictionary meaning understood as the closures commonly and typically associated with a tag but these associated closures do not exclude many others – sometimes even contradictory ones – nor is there a stable core which is safe from revision. If the salesman's use of the tag became widespread in the community so that 'car' was only used to refer to vehicles that had a significant resale value, its dictionary or core meaning would have shifted accordingly. The apparently fixed tags of cultural space can be regarded therefore as consisting of imperceptibly moving points of similarity found between the personal spaces of the individuals of a society, which are themselves undergoing continuous change. Each language is a constantly shifting system with a myriad eddies and currents, resulting from the changing and diverse personal spaces of the speakers of the language and their interaction.

Although the personal space of each speaker is in fact divergent and although linguistic marks are, as a result, associated with different closures on each occasion of their use, speakers of a language operate on the contrary assumption that linguistic marks in general generate the same closures with the same tags. This assumption is a requirement of language if it is to function as means of communication. The assumption that in general linguistic marks are associated with the same closure or set of closures provides a framework of stability in what would otherwise be a space of unlimited flux. Such a framework of linguistic stability is required for otherwise there would be no means of successfully identifying a particular closure from a set of tags. In this respect linguistic closure does not differ from closure in general. In order for closure of any type to be realised, and material and texture provided, the process of closure needs to operate on the basis that the closure in question can be completed, even though such a completion cannot be achieved. Without this requirement closure could not get off the ground. The same principle applied to linguistic closure has the consequence that speakers of a language operate on the basis that tags generate the same closures for each individual even though this is not the case, nor could it be the case.

While therefore the function of language in communication requires the containment of diversity in the associations that individuals make with individual tags and particular linguistic closures, this containment of diversity can be seen to be a continuance of the process of closure. Closure, in general, is the containment of that which is diverse, the holding of that which is different as the same. As a result even within the personal space of a particular individual the marks of language are, in general, associated with the same linguistic closures, and are understood to have the same meanings when used in a similar manner. Moreover it is the association of a mark with a single meaning, and thus its identification as a tag, which is the realisation of linguistic closure. Although the requirement of linguistic stability can be thought to be the mechanism by which the potential diversity of linguistic closures across speakers of the language is contained, such a containment of diversity is also an extension of the process of closure on the part of each individual. The linguistic closures of personal space are thus inherently social and the relative stability of the closures of cultural space is itself driven by the process of closure within each individual. It is in this manner that the process of closure as it occurs within the personal space of each individual determines the relative stability of cultural space as a whole.

In addition to ensuring the relative stability of cultural space, the character of the process of closure within individuals has the consequence that the closures of personal space and cultural space are inherently political. The closures of personal space are at once political for in addition to providing a means by which to hold openness they are also a means by which to intervene. The interconnectedness of personal and cultural space and the adoption therefore of the closures of others has the consequence not only that we share a common way of seeing but a common means of intervention. Through the sharing of closures we have the possibility to intervene in certain ways, and at the same time are denied the possibility of intervening in other ways. In the same way that by seeing a face in a page of dots we are in a position to cut it out, or point to it, or to seek a similar face in another location, while at the same time the identification of the face has the consequence that at least for the time being alternative closures that might have been realised have been obscured, with the further consequence that the means of intervention that might have followed from such closures is also unavailable to us.

The closures of cultural space, as the product of myriad trials of closure by individuals and institutions over time, reflect the capacity of those closures to hold openness in a manner that has been regarded as desirable by those individuals and institutions. These closures have been retained because the manner of intervention that has been made possible has been deemed valuable by those individuals or institutions who have been in a position to maintain the closure. So, for example, by dividing the world into things and assigning these things to some individual or group of individuals through the notion of ownership it becomes possible to devise rules to govern the interaction of individuals and property. These closures, and associated sets of closures on which they rely and which they spawn, have profound consequences for the nature and structure of society. Such sets of closures allow the creation of poverty and wealth; allow the possibility of property crime and the identification of thieves; allow the formation of laws that seek to constrain behaviour between individuals regarding property and maintain within certain limitations the current allocation of goods. It will be apparent therefore that the adoption of the closures of property and ownership, notions that are embedded deep in the framework of the closures of our cultural space, carry with them the capacity not only to hold openness in a particular manner, but also to enable intervention to certain effect. More importantly perhaps these closures encourage certain forms of social organisation and allow the formation of certain types of society. The realisation of these closures is a complex product of the desires of individuals and institutions within society, and reflects the value attached to this manner of holding the world and the opportunities it offers for intervention. It can be seen therefore that the adoption of the closures of others through the marks of language results in the inheritance not only of a way of holding the world but certain means of intervention which make possible a particular structure and organisation of society. The interconnectedness of personal, institutional and cultural space has the consequence that the structure of society flows from the closures that make up the personal space of the individuals within that society, but at the same time the closures of personal space are influenced by that social organisation. In adopting the closures of others we thus acquire the historical relationships and power relations that have applied between individuals as well as the way those individuals have held the world.

18

STORIES OF DESIRE

> A set of closures provides a means of holding the world, a
> story of understanding but also a means of intervening to
> achieve an end, a story of desire; and as the stories that make
> up personal space change so do our desires.

If we are to understand the manner in which the structure of closure
influences the relations between individuals and the exercise of power we
must first clarify the nature of desire. For the exercise of power involves
the capacity of one individual or institution to change the behaviour or
beliefs of another, and that capacity in part depends on the desires of the
individual concerned. If an individual is capable of changing the desires of
another they are capable also of influencing behaviour and thus can be said
to have some power over the individual concerned. It is important therefore to address the origin of desire and its role in personal space before an
attempt is made to elaborate the role of closure in determining the power
relations between individuals or institutions.

It will be argued that desires are the product of the process of
closure rather than its motor. The perceptual world provided through
the material and texture of our personal space enables us to manipulate
our environment in order, at an elemental level, to ensure survival and
to satisfy complex desires whose fulfilment can be experienced as pleasure. While it may be superficially tempting to propose therefore that
we generate closures for the express purpose of satisfying desire, such a
conclusion would have the consequence that desire is both prior to, and
of a different nature from, closure. There are many circumstances in
which desire can be seen to be the motivation for the realisation of
closure, but it is not necessary to propose that closure is in general
driven by some pre-existent will or desire. The experience of desire is
instead the product of a sophisticated system of closure which has
enabled subjectivity.

While the machinery of the body needs to carry out a range of activities in order to survive, such activities need not be considered to be the consequence of a will or desire on the part of the individual any more than it is necessary to suppose that an amoeba divides because it desires to do so. The functioning of the body is a result of the operation of its machinery. In the context of this mechanistic account of the body, the process of closure is part of the operation of the machinery of the body and does not require any further motor or direction. We generate visual sensory closures because that is how we operate, not because we desire or choose to see. Similarly, an infant seeks linguistic closure because it is set up to do so, not because it requires understanding in order to be able to fulfil its desires. Through sensory and linguistic closure we are able to function, but it need not be supposed that we engage in sensory and linguistic closure so that we can function. Pleasure and pain can in this context be regarded as feedback mechanisms that encourage the system to repeat certain actions and avoid others. As such they serve to influence behaviour but such mechanisms do not require desire on the part of the individuals in order to function.

Although without closure there could be no desire, it is not to be supposed that closure realises the desire from nothing, any more than closure realises a visual field from nothing. That there is desire is a function of the workings of the body, but its character owes to closure. We can for example only desire to eat an apple if we have the closure 'apple', but the closure alone does not instigate the desire; in the same way that our visual sensory closures enable us to realise the experience of seeing an apple, but that experience is not wholly created by the closure alone but is also the outcome of openness. Closure does not create the world, but realises the world from openness. Similarly closure does not create desire but enables the experience of desire. Without the machinery of the body there would be no desire, but the character of desire is a product of the closure.

In referring to the machinery of the body and the process of closure it is not intended that an ontological priority should be given to mechanism over closure – as if in the physicality of process there is metaphysical bedrock. As I have suggested previously 'the machinery of the body' and the 'process of closure' are themselves closures, they are not descriptions of openness. The mechanical metaphor is close to the familiar closures of realism but it is not proposed that closure is the outcome of machinery in a realist sense, and that as a consequence the machinery of the body and indeed of the world is prior to closure. The theory of closure can be dropped into a mechanical account of the universe; as such it might be placed in the scientific story as part of the evolution of the brain and the emergence of consciousness. So long as such an account is recognised to be itself a set of closures it can be endorsed. It is to be taken therefore as a story that provides an account of the manner in which experience is generated but it is nevertheless a set of closures. As such it is no more a description of openness than any other closure or set of closures.

To return to the matter in hand: it has been argued that desire is the outcome of subjectivity, itself the product of a process of closure that has provided personal space, whose closures can be used to engage in directed intervention and thus support the functioning of the body. It is through the provision of material that we are able to distinguish one object from another and thus are able to desire one object rather than another. If we have not realised a closure and generated the corresponding material we cannot desire that thing. Furthermore although some desires are associated with simple closures, the majority of adult desires are a complex product of personal space. The desire to be on time or the lack of such a desire, the desire to go to church or the desire not to go to church, will be dependent on the detail of the organisation of tags that make up personal space. In such examples the desire is a consequence of a system of belief, a set of closures that together provide an overall account, a story, that is used by the individual to determine action. Desire can be seen therefore to be the outcome of closure. The precise character of personal space, the stories that it maintains and the manner in which they are interconnected, determine the extent and priority of our desires.

A football supporter while watching the final of the World Cup is likely to desire to follow the game and little else. This desire is a consequence of a whole framework of closure involving knowledge of the game and of the teams and the structure of the competition. Without these closures and the stories thereby made possible, there would be no such desire. When we do not share the closures of others their actions can appear inexplicable: the behaviour of the football supporter or the trainspotter can to the outsider look bizarre. Only from the inside of the story do the actions of the individuals make sense, for the desire to act in this manner is an outcome of the closures of the personal space of the individuals concerned. Part of the personal space of the football supporter, for example, will consist of an evolving set of closures about football. The process of closure will in a routine manner seek to extend the 'football story' so long as it offers new closures. A point may come where the individual becomes uninterested in the story perhaps because the closures feel repetitive and familiar, or because alternative stories have taken its place. In any event the changing character of personal space remains responsible for determining desire.

As it was argued in earlier chapters, the closures of personal space can be seen to be organised into stories, combinations or sets of closures that provide a sense of completion to an otherwise haphazard jumble. These stories, the product of the process of closure, provide us with an understanding of reality and a strategy for directed intervention in openness. We thus operate with stories that give each moment a purpose – we know at all points 'what is going on', we are reading a book, we are sitting in a room, we are about to do such and such. In the event that we lack an overall story to account for our current experience we feel ill at ease, we are lost. So it is that we have stories of what we have done and what we are doing and what we are going to do, that range in scale from the current moment to our lifetime. Each of these collections provides a means of holding the world: a story of understanding certainly but also a means of intervening in order to achieve an end: a story of desire, and as the stories which make up personal space change so do our desires. Each story generates its own desires, and as personal space adopts new closures and new stories, so it also generates new desires.

Robinson Crusoe's desires on being shipwrecked on his desert island would for example be a consequence of his personal space, and divergent from the desires of one who had always lived on the island. We can only desire that for which we already have closures, for those things that we already know, or that we have been told about. Of course Crusoe in his first moments after the shipwreck can try to imagine what it will be like on the island from his current space, but he cannot know in advance how his desires will change. At first he may have a desire to return to his home country, to fulfil the desires offered by the stories of his culture, to live in comfort, to eat with ease, to listen to music, to read its texts, to achieve recognition. Over time his personal space will change as the result of being on the island. New closures, and new stories will gradually replace the old, and with them new desires. Even to the extent that we can imagine that Crusoe might finally no longer desire to return to his home country, for in the process he would lose the ability to satisfy these new desires. The manner of the change in his desire is not predictable from his current space, for it depends on closures not yet realised. It can be seen therefore that personal space accrues as the result of chance events and the response to those events is dependent on the particular structure of personal space at the time. It can be seen therefore that personal space does not develop in a linear rational fashion. As a consequence our future desires are no more determinable than our future closures or our future perceptual world.

Since the stories of personal space enable intervention they also are the basis for potential desire. As a result at any given point each individual has many potential desires, and these desires may well conflict. This conflict is not always apparent because we do not experience all of our possible desires at once, any more than we realise all of the closures of personal space at once. At any point in conscious experience we are in a sense lost to the closures of the moment, and thus also to the desires of the moment. The nature of our desire in any particular circumstance is therefore a complex product of the immediate closure or sets of closures in hand and the hierarchy of desires that is the outcome of the organisation of personal space. This can be illustrated with the aid of an example. If Crusoe begins to build a hut, his desires at this point will in large measure flow from this particular story. He is for example likely to look for and desire building materials such as wood or bamboo; no doubt a more general desire might consist in the character of the hut itself, that for example it will be strong and the location prove to be a good one. So it is that the activity of building a hut and the story with which it is associated generates its own subsequent closures and subsequent desires. No doubt Crusoe could also desire to make a fire, or lie in the sun, but when wholly engaged in the building of the hut these other desires are not so much temporarily forgotten as not realised. At some future point they may or may not be realised dependent on the accident of circumstance and the character of space.

It would be a dangerous survival strategy if we were so engaged in any one set of closures that all others were ignored. As a result human experience can be regarded as operating a number of stories in parallel each of which generates its own desires. At an everyday level we are aware that the greater attention we give to any one story the more it is inclined to crowd out the others. So for example if we are engaged in a conversation at the same time as driving a car at various points one or other of these stories may come to dominate. Either set of closures may obscure the other. An intense conversation may lead to a lack of care in driving. A complex situation on the road and the conversation is liable to be dropped. Meanwhile in the background there will be other stories operating: a geographical story perhaps, or a story about the relationship between the individuals in the car, and so forth. To use the almost ubiquitous contemporary analogy of the computer, it is as if we run a number of programs at the same time but at any one moment we operate in only one. In the context of this metaphor it can be seen that desires are generated as the result of each program and our actions are an outcome of these combined desires.

The impact of changes in personal space on the character of desire can be further illustrated with the example of Crusoe. When planning the building of his hut Robinson Crusoe will apply the current closures of his personal space to the matter. He will choose the location and the size of the hut accordingly. No doubt there will be conflicts of desire. He may consider a large hut to be more comfortable, but a smaller hut to be more secure. His search for materials will similarly be based on the way that his current closures determine his choices. If the following year he came to build another hut his closures are likely to have changed in a myriad of ways. If he has encountered no threat, he may no longer think the defence of the hut as significant. He may conclude that using different materials would make it easier to build. He may wish to change the shape and design of hut as the result of the closures generated from living in the previous one. We usually call such shifts in behaviour 'learning from experience', but the process is less linear and less progressive than implied by this description. For it can be seen that there is no ideal hut towards which these changes in action tend. This is because what Crusoe at any point will consider an ideal hut depends on his personal space at that juncture. As a result over time his notion of an ideal hut will alter because the character of his personal space will have altered. Similarly all of the subsidiary closures involving the materials and the construction of the hut will change according to his space and so also will his actions in seeking them out and building the hut.

At any point the actions taken by Crusoe may be more or less successful in delivering his desires at that point, and in this respect progress can be made. It does not however follow that the interaction of space with circumstance is automatically progressive. After ten years on his desert island Crusoe may be constructing a hut yet again having previously made numerous huts. Throughout this period as the character of his personal space developed so would the way it interacted with circumstance with consequential further changes to his personal space. At each stage his desires and resulting behaviour would engage in subtle shifts. Although on the first few occasions Crusoe may, for example, have altered his actions to achieve a hut closer to a desire which it can be supposed remained roughly stable – and thus might be said to have learnt from experience and could in this respect be seen to have made progress – as time passed the process of closure would have greater impact as the realisation of closure progressively shifts the organisation and character of personal space with a resulting change in desires. It can be seen therefore that desire is no less plastic than the rest of personal space with the consequence that behaviour is equally changeable. Crusoe's closures concerning the appearance of the hut in this case might develop with their own sub-closures concerning for example the type of wood used and the precise positioning of the doorway or the angle of the roof. These new stories would generate their own desires which might in time come to dominate over previous desires concerning supposedly more basic matters such as the size of the hut and its location. This is presumably the mechanism that influences the formation of many artefacts and is the reason that seemingly irrelevant issues of design and fashion can have a greater role in determining the outcome of behaviour than supposedly more salient matters. For Crusoe, the hut finally built will depend on his desires at that point and his capacity to execute those desires. These desires in turn depend on what he regards as improvements to his current abode, which again depends on the various operating stories of his space, itself the result of the history of the interaction of space and his circumstance. Crusoe's experiential world and Crusoe's desires can be seen therefore to be the outcome of the play of closure and openness. In this respect therefore he is as much at the mercy of his closures as at the mercy of his circumstance.

Since desire is a product of personal space, and since there is ultimately no limit to the plasticity of space, it follows that there are no desires that cannot in principle be abandoned, although psychologically this may not be possible for a given individual. Our elementary desires can be interpreted as being closely allied to the mechanical feedback mechanisms of pleasure and pain and which result therefore in what are sometimes referred to as basic instincts. These so-called instincts have a central place in the hierarchy of personal space, but it can be seen that even these desires that can be regarded as stemming from the initial character of the machinery of the body can be altered by the changing character of closure. For example, individuals on hunger strike or those suffering from anorexia can in some cases no longer have any residual desire for food.[1] Such an outcome might be interpreted as the individual will having overcome an innate response, but in the context of closure it can be seen not as a battle between instinct and intellectual understanding, but as the product of a personal space that has developed to include closures that result in desires which conflict with previous desires and which threaten the existence of the organism. It is testament to the plasticity of personal space, that we can come to cease to desire those things which we require for survival.

The experience of fighting to impose one's will over seemingly basic instincts can in this context be seen as the outcome of a tension between competing stories within our personal space that generate different desires and thus different courses of action. With the emergence of a new desire that follows a reorganisation of personal space or the realisation of new closure, we can experience a conflict of desire. An individual may feel torn between what is regarded as an 'innate' desire and a more complex desire that follows from the later organisation of some part of personal space. As a consequence an individual may feel unsure of what to do and can oscil-late from one story to the other. As the structure of personal space continues to evolve a point may come where the initial or innate desire is lost. Alternatively the internal contradictions may encourage the abandon-ment of the new closures which threaten the earlier desire.

It can be seen therefore that human desires do not precede closure, nor are they fixed. Desire is a product of a system of closure and in being a product is itself plastic. We have the impression that our desires are inde-pendent of our understanding. It is as if what we think about the world is separate from our feelings and desires. While the initial feedback mecha-nisms of pleasure and pain can be regarded as being part of the machinery of the body, with the emergence of experience and subjectivity, the organi-sation of personal space has a direct impact on desire and is potentially able to overcome any supposedly innate desires. Thus while behaviour is partly driven by desire, desire is not prior to closure but its outcome, and in due course, even the most elementary desires can be abandoned as the closures of personal space generate new desires that replace them. Even in Crusoe's world, a world without interaction with others, the character of personal space shifts with resulting alterations in desire and changes in behaviour. These changes in behaviour in turn generate new closures with further consequent effects on personal space. Unlike Crusoe not only are our desires changing as the result of the play of our own space with open-ness, but our interaction with cultural space and the closures of others provides a further impetus to the shifting character of our own personal space. As a result our desires and the hierarchy of those desires is under-going continuous change.

19

THE FIRST POWER RELATIONSHIP

It is because desire is the outcome of personal space that a
parent is able to influence the behaviour of a child and thus
exercise power over the child through the manipulation of
the child's personal space.

We are now in a position to examine the role of closure, and in particular
linguistic closure, in determining power relations between individuals.
Directed intervention, made possible through closure, enables individuals
to alter the world in a desired manner, and thus to control their circumstances. It follows therefore that when a person adopts similar linguistic
closures to another the person acquires the ability to intervene in a similar
way. If the closures are effective their power over the world has been
extended. It is also the case that at the same time the person's behaviour
has been influenced by another, and that as a result the other individual
can be said to have power over the person in this respect. The same principle applies to groups of individuals and institutions. If a group of
individuals is able to add to or alter the linguistic closures of other individuals or groups they change the personal space and sometimes
institutional space within which the others operate. As a result the group
of individuals change how the others are able to intervene, and have in
this respect exercised power over them. It is for this reason that the imposition of linguistic closure between individuals or groups of individuals,
be it overt or covert, welcomed or opposed, is at once an exercise of
power.

In the ensuing analysis power will be specifically and narrowly defined
as the capacity to alter the behaviour of another without physical
constraint or the threat of such constraint. In this sense it is intended to be
distinguished from control which is understood to require constraint. In
the context of these definitions, power enables an individual to change the
behaviour of another through the capacity to impose or enable closure on

the part of the individual over whom power is exercised; while control achieves a change in behaviour through constraint. A farmer can in this sense be said to have power over a trained sheep dog but exercises control over the sheep through constraint. One of the consequences of this definition is that it is only through the ability to get others to adopt closure that an individual has power. It is usually the case that the greater the power of an individual over another the less requirement there is for control. For the greater the power the easier it is for one individual to change the desired behaviour of another. While the greater the constraint that one individual uses in order to control the behaviour of another the less influence they are likely to exert over the desired behaviour of the person concerned.

The intention of the remainder of this chapter and the one following is to elaborate the role of linguistic closure in determining the manner in which individuals are able to exercise power over each other. It will be shown that an individual can change the personal space of another simply through the expression and repetition of linguistic marks and that through this means it is possible for one person to gain some power over another through the realisation of similar associated linguistic closures. In addition personal space of another can also be altered by interaction. Furthermore it will be shown that an individual can by these means alone encourage the adoption of a set of closures in another that will be referred to as 'the closures of authority'. These closures of authority play a key role in the exercise of power and as I will later argue form the basis of institutional space.

To illustrate the manner in which the character of closure influences the relations between individuals I shall begin with the relationship between parent and child: an interaction that can be considered to be the first power relationship. Since a child in the early stages of childhood usually relies on the closures of parents, the consequences of adopting these closures is more easily laid bare than in adult society where the closures of different individuals compete for dominance. With the benefit of this contained example in mind it will then be easier to turn to the question of how individuals in general, and subsequently individuals in groups and institutions, employ linguistic closure in order to exercise power over each other.

In earlier chapters the way in which the acquisition of closure drives the development of a child's personal space was described. This description was largely offered from the perspective of the child, in so far as it sought to demonstrate how the accretion of closure both extended understanding and the capacity to intervene. While it is the case that from the child's perspective the process of closure drives the development of personal space, from the parent's perspective the aim is to impose the closures of the parent and thereby, either implicitly or explicitly, the closures also of cultural space. Parents observe the progress of the child by comparison with the rest of society, while the child is simply engaged in the development of his or her own system of closure. The imposition of social closures on a child is not of course onerous. It is valuable and more than that it is essential. Without it the child would not be able to communicate and would not benefit from the cultural history of the realisation of closures and the accumulated experience of their outcome. Nevertheless, the first acquisitions of linguistic closure are at the same time the imposition of closure and can be regarded therefore as an exercise of power. Other non-linguistic intersensory closures adopted prior to the acquisition of language may also involve an exercise of power – the behavioural encouragement of a certain object-orientated reality, or a certain notion of self or gender for example. These non-linguistic intersensory closures however are with difficulty identified and as a result the politics of closure will be regarded as starting with the inception of language.

The social character of linguistic closure enables parents to intervene in and to a considerable extent form the personal space of their child. At the outset of the adoption of linguistic closure, this occurs simply through the expression and repetition of the marks of language. The adoption of closure between parent and child is in this respect largely a one-way process. The parent seeks to impose closure through expression and looks for such closures to be echoed in return. In contrast, the parent does not in general seek to realise linguistic marks uttered by the infant, unless they already conform to the framework of closure that makes up cultural space. The linguistic responses of the infant that fall outside the combination of tags identifiable from the parent's space are thus largely ignored, while an infant's response that is close enough to those of cultural space will be encouraged by repetition. A baby that calls a stranger 'mama' is therefore unlikely to get the attention that usually follows when the mother is so addressed. As the rudiments of language are acquired, the parent refines the infant's use of linguistic marks by ignoring closures that are difficult for the parent to realise, and that are in this sense not correct. An infant that calls all objects with wheels 'cars' may simply be told 'it's not a car, it's a bus', or 'a bicycle', or whatever closure would normally apply. Such assertions result in the gradual imposition of the parent's closures and thus the closures of cultural space. The adoption of the marks of linguistic closure takes place therefore simply by the expression and repetition of these marks. In principle the child could refuse to realise closure from these tags – for they are merely cultural accidents – but the child does not have a sufficiently developed alternative space to mount such a denial, and even if the child did have access to such a space it would not be understood.

In due course therefore children gradually adopt both the tags of cultural space and the organisation of those tags, thus acquiring the closures of cultural space: the combination of tags that are regarded as being capable of being realised in all circumstances. As a result the reality that children inhabit converges on that of those around them. The adoption of the closures of cultural space enables children to communicate with other speakers of the language but it also provides a means by which a child's own behaviour can be influenced. The behaviour of a baby may for example be in conflict with the desires of the parents but that behaviour cannot be changed through linguistic closure. Telling a baby to stop crying or to wait for food is unlikely to be a successful strategy. With the acquisition of linguistic closure the parent has a powerful means to intervene in the behaviour of the child without recourse to constraint, distraction, or overt reward.

A specific example may serve to illustrate the mechanism of intervention in more detail. A child is playing a drum and desires to continue to play with the drum. At this juncture the child is engaged in a set of closures, a story, about playing a drum. This may well involve experimentation with the force of the physical movement and the sound produced, or with combinations of different sounds from different parts of the drum. The child's desire is driven by this story which opens up various possibilities which can be explored, and in the process adding to the child's stock of closures. Most children will in due course tire of this story as it becomes more difficult to generate new closure and the activity becomes repetitive. A few will find ways to provide new closure almost indefinitely, by varying for example the speed of the beat and the type of rhythm, and as a consequence will become engrossed for a longer period of time. In this context a professional drummer can be seen as someone who is likely to have found in the drum story the capacity to consistently generate new closure and who as a consequence has become fascinated with the activity.

Let us suppose that while the child is enveloped in the story about the drum, the parent is reading a novel. The parent is also engaged in a set of closures, which in a similar manner generates desire: the desire for new closure or for a completion of closure. Assuming that the author of the book set out to be entertaining, the book was specifically designed to offer closures which are desired, first through accretion and later through completion. Given such a situation, it is not unlikely that the parent is faced with a threat to his or her desire – the seemingly haphazard drum beats are making it difficult to read. As a consequence the parent wishes to intervene and change the behaviour of the child.

The desire of the parent could be achieved by force: for example by removing the drum from the child, or removing the child from the room. Such a strategy however is unlikely to succeed in changing the child's current story. The child's desire to play the drum in the same room as the parent may well remain undiminished. Furthermore the enforced curtailment of the story may even result in increased desire. For in the same way that the parent will wish to continue to read the book until the story ceases to offer new closures – which could be for example that boredom has set in or because there is a sense of completion as the result of finishing a chapter or the book as a whole – so the child is also likely to wish to continue the story of playing the drum so long as it generates new closure. In addition, one of the new closures involved might be the attention generated by playing the drum, and the resultant effect on the behaviour of those nearby. An enforced abandonment of the drum story is likely in such a context to highlight the absence of new closure that was within grasp. The child is likely therefore to attempt to play the drum unless there is a continuing reason for not doing so. If force is the chosen means to alter the child's behaviour a single intervention by the parent is therefore unlikely to be sufficient. The continuing use of force, or threat of force, is probably going to be required. All of which is unlikely to place the parent in an ideal position to continue with the book.

So a possibly more effective parental strategy is to change the set of closures the child is utilising so that the child's desire is also changed with the resultant modification of behaviour. The parent may attempt to achieve this outcome by physical intervention. The child is given another toy, or distracted with a game, thereby offering a new set of circumstances with the potential for a new set of closures. The disadvantage of this strategy is that the parent is forced to temporarily abandon the book. There is also the risk that the new circumstances provided by the toy or game and the alternative set of closures thereby offered will not prove more desirable to the child than the drum story in which case further action will be required.

Aside from such physical interventions the parent can seek to alter the child's behaviour through linguistic closure alone. The aim is to get the child to abandon the drum story in favour of a different set of closures, and thereby change the manner in which the child chooses to intervene in the world. The parent seeks to encourage the realisation of a different set of linguistic closures, with it is hoped a different set of desires and, as a consequence, a different behaviour pattern.

To understand the various strategies open to a parent for imposing linguistic closure we must first clarify the relationship between linguistic closure and behaviour. For up until this point our consideration of both practical and formal linguistic closure has for the most part been restricted to descriptions of reality. In the context of human behaviour however many other forms of linguistic use are employed: commands, promises, and threats for example. An account of how these forms of communication function has not yet been provided and a preliminary indication is needed if we are to understand the role of closure in the exercise of power.

To understand the way that commands function we need to bear in mind the three levels of response of competent speakers to a linguistic utterance. These are the identification of the proposed closure; the realisation of the formal closure; and the realisation of practical closure. Assuming that an utterance contains previously identified marks which are therefore taken as tags, the order of the combination of tags indicates how closure is to be attempted. In such circumstances, on hearing an utterance a competent speaker knows the closure that is proposed. It is in this sense therefore that competent speakers can be said to know the proposed meaning of a sentence. At this stage the speaker does not need to have actually realised closure, either formal or practical, merely to have identified what would be involved if closure was to be realised. Some of the implications of this distinction became clear in our earlier discussion of truth in the context of the claim 'London is the capital of France'[1]. As it was shown, although competent speakers will at once be able to identify the proposed closure in this example, most speakers will be unable to realise either formal or practical closure. The identification of the proposed closure – knowing what is proposed if closure is to be realised – is central to the great majority of language use since whenever language involves the use of known tags the identification of the proposed closure either precedes or is concurrent with formal closure. So it is that in the case of competent speakers of a language a significant part of linguistic communication functions solely at this first level where closures are not in fact realised at all.

Having identified the proposed closure an individual can then attempt to realise formal closure, although the individual may not in all circumstances actually choose to do so. The realisation of formal closure ensures, as we have previously shown, the existential truth of the utterance as far as the individual realising the closure is concerned. The realisation of formal closure changes the individual's personal space and in relevant circumstances will therefore change the practical closures available. In the event that practical closure is not capable of being realised the formal closure may be abandoned or revised. In this manner formal closure often precedes practical closure but there are, as we have seen, also cases in which practical closure precedes formal closure. For example in circumstances where practical closure through the realisation of cues provides meaning which can then through formal closure be held as tags.

In general the desires of an individual, as I argued in the last chapter, flow from the character of personal space and thus from the closures available to the individual. A change in the character of personal space results therefore in a change in the potential desires of the individual. However, the addition of a closure does not necessarily result in a change in desire and therefore does not necessarily result in a change in behaviour. We can be told that food is on the table, we can realise the closure, but it does not necessarily result in a desire to eat. There is therefore no direct correlation between the realisation of closure and the immediate behaviour of an individual. Since communication often involves attempts to change the behaviour of others to achieve certain outcomes, many aspects of linguistic communication seek to overcome this problem the most obvious being the use of commands or orders.

The case will be made that commands are disguised future descriptions. They propose a course of action the result of which is to make a practical closure possible. The command 'Run!' for example can be taken to be a short form for the utterance 'Act in such a way that I will be able to realise the practical closure associated with the phrase "You are running".' Since we are not in a position to know whether the individual issuing a command has realised the relevant practical closure as the result of our activity we often interpret a command such as 'Run!' to mean 'Act in such a way that you are able to realise the practical closure associated with the phrase "I am running."' On hearing a command a competent speaker through their acquaintance with the tags involved will usually be able to understand the proposed meaning of the utterance. In the event that it is not apparent to the person to whom the command is addressed the nature of the specific change in behaviour that is being proposed formal closure may not go through. So for example, if while driving a car a passenger says to us 'Run!' we can be said to know the proposed meaning of the command but be unable to realise formal closure because it is not apparent to us what form of behaviour on our part would enable the realisation of the practical closure 'You are running.' However, in the majority of circumstances, knowledge of the proposed meaning is accompanied by the realisation of formal closure.

It is of course quite possible to understand a command, and to understand what is proposed in terms of one's own behaviour, and yet to choose not to act in this manner. We identify the proposed meaning of the command, we realise formal closure, but we choose not to obey the command. In such circumstances formal closure has been realised but we act in such a way that the practical closure that is proposed cannot easily be realised. Commands are similar in this respect to statements about reality which refer to a future time or to a different location than the one we are in. 'John will arrive in an hour'; or 'round the next bend in the road there is a church' can only be realised as a formal closure at the time of the utterance. If we are present in an hour, or venture along the road, the practical closure can be realised if circumstances allow. Commands in a similar manner put forward a future circumstance in which a practical closure is to be realised. It is for this reason that commands can be regarded as disguised future descriptions. So that the command 'Run!' is equivalent to the statement 'You will run'. On hearing the command an individual can realise formal closure and as such understand the command yet not act in such a way that the practical closure can in fact be realised. When a command is obeyed, the formal closure is realised and acted upon in in a such a way that the practical closure is also capable of being realised. An individual cannot know that the person issuing the command has realised the practical closure, so as we have identified, the command is often understood to mean that the person to whom the command is directed is able to realise the practical closure. It is in this sense that the obeying of commands can be seen to be the adoption of the closures proposed by others, for the individual acts in such a way that the individual can realise the appropriate practical closure.

Returning therefore to the consideration of the strategies open to a parent in an attempt to change the behaviour of a child through the imposition of linguistic closure, three distinct approaches can be identified which can achieve the parent's goal. All of these stem from the social and political character of linguistic closure, which as we have seen itself flows from the underlying structure of closure in the context of the human organism. The first of these strategies has already been encountered: expression. Through expression the parent can offer a set of tags which if realised by the child will result in the abandonment of currently realised closures that are relevant to the behaviour in question. The set of tags expressed by the parent may seek to distract the child with other desires, or may seek to express the parent's dislike of the drum, or may simply tell the child the end that is desired by the parent. If the parent is to be successful the child needs to realise the closure offered. In order to do so the child must at least consider the set of tags sufficiently to contemplate their realisation. To this end the parent may therefore seek to attract the child's attention – perhaps by calling their name, and will seek to express the closure in a way that makes it as compelling as possible. If the child identifies that the parent is seeking to impose closure this may in itself make realisation less likely. To avoid this outcome the parent may seek to disguise the desire to impose closure. For example the imposition of closure could be disguised as a question 'Didn't you want to play with your new football?' The parent may use humour, or irony, the parent can be light-hearted or stern, forceful or determined. Each of these strategies however can be regarded as a means to encourage the realisation of a set of tags which in turn leads to new desires and a consequent shift of behaviour.

A second means of imposing closure can be brought into play if the closure suggested through the expression of linguistic marks has not been realised, or if the realisation of the closures in question has not resulted in the appropriate change in behaviour. The parent can seek to encourage the realisation of the suggested closure through interaction with the child's organisation of personal space. The purpose of this interaction is to encourage realisation of the combination of tags in question by demonstrating connections between the proposed closure and other closures which the child is prepared to realise. Through interaction the links between the closures of personal space are examined, making it possible to draw attention to consequences of a closure that had not been recognised or to identify associated closures that might have to be abandoned, or modified. Interaction with the child's space may thus enable the parent to shift the hierarchy of desire and thereby to alter the child's behaviour.

The effectiveness of expression or interaction will in turn depend on the current personal space of the child. An understanding of the child's space and its organisation is therefore an important part in the successful attempt to impose closure. Parents are in a potentially strong position for it is likely that they have played a significant role in the formation of the child's space. Through a knowledge of the child's space, parents are more likely to be able to choose combinations of tags that will be realised as closures by the child, with the consequent and desired effect on behaviour. In addition the understanding of the structure and organisation of the child's space will make it easier for the parent to interact with the child to change the child's current hierarchy of desire. Outsiders, as a result, are at first unlikely to be as successful in changing a young child's behaviour, unless they are operating in a context that provides closures of authority – a circumstance that will be considered shortly. Over time an outsider may be able to become more successful in directing the child's behaviour than the parent but this will be a consequence of gradually changing the child's space through the extended use of expression and interaction.

The third means by which the parent can seek to impose closure is also potentially the most effective. It involves the acceptance by the child that the child seeks to realise not only the formal closure proposed by a parent's command but to act in such a way that the practical closure can also be realised. Such behaviour is the outcome of a framework of closure that will be referred to as 'the closures of authority'. The closures of authority do not refer to a specific set of closures as if the parent needs to inculcate certain key closures that might be written down and referred to in a parental manual, but rather identifies any combination of closures that results in the child agreeing to seek to act in the manner proposed by the parent. The adoption of closures of authority has a radical impact on the capacity of a parent to influence the behaviour of a child. The structure of personal space is therefore critical to the realisation of any particular closure and the ease with which one individual can intervene in the behaviour of another. The parent over time may be able to influence the structure and organisation of a child's space so that closures of parental authority are adopted. As a result the parent may be in a position merely to ask the child to stop playing the drum for them to do so. In this respect closures of authority can be considered as a variable set of meta-closures which once realised enable an individual to intervene in the behaviour of another with relative ease. They are thus a short cut to the imposition of closure and it will be argued therefore that as a consequence they play a pre-eminent role in the exercise of power.

In this context it can be seen that commands often rely on a structure of personal space that incorporates a network of closures that in combination can be regarded as closures of authority in this particular respect. The commands do not of course by themselves entail such closures. A parent may seek to alter the behaviour of a child by issuing a command. For example, in the circumstances that have been described: 'Stop playing the drum.' Even if the child hears the command and correctly identifies the proposed linguistic closure, there is no requirement for a child to act in such a way that the practical closure 'I am not playing a drum' could be realised. The command may simply be ignored. Alternatively a child may realise formal closure but not act on this formal closure to realise the appropriate practical closures. In defence of such an action the child may seek to defend his or her current set of closures and desires by responding with a simple 'why?', or by asserting the value of the currently held closures by saying 'but I like playing my drum'. In response the parent can seek through expression and interaction to offer closures which if realised by the child are liable to result in the requisite change in the child's behaviour. This may consist in seeking to draw the child's attention to other desires in the child's personal space, or to demonstrate the strength of the desire of the parent and the likely consequences for the medium-term desires of the child. However, unlike expression and interaction, an appeal to closures of authority does not require further explanation so long as the relevant closures of authority are realised. As a result closures of authority have the capacity if realised to bring further discussion to a halt.

An attempt to invoke the realisation of closure on the basis of authority alone might at first consist of the crude response 'because I say so', but these tags are unlikely to be realised unless they are part of a wider framework of closure. This wider framework associates parental remarks with the adoption of the appropriate practical closure, perhaps because the realisation of parental closure has become associated with a positive outcome, or the non-realisation with a negative one, or because of the realisation of a network of other linguistic closures involving tags such as 'obedience' or 'good behaviour'. The closures of parental authority do not consist therefore in a particular set of closures for there are many different, not to say countless, closures that in combination can lead to an acceptance of the authority of the parent. However, the adoption of any such network of closures has the consequence that the child seeks to realise the practical closures proposed by the parent even though these are potentially at odds with current desires.

A network of closures that results in the acceptance of parental authority on the part of the child in addition to being non-specific is also not adopted in a once and for all decision. The closures of parental authority may be adopted with regard to the imposition of certain types of closure but not others, and closures of authority can be abandoned or modified at any point. In order to maintain closures of authority the parent will need to continue to influence the child's personal space in such a way that closures of authority continue to be endorsed. This may involve seeking to support the child's current closures but it may also involve providing alternative closures that also function as closures of authority. If closures of authority were re-examined with each attempt to impose closure they would of course have little purpose. While therefore the realisation of closures of parental authority may well be encouraged by a recognition of benefits, once closures of parental authority in certain respects are realised the child will in general seek to carry out the commands of the parent in the relevant respect independently of the consequences. A major conflict between the current desires of the child and the command may in due course, and combined with other such conflicts, lead to a reconsideration of the closures of parental authority. The issue from the parental perspective – assuming the aim is to invoke closures of authority – is how to persuade the child to realise such closures and to maintain them so that they become a permanent part of the child's space. For once the story of parental authority has been accepted by the child it is going to be much easier for the parent to intervene in the child's behaviour.

Persuading a child to adopt a structure of space which entails closures of parental authority is not different from the means that can be used to persuade the child of any closure: the consistent use of expression and interaction to encourage the organisation of the child's space in such a manner that closures of authority are realised. Until a rudimentary element of the closures of cultural space has been adopted by the child, so that the child shares the linguistic closures of the parents to some degree, it is not going to be possible for the parent to influence desire through linguistic closure. Closures of authority involve a further level of sophistication in the structure of personal space. It will not be possible therefore for a very young child to realise closures of authority since they will not be accessible from the child's personal space. In due course, however, consistent application of the tags associated with the story of authority is likely to result in their adoption by the child on the basis of expression, in the same way that the child was encouraged to acquire the initial framework of language. As the child's space develops, these closures will come into conflict with the child's own desires consequent upon other regions of the child's personal space. For the child to continue to realise the practical closures proposed by parental commands when they are apparently in conflict with their own desires, the child's space will need to be organised so that the hierarchy of desire places the realisation of the proposed closures of the parent at a high level. This is made easier so long as the parent retains control over the daily details of the child's life. For the parent can seek to demonstrate that the realisation of the practical closures proposed in the parents' commands results in an outcome which in the medium term fulfils the child's desires as well. As the child's space matures and develops its own character and desires, and as parental control over the minutiae of everyday life diminishes, the story of parental authority is less easily maintained. If the child is to maintain the story of parental authority, the set of closures that make up the child's closures of parental authority will need to be linked into the remainder of space in such a manner that they are still seen to satisfy the desires of the child. Since the realisation of practical closures proposed by others, the central characteristic of authority, would on the surface of it lessen the opportunity to accomplish one's own desires, a complex web of closures is required to hold such a story in place.

The typical pattern of a child's response to parental desire is therefore an outcome of the manner in which personal space develops and the manner in which closure is acquired from others. When the child is very young closures of authority are not capable of being realised and therefore the ability of the parent to intervene through linguistic closure is limited. The parent has frequently to rely on non-linguistic means of control. Once the initial framework of closure is in place it becomes possible for the parent to impose closures of authority, and the desire of the child to accrete closure encourages this outcome. As the child's space becomes more sophisticated closures of authority come under increasing pressure from other closures and desires. As a consequence the parent may find it more difficult to intervene in the behaviour of the child.

The continued realisation of closures of parental authority depend on the acceptance by the child that the realisation of the practical closures implied by the parent's commands are in the interests of the fulfilment of the child's own desires. If desire was independent of closure the parent's capacity to intervene in the behaviour of the child would be limited. It is because desire is itself a function of the structure of personal space that the parent is potentially in a position to manipulate the child's space in order to exercise power. The child may continue to realise the closures of parental authority because the parent is perceived to be acting in their best interests, or because the child perceives the greater experience of the parent and thus the likelihood that the parent's closures will be more effective in operating with openness, or because the child wishes to avoid the negative consequences of going against the parent's wishes. These perceptions of the child are themselves dependent on closures over which the parent may have had considerable influence. There can, for example, be no independent notion of the 'best interests' of the child. A religious fundamentalist will see the best interests of the child in a wholly different light to an atheist liberal. The child's space will probably have adopted in large part such parental closures and the desires of the child will alter accordingly. As a result of this interplay of closure the parent may be able to maintain the realisation of closures of authority despite circumstances in which the child's immediate desire is apparently at odds with the adoption of the closure advocated by the parent.

As we have seen, personal space builds on its initial closures, with the consequence that later closures are dependent on the structure of closures realised in the early stages of development. Although in principle personal space has unlimited plasticity,[2] in practice the removal of closures embedded in the initial structure can prove difficult since the abandonment of such closures may have the potential to destabilise the space as a whole. Closures of parental authority are likely to have been realised relatively early and are therefore embedded into the structure of personal space. A child may therefore continue to maintain closures of parental authority even though conflicts arise between the realisation of parental closure and desires generated by other parts of the child's personal space. Furthermore the parent may well be in a position to encourage closures in the child which maintain the story that the parent is acting in the best interests of the child and thereby support closures of parental authority. As a result it is possible for an individual to maintain the story of parental authority in circumstances where parental attempts to impose closure are in opposition to many of the desires and the interests of the individual concerned.

In addition to calling upon closures of parental authority a parent can seek to intervene in the behaviour of a child by calling upon other forms of authority. These closures are a consequence of social organisation and the interaction of individuals and institutions, for example the control of order in a society in the form of the police and the legal system. The manner in which such closures flow from the structure of society will be examined later, but there is one type of authority which due to its importance in supporting and defending closures of parental authority it is worth briefly considering at this point. These are the closures of morality.

As with closures of parental authority, closures of morality provide a framework which aid the imposition of closure and the ability therefore to alter the behaviour of another in certain respects. To say to a child 'you should do such-an-such' is to call upon an additional authority that is apparently outside and beyond the interests of the individuals involved in support of the child acting in a manner that would enable the realisation of the practical closure implied in the command. Unlike the closures of parental authority which rely on the child's acceptance that these are in the long-term interests of the child's desires − either positively by enabling fulfilment of the desires or negatively by avoiding a challenge to those desires − the closures of morality can be seen as an attempt to find an ultimate requirement for the realisation of closure that does not entail some further explanation.

In this context it can be seen that moral closures to be fully effective require a framework of unchallengeable authority. It should not be surprising therefore that moral closures have often been linked into a network of religious closure, with God as ultimate authority. In a secular culture such unchallengeable authority is more difficult to sustain but the authority of moral closures is often still sought. In either circumstance the adoption of moral closures can be seen to be encouraged by notions of long-term benefit and threat. In a religious framework the adoption of moral closures is encouraged by the benefit and threat that acting according to or against God's will has respectively positive and negative consequences. The notion of heaven and hell is a particularly virulent form of this moral story. In a secular culture the benefit and threat are likely to be accounted for in terms of the impact on society as a whole and thus in due course on the individual concerned. Although closures of morality are encouraged by a framework of benefit and threat in a similar manner to closures of parental authority, they aim to provide an ultimate reason for acting in a manner that would enable the appropriate realisation of practical closure which is less easily challenged for it in principle applies equally to the parent and the child. As with the adoption of closures of parental authority the adoption of a framework of right and wrong enables intervention in the behaviour of the child. The apparent independence of closures of morality from the interests of the parent has the advantage that they may be more effective in influencing the behaviour of the child. Parents can seek to employ closures of morality as a higher authority to encourage closures of parental authority as in the claim: 'children should obey their parents'. In this respect parents may seek to use non-parental closures of authority to pursue their own attempts to impose closure. This outcome, I shall later argue, is not dissimilar from the use of morality by many other institutions in society.

In summary therefore an attempt has been made to give a description of the means by which an adult can seek to influence the linguistic closures and thus the behaviour of a child so that the desires of the parent are fulfilled. The primary mechanism is expression, which succeeds in the first instance because the process of closure results in a desire on the part of the child to realise linguistic marks. This enables the parent to instil a framework of tags, and in due course the realisation of the dominant closures of cultural space. The parent can then seek to overcome conflicts of desire with the child by further interaction with the child's space. Then over time the mechanisms of expression and interaction can be used progressively to enable the imposition of a set of closures which in combination realise parental authority which shortcuts the procedure for imposing closure more generally and which encourages the child to act in a manner desired by the parent. Finally, these closures of parental authority can in addition be supplemented by other types of authority of which morality is perhaps the most significant.

20

SOCIAL POWER RELATIONS
Repetition and rationality

Repetition and rationality as forms of expression and interaction play an important role in the imposition of closure
between the individuals of adult society.

The power relationship between parent and child provides a simplified
example of power relationships in general. Simplified because the assumption has been made that the child's space is primarily constructed through
interaction with the parents and as a result the child is not able to compete
– at least in the first instance – with the closures of the parent. As the child
grows up the relationship becomes more complex and more akin to that
between individuals in general, and it is to these relationships that I shall
now turn.

Adult members of society have already in large part acquired the
closures of cultural space – and indeed if they have not done so are not
treated as full members of society. As a consequence they have acquired
closures of authority already embedded in the closures of cultural space.
These closures of authority both flow from the structure of social organisation and are responsible for that structure for they are a prerequisite of
institutional space. An examination of the formation and function of the
closures of authority will therefore be central to understanding the role of
closure in determining the structure of society. Before this is attempted
however I will consider the more elementary mechanisms of expression
and interaction and their role in influencing the behaviour of individuals
in society as a whole.

Communication it has been argued is a product of the process of
closure. Through communication individuals are able to extend their
framework of closure and their capacity to intervene in the world. In an act
of spoken communication both the speaker and the hearer, the individual
proposing closure and the individual realising closure, seek to extend their
capacity to intervene and thus their power. From the perspective of the

individual expressing closure, a closure is offered which if realised will alter the personal space of the hearer and thereby influence future behaviour. As a result the speaker will have intervened in the behaviour of the other and in this sense can be seen to have exercised some power. The individual realising closure on the other hand has through communication thereby acquired the closure and thus the capacity of that closure to enable intervention.

In a conversation between two parties the roles of speaker and hearer are exchanged, so that both are engaged in an extension of personal space and in the intervention in the space of another. These two aspects of communication can be seen therefore to be primarily responsible for encouraging us to engage in the activity of talking to others. Contrastingly a lack of desire to communicate is liable to follow from a belief that we are unable either to extend our space, or intervene in the space of another to encourage a desired outcome, or both. While in the case of a young child, expression on the part of an adult is often sufficient to result in the imposition of closure, amongst adults there is a greater likelihood that expression will not result in realisation. An adult will already be attached to a complex array of stories and their interrelation. So long as the realisation of closure serves to satisfy the desires of the individual it is likely to go through, but there will be many occasions when the realisation of closure is rejected on the grounds that it conflicts with other closures and thus potentially with other desires. As the desire for the accretion of closure is increasingly replaced by defence of personal space[1] in a manner that was outlined in an earlier chapter so there is a greater likelihood that realisation will not satisfy the desires of the individual, and that the expression will not result in formal closure and that even if this occurs it will not result in the adoption of behaviour which would enable the appropriate practical closures to be realised.

Although the satisfaction of desire makes realisation more likely, the pursuit of closure has the consequence that the satisfaction of desire need not itself be a requirement for realisation. Nor do conflicts with other closures necessarily result in a failure to realise the closure in question. In an adult individual's personal space many combinations of tags can be seen to continue to be realised as closures on the basis of accretion alone even though they may have no direct bearing on the satisfaction of desire other than the desire for closure itself. In addition, the complexity of personal space is such that a closure can conflict with desires generated by stories within other regions of personal space without either the closure or the desires being abandoned. For it may be that the individual is not aware of the threat to current closures from a new closure, or unaware of how in conjunction with other closures it may be capable of generating a story which would be a threat to the closures of current space.

If an individual does not realise a closure from a combination of tags expressed, one means open to the speaker to encourage realisation is the repetition of the expression. Through repetition the speaker can be regarded as seeking to encourage the realisation of closure without concern for the organisation of tags and the structure of personal space of the individual in question. Repetition can result in realisation because factors which might initially have undermined closure are either overcome or ignored. These factors might include doubts as to the speaker's intentions; an inability to realise closure from the tags despite an attempt to do so; or a perceived conflict with currently held closures or desires consequent upon those closures. It is because of the inherent complexity of personal space that expression which in one instance fails to lead to realisation can do so on another. Changed circumstances may lead to a preparedness to abandon previously held closures, but it may also simply be due to overlooking conflict, and in the end since in the limit closure fails, ways can be found to make any two linguistic closures compatible.

Repetition is used therefore to encourage realisation. In an attempt to change behaviour individuals will 'remind' others of their opinion, they will seek to 'wear them down'. An example may prove helpful in illustrating the mechanism. If an individual smokes cigarettes aware of the health risks involved someone close to them may choose to continually remind them of the risks being undertaken. In some cases this will be a successful strategy, in others it will be seen as an irritating attempt to impede the individual's knowing desire. In this later case there may well be recognition on the part of the cigarette smoker that the repetition of the relevant linguistic marks is precisely aimed at the imposition of closure. Indeed, it is one of the weaknesses of repetition as a strategy that it at once flags intent to impose closure and thus alter the personal space of the individual involved.

Despite its crudity as a means of imposing closure, repetition is still capable of being a powerful means of encouraging closure. It often plays for example an important role in organisations and institutions. Many organisations seek to inculcate new recruits with a range of beliefs at the point of entry into the organisation, and then aim to maintain these closures through repetition. In time individuals may come to challenge these closures but in the first instance they are likely to be adopted on the basis of accretion rather as we adopt the rules of a game in advance of playing. Only after a period of time is it likely that the rules will be challenged, and repetition plays an important part in avoiding such a challenge. The behaviour and belief system of those in the army or the church can in this context be seen to be aided by the repetition of a whole system of closures, a linked set of stories. The continuous repetition of rank and the requirement for salutes, for example, in the army can be regarded as functioning to maintain closures of authority. Similarly the routine of the church service, with the same prayers with the same phrases, serves to encourage and maintain the closures of belief. Such mechanisms are to be found widely throughout social organisation. In the same way that a child learns the closures of cultural space through expression and repetition, so organisations can seek to inculcate a set of closures by the same means.

Repetition, in the form of routine, is used therefore by organisations to encourage the adoption, and maintenance, of a set of closures by its members and thereby cement the internal power relations and ethos of an institution; but it can also be seen to be used, in the form of advertising or propaganda, to encourage and maintain closure in those who are not members of an organisation and are external to the closures of its institutional space. Advertising and propaganda aim to change or reinforce the behaviour of individuals, and the means for achieving this end is to encourage closure through expression and repetition alone. An advertisement by its very nature usually indicates the desire to impose closure and thus is inclined to excite the defences of the individuals to whom it is directed. As a result the advertisement often offers closures that satisfy unrelated desires: images and stories that entertain, amuse, surprise, and so forth. The resistance to the message of the advertisement is thus lowered by its association with a set of closures whose realisation is pleasurable. Through repetition, the realisation of the set of closures involved in the advertisement is thereby encouraged and maintained. Unlike the routine of an organisation, however, the advertiser is not in a position to insist that an individual watches an advertisement. As a consequence, the form of the advertisement can be seen to require change if it is to continue to be realised, since its capacity to satisfy desire will decline once its closures are known.

Although repetition is a simple strategy for encouraging the realisation of closure, the power of advertising testifies to its influence. Unlike interaction, repetition does not offer any other reason for adopting closure, but as a consequence the resulting closure if realised is in some respects less open to challenge. Closures adopted through advertising can as a result be seen to be remarkably stable. Although we choose the products with which we surround ourselves we also in part become them, for the closures encouraged by the advertising of the product become part of our space. It is for this reason that the advertising of products from cars to cigarettes relies on the presentation not of the product but of a lifestyle, the purpose of which is to offer an associated closure that we wish to realise which in turn encourages the realisation of the related message to buy the product. The acquisition of the product is also likely therefore to be the acquisition of an integrated set of closures, instilled through expression and repetition, that take their place in personal space. Those who, for example, smoke a particular brand of cigarettes are likely to have acquired an elaborate set of associated closures which have become part of their personal space. These associated closures, adopted as the result of the expression and repetition of the closures in the form of advertising, serve at first to encourage the desire to acquire the particular brand of cigarettes but in due course are embedded in the personal space of the individual and thus in part make up the individual; with the consequence that the acquisition of the brand of cigarettes can cease to be a desire, and become a self-defining aspect of the individual, so that the individual identifies his or herself with the image, or the type of person, associated with the cigarette brand. The most successful advertising does not therefore stimulate desire alone but by adding a set of closures to an individual's space plays a role in determining the individual's personality. As a result it can be seen that a society where advertising is successful in the sale of consumer goods will be one in which individuals increasingly define themselves in terms of the goods acquired, with consequential impact on the perception and the desires of the individual concerned. The expression and repetition of closure, without recourse to interaction or authority, remains therefore a persuasive means of imposing closure even amongst adult members of a society, and plays an important role in exercising power, not only changing the behaviour of the individuals within the society but in determining who they are.

In the event that expression and repetition do not result in the realisation of closure, a further means to persuade an individual to adopt closure is an extended interaction with the personal space of the individual concerned. As we have seen in the case of the parent–child relationship, interaction seeks to change the personal space of the individual in such a manner that the realisation of the closure becomes possible. This can be achieved either by offering further associated closures which might encourage realisation of the closure in question or by drawing attention to aspects of the personal space of the individual concerned which have so far been overlooked. Interaction relies on expression and repetition but, in addition to the simple accretion of closure, interaction operates through the reorganisation of the closures of personal space. In order to change the personal space of the individual concerned, it will of course be necessary for the individual to realise some closures on the basis of tags offered by the speaker, otherwise no changes could take place in the organisation of space. Interaction is therefore of no value if an individual refuses to realise any tags offered by a speaker. Such a circumstance is not uncommon where the attempted communication is between individuals whose desires are perceived, by one or both parties, to diverge greatly.

Interaction is an extension of expression, and as with expression comes in a host of forms. We can, for example, seek to encourage realisation through humour, or sympathy, through fear or boredom, but there is one particular interactive mechanism for encouraging the realisation of closure which plays a special role. This form of interaction is found in discussion, debate, and argument and can be regarded as consisting in the application of what is commonly described as rationality. The everyday notion of rationality carries with it many associated connotations, for as with moral closures the identification of a closure as rational is linked to the attempt to impose the closure on others – as if the closure was unavoidable. Since it has been demonstrated that it is not possible to identify a closure that is realisable in all circumstances, such associated connotations will be jettisoned.

Since rationality as a term has many uses, it is necessary to refine the manner in which it is to be used in this case. Rationality will be used to mean the identification of connections between closures and the avoidance of conflicting closures. In a circumstance where the relations between the closures of personal space were at once evident, a strategy of rationality in this sense could not succeed, for the connections between the closures would already be known by the individual concerned. It is however because of the complexity of personal space, and the limitless potential for further closure, that the strategy of rational interaction can seek to identify connections that have gone unnoticed or unrealised.

While rationality is often portrayed as the identification of necessary links between fixed closures, its success as a strategy is in contrast largely due to the lack of fixity of the closures of personal space. It is the plasticity of closure that results in the potential to identify links within the personal space of someone else which have not yet been realised. In contrast to the normal understanding of rationality, it can be seen therefore that it is because the relations between closures are not wholly logical in character that it is possible for one individual to alert another to connections or conflicts within personal space and thereby encourage either a reorganisation of that space or a change in presently realised closures.

In the light of this account of rationality it is not surprising that the employment of rationality is far from a foolproof method for the imposition of closure. For what is rational for one individual will not be rational for another. A divergence between individuals in the organisation of space will result in different closures being rational outcomes. Similarly whether two closures are in conflict will also be dependent on the nature of personal space: what is evidence of conflict will vary from one individual to another. Furthermore, as I have previously identified in an earlier chapter,[2] even within the context of a particular space it is always possible to deny that one closure follows another and thereby defeat the supposed deductions of rationality. This will require some agility in the operation of current closures in order to find the point at which the failure of closure allows two seemingly contradictory closures to reside together but it is in principle always possible. In addition, even if connections or conflict between closures are accepted, the individual can still refuse to acknowledge the apparent consequences of these connections. Individuals do not have to be rational; rationality is itself only a strategy. For example, an individual's space does not have to exclude contradictory closures. It is not difficult to find circumstances in which some individuals find it convenient and possibly effective, if anti-social, to adopt contradictory closures – for example in the conduct of sexual relations. As a result, even if through rational argument a connection or conflict is demonstrated between the closure in question and other closures within an individual's personal space, the individual may still refuse to take the appropriate action.

Since we are used to the notion that rationalism is always a positive characteristic or strategy, I will endeavour to provide an example that illustrates one of the weaknesses of rationalism before considering the reasons for its success. In order to identify conflicts in closure and thereby eradicate them, one of the familiar aspects of a so-called rational approach is to identify closures precisely in order to determine whether conflict is present. This is sometimes described as the defining of terms. In so far as this process assumes a completion of closure which cannot in fact take place it can be seen to be liable to error.

The rational strategy of the defining of terms can be applied to any practical, mechanical, or technical problem. If a machine is broken an attempt can be made to find the fault by defining each part of the machine and determining whether it is operating satisfactorily. So, from a rational perspective, a fault in a computer, for example, can be taken to be either caused by a software problem or not be caused by a software problem. This segmenting of reality into a series of separate functions, which can then be deemed either to be operational or not operational, has potential to aid the process by which the fault might be discovered. Rationality in such a situation encourages an orderly approach in which different items are examined for faults. It is not difficult however to imagine certain circumstances in which a strict adherence to such a strategy might impede rather than aid the solution to the problem. Since a rational strategy seeks to avoid conflict between closures it is assumed that a thing, in this case 'software', is either responsible for the fault or not. Thus it is assumed that the tag 'software fault' provides a closure that is discrete and could be completed, and that as a consequence all possibilities are exhausted by s v ~s, s or not s, where s is 'software is the cause of the fault'. The limits of the closure 'software', which we take for granted, are however not ultimately definable,[3] and this assumption of completion can lead to error. For example, suppose the fault is due to an incompatibility between software and hardware. In such a circumstance the software could be examined and no fault found. The conclusion: 'software is not the cause of the fault' would in one sense be an accurate description of the circumstances, but at the same time it might obscure the cause of the problem which is only with difficulty assigned within the simple opposition 'software is the cause of the fault' and 'software is not the cause of the fault'.

It can be seen therefore that the employment of a rational strategy to the finding of a fault with the computer might lead to an examination of each part of the machine only to find that none of the parts were faulty. Such an outcome would not have shown the rational strategy to be wrong for it will always be possible to introduce further closure in an attempt to complete this particular story of events. In this case, it might consist in distinguishing for example between a part and the way in which the part is combined with other parts, or in redefining software to include the way it interacts with hardware. However not only is the introduction of further closure a potentially limitless exercise but there may be more effective strategies for solving the problem. The most obvious is to find someone who has experience of similar problems and who is able to identify the fault not by an attempt to understand each part of the machine but by seeking to identify similarities with previous examples of failure. In the same way that a farmer who grows apples will choose to pick them on the basis of previous experience rather than a supposedly rational attempt to analyse the weather, the size of the apples, the force of gravity and so forth.

It is often supposed that the success of rationalism is due to its capacity to result in an accurate reflection of the character of the world. Openness is not however logical any more than it is illogical, it neither excludes contradiction nor contains it. So if a rational strategy is not successful because it enables an accurate description of the world, why is it successful? Rationality is embedded in the process of closure. In order for a closure to be held, in order that it realise material, openness must be held as this thing and not as something else. If closures that are in conflict are realised at the same point a choice between the closures will be required. We see a duck or a rabbit but not both at the same time. An avoidance of conflict is therefore part of the process of closure from the outset. Moreover, the reality within which individuals operate is largely the outcome of the closures of cultural space. It is through the successive application of a rational strategy to weed out conflict between these closures that a relatively fixed and agreed framework has been adopted which enables effective communication between individuals. Without this stable framework communication would break down and so therefore would the capacity to enable effective intervention. Rationalism is thus both a requirement of the individual process of closure and of cultural space. Although the world is open and although each thing could be regarded as something else, if at the individual level we attempt to hold something as many things we fail to provide material at all, and if collectively we were all to realise different closures in a haphazard fashion there would be no way of communicating, and no means of coordinating intervention. The process of closure and the structure of language and cultural space therefore entails an element of rationality.

It is because personal and cultural space require a level of rationality that the identification of conflicts in the closures of others, or the identification of connections that have been overlooked or forgotten, is a means of encouraging the imposition of closure. However, as I have sought to demonstrate at many different points, the pursuit of rationality does not result in a gradual approximation to an accurate description of openness. It is part of the means by which we extend the process of closure, and as a result it is capable of increasing our capacity to intervene. It does not however necessarily aid intervention, and any particular supposedly rational conclusion can be evaded by an individual or a culture.

The identification of conflicts and connections in the closures within the personal space of another, and thus the application of rationality through discussion, debate, and argument, is a vital part of human interaction and the formation of cultural space. As with other forms of interaction, however, the successful imposition of closure will in addition depend on an individual's knowledge of the space of the person or persons whom they are seeking to convince. Interaction that consists merely in the stating of connections between the closures of one's own space and the identification of apparent points of conflict in another's space is only by chance going to prove successful. The trading of alternative points of view between individuals whose space has little in common is as a consequence only rarely effective for either party. If an individual does not see the force of a closure or set of closures, or wishes to retain a closure, only a detailed knowledge of their space will put another individual in a position to either encourage the closure or undermine it through interaction. With individuals we know well this may be possible, using a variety of subsidiary strategies – entertainment, humour, anger, irony, enticement, and so forth, along with challenge and debate. For those we do not know it is likely to be less successful. All forms of interaction, both rational and otherwise, are therefore most commonly applied amongst groups of individuals who know each other well, or who share a similar background and experience and thus share similar desires. It is less effective for deciding between desires or for shifting the closures of those whose experience is disparate and who do not have knowledge of each other's space. In principle therefore it can be seen that the wider and more general the debate on an issue the less interactive and rational the debate is likely to be and the more it will rely upon the trading of different closures in the form of expression and repetition. This outcome is perhaps particularly apparent in the way politics is conducted in the mass democracies of Western culture.

While rational interaction is an important means of seeking to impose closure on others, and plays a central part in the formation of cultural space, its role in determining the structure of society is minor by comparison with closures of authority to which I shall now turn.

21

THE STRUCTURE OF SOCIAL CONTROL

Closures of authority

> The realisation of closures of authority creates an institution
> and with it a network of rights and responsibilities that in
> combination is responsible for the structure of society.

The structure of closure has the consequence that high-level intersensory closure, in particular linguistic closure, is both social and political. As a result it is possible for one individual to have power over another through the imposition of linguistic closure. In addition the social and political character of personal space has the consequence that individuals can manipulate the personal space of others so as to allow the imposition of closures of authority. In the context of a group of individuals the closures of authority initiate an institution for they allow the development of institutional space. The possibility of institutional space is therefore the consequence of the social and political character of each individual's personal space and the consequence of the structure of closure. In this chapter I will be concerned to outline the nature of institutions and the role of closures of authority in forming those institutions, and thereby cast light on the mechanism by which power is exercised in society.

Closures of authority help one individual, or group of individuals, to impose closure on another. Unlike repetition and rationality, which encourage the adoption of closure, closures of authority seek to guarantee its imposition and to ensure that the behaviour of the individual or group of individuals is such as to enable the appropriate practical closures. Closures of authority are not merely an extension of repetition and rationality as if they had the character of an irrefutable argument providing an ultimate rational hammer blow, but are different in kind. The power that stems from the adoption of closure as the result of expression and interaction is specific to the closure in hand. Rational interaction may encourage the adoption of a closure but it does not thereby influence the adoption of future closures, other than as a consequence of the impact of the closure in question on personal space. The power that results from closures of authority is of a different and institutional form. While the successful application of expression and interaction give one person power over another, closures of authority institutionalise power and are thereby responsible for the structure of society.

In order to examine the operation of closures of authority a distinction will be made between the individual or individuals who seek to impose closure, who will be referred to as 'the author', and those on whom closure is imposed, 'the realiser'. Within this framework the closures of authority can be seen as a variable set of closures adopted by the realiser that result in closures of the author being realised. The closures of authority are always constrained to some degree. It is possible to distinguish two primary ways in which these constraints limit the capacity of the author to impose closure: the imposition of closure is limited to an aspect of personal or cultural space, and the imposition of closure is limited to those individuals over whom authority can be exercised. These limitations are a consequence of the nature of the closures of authority and are liable to variation. Authority is not therefore a black and white affair. An individual or individuals have authority in certain respects and not in others. One consequence of this is that the power relationship between individuals need not therefore be only in one direction. Two individuals may each have authority in certain respects over each other.

It will also be apparent from this account that the character of the power bestowed by closures of authority is a result of the particular form of those closures, and is only consequent upon the adoption of closures of authority by the realiser. It is therefore only the realiser who is capable of realising the closures of authority. Closures of authority cannot be insisted upon by the author alone, since these closures must be part of the space of the realiser and this can only occur if the individual in question does actually choose to adopt them. In short, power is dependent on the realiser adopting closures of authority and these cannot be imposed by diktat. Of course an individual or a group of individuals can seek to impose their will on others, through force or the threat of force, and in doing so may be said to have control over others, but they have not thereby put in place closures of authority for these have not been realised by those others over whom control is exercised. As a result, as it will be argued, there is no institution formed nor is institutional space created.

In any particular instance, the adoption of closures of authority, as with communication in general, is the outcome of the process of closure. Through closures of authority the author is able to intervene in the behaviour of the realiser. As a result the author is able to extend the

capacity to intervene and thus in principle to satisfy desire. It is not difficult to see why therefore the author might wish to impose closures of
authority. It is perhaps less immediately apparent why the realiser should
adopt closures of authority. For closures of authority not only have the
consequence that the actions of the realiser are constrained by another but
that the self is partly abandoned to another. For to agree to adopt the
closures of another in some respect is to agree to become them in that
respect. There are two issues therefore which will be addressed: how it is
that closures of authority are adopted by the realiser, and the consequences
of adopting closures of authority.

Closures of authority may be adopted by the realiser because they are
part of the closures of cultural space, and may in such circumstances be
adopted on the basis of expression alone. Since cultural space is itself the
outcome of previous social organisation, it is first necessary to examine
how closures of authority are adopted independently of the particular
social circumstance if we are to understand how closures of authority are
themselves responsible for that social organisation. I shall return to the role
of closures of authority in cultural space once the basis for their adoption
independently of cultural space has been considered.

Many outcomes desired by an individual are more easily achieved
through joint action with others, some are only achievable as the result of
such action. Accordingly, an individual may be prepared to accept the
authority of another, and thereby adopt the closures of another, in order to
see this desire fulfilled. It is possible to act jointly in pursuit of an agreed
end without a framework of authority – such a modus operandi may apply
between two friends engaged in joint task. In such circumstances a decision over closure is based on an exploratory adoption of suggestions for
closure from either party. So long as the individuals have similar desires
and the adoption of closure is independent of the person who initiated the
closure, such a strategy is effective since it utilises the framework of
closures available to both parties, and thus the ability of each space to
enable directed intervention in the handling of activity. The weakness in
the strategy is the time taken to agree each closure involved in the task at
hand. A long-term task may be suited to such an approach if decisions are
not frequently required to determine the behaviour of the relevant individuals. If however a task requires rapid decision-making the need for
authority will become more apparent.

Sailing a boat in a storm without having one person in authority may
prove to be more dangerous than agreeing a leader even if the combined
closures of the individuals would in time produce a more effective intervention than the single individual. The identification of a leader will

involve the adoption of closures of authority by author and realisers. When speed of response is important, adoption of closures of authority may therefore prove beneficial even if desires are in common between the individuals. In the event that subsidiary desires are in conflict, the adoption of closures of authority may be essential if the primary desire is to be satisfied, since the difficulty of agreeing joint closures may be sufficient to make the completion of the overall task either tiresome or impossible. As a result it can be seen that it is usually the case that the greater the convergence in personal space between individuals, and the fewer individuals involved, the less requirement there is for closures of authority, and the more effective is a reliance upon agreement of closure in pursuit of what is recognised to be a mutually held desire. Conversely the greater the number of individuals and the more divergent the personal space of the individuals the more important will be the adoption of closures of authority in pursuit of the primary desire. If a task requires more than a few persons, unless there is an unusual degree of convergence of personal space, some form of authority is likely to aid the completion of the task. In the first instance therefore, closures of authority are adopted for this reason, thereby giving one or more of those involved in a task the capacity to impose closure in certain respects and thereby aid the completion of the task. The case will be made that the adoption of closures of authority on these grounds, and the means of containing the authority once it has been agreed, provides the initial basis of social organisation.

The realisation of closures of authority does more than enable the imposition of closure for it results in the creation of an institution, an institution which can be seen to be distinct from any of the individual parties involved. Closures of authority create an institution by giving rights to the author to impose closure in certain respects on those individuals who are members of the institution – including the realisation of practical closure that is enabled by appropriate behaviour. These rights are not moral or ethical but stem from the closures of authority themselves. It is therefore in the nature of closures of authority that their adoption by realisers can be said to give rights to the author to impose closure in certain respects so long as closures of authority are maintained. The author can be seen in the first instance to be granted these rights by realisers in the belief that there is as a consequence a greater likelihood of successful intervention in pursuit of the relevant desire or desires.

The formation of an institution through closures of authority not only grants rights to the author to impose closure but also generates responsibilities for the author. If closures of authority are to be retained the author will need to maintain the realisation of closures of authority on the part of the realiser, either by actually satisfying the relevant desires of the realiser, by maintaining the possibility of such a fulfilment of desire, or by the threat to deny these desires in the event that the closures of authority are abandoned. It is the need to satisfy the desires of the realisers in certain respects which has the consequence of placing the author in a position of responsibility. This consequence only applies of course so long as the author desires to maintain the closures of authority. In general it can be seen that it is usually the case that the more extensive the rights to impose closure the greater the responsibility to satisfy desire. While the more effective the author is in satisfying the desire of the realiser the more firmly will be held closures of authority. The adoption of a set of closures of authority results therefore in an institution consisting of the individuals involved, linked by rights and responsibilities.

It is important to note that the rights and responsibilities that go with closures of authority flow purely from the structure of the institution and the relationships between the individuals. These rights and responsibilities have no higher authority. Higher authorities, such as the law, morality, or religion, may be invoked in an attempt to encourage the realisation of closures of authority but they are neither a necessary nor a sufficient condition for the creation of a network of rights and responsibilities on those within the institution. As a result, the rights and responsibilities that go with closures of authority are dependent on the continued realisation of the closures of authority concerned. There is no right to impose closure independent of closures of authority nor is there a responsibility independent of the requirement to maintain those closures. If the author does not carry out the responsibilities involved in closures of authority the closures are likely in due course to be abandoned and the right to impose closure will lapse. Responsibility is therefore not to be taken as an abstract goal independent of closures of authority as if an author takes on responsibility out of a higher notion of justice or morality; in the same way that the rights of the author to impose closure is not driven by a moral or political sanction. In both cases the rights and responsibilities are the single and immediate consequence of closures of authority. If the responsibilities are not carried out the authority is placed in jeopardy, and if closures of authority are abandoned the right to impose closure at once ceases.

This argument is not putting forward a moral claim. The assertion that rights and responsibilities are the outcome of the realisation of the closures of authority, and thus cease to operate if the closures of authority are no longer realised, is not intended as an endorsement of a certain social order. It is instead simply a description of the way institutions are formed and the nature of the institutional space that is created. The notion of rights and responsibilities is often thought to be at once moral, as if the assertion of certain rights is an optional matter. Instead I wish to argue that rights and responsibilities are part of the logic of institutional space and thus describe the way the individuals concerned have, by their realisation of a particular set of closures of authority, agreed to behave.

While the rights and the responsibilities generated by closures of authority provide the structure of the institution, the content of the institution is made up of the closures imposed by the author. These closures constitute institutional space. They are, of course, part of the personal space of the realisers – for this is a consequence of their adoption of closures of authority, and are likely, but not necessarily, to be part of the personal space of the author. The institutional space that results from one institution has its own dynamic and interacts with the institutional space from others. The combined interaction of institutional spaces plays an important role in the provision of cultural space. Each set of closures of authority therefore generates an institution which takes its place alongside, or within, other institutions. The structure of society is as a consequence the product of institutional space as a whole, and this structure is incorporated into the closures of cultural space along also with the closures of authority which it embodies.

In order to follow the process whereby closures of authority are first adopted and then generate an institutional framework, it is helpful to consider circumstances in which the closures of authority embedded in cultural space play a minor role. To this end, take once again the circumstances of Crusoe, only on this occasion suppose that he is wrecked on a desert island along with a number of others who did not previously know each other and who are not distinguishable in terms of their prior status in society. Suppose also that there is a mutually held desire amongst the shipwrecked islanders to build a defensive camp. The adoption of closures of authority will aid the building of the camp, so long as the individual chosen is capable. The realisation of closures of authority need not involve giving a single person rights to impose closure over all matters concerning the camp. The more extensive the closures of authority however the greater the likelihood that a camp will be completed, assuming the capability of the individual concerned. In such a circumstance, the decision to adopt closures of authority by those involved will be the outcome of the hierarchy of desire resultant from their personal space. If the primary desire is to build a camp at speed, subsidiary desires such as the details of the construction of the camp and an individual's own role in the construction may be overridden and closures of authority adopted which give extensive rights to an individual, or individuals, to direct the construction of the camp through the imposition of closure. Considering such a circumstance it will be apparent that an individual is only likely to adopt closures of authority with regard to an author who it is believed will effectively satisfy the relevant primary desire, or has the best chance of satisfying such a desire.

It is possible that in adopting closures of authority the realisers specify limits to that authority, for example by placing a time limit on the authority or by making closures of authority dependent on the satisfaction of additional subsidiary desires. It is often the case however that closures of authority are ill-defined in advance and it is only in practice that the limits of authority are determined. Nevertheless in the context of the circumstances supposed, as the result of the closures of authority an institution has been created: a camp-building institution in which the author has the responsibility to get the camp built and the right to impose closure in order to do so. It can be seen that the institution is dependent on the maintenance of the closures of authority. In the event that the realisers abandon the closures of authority the author's power is lost and the institution folds. In order to satisfy the primary desire therefore the author must ensure that the realisers continue to adopt the closures of authority.

In this hypothetical circumstance the author is likely to be constrained in the attempt to impose closure. At any point, one of the members of the institution may refuse to accept the imposition of a particular closure. This is likely to result in a redefining of the closures of authority and the limits of its application and may thereby threaten the closures of authority in general. The position of power initially held by the author is therefore precarious and such an outcome can be seen to be commonly the case in the early stages of an institution. In the case of the ship-wrecked islanders it can be supposed that the closures of authority are likely to be constantly renegotiated in the preliminary stages, thereby setting out the author's limits of authority. The urgency with which each realiser desires to build the camp and the alternatives available will play an important part in the extent to which the realisers are prepared to adopt closures of authority.

Since closures of authority are adopted on the basis of a hierarchy of desire within the personal space of the realiser, any change to that hierarchy or any threat to subsidiary desires may bring the adoption of the closures of authority into question. As a result, in order to maintain closures of authority the author may seek to sustain the primary desire and avoid conflict with subsidiary desires. In this example, the author may seek therefore to stress the need and urgency of building a defensive camp in order to maintain the primary desire – for example by stressing the potential danger from warlike tribesmen or wild animals. In addition the author may at the same time seek to avoid imposing closure in a manner that will cause conflict with other desires held by the realisers, for being attentive to the subsidiary desires of each individual may be necessary if the author is to maintain the attachment of each realiser to the closures of authority. As a result, in the early stages of a project where closures of authority are based on jointly held desire, not only are the closures of authority precarious but the power held by the author is likely also to be highly constrained.

Although an author's power is initially constrained, not infrequently in the early stages leading to the collapse of the institution altogether, closures of authority can be strengthened and extended as a result of satisfying either the primary or subsidiary desires of the realisers. Having for example successfully built the defensive camp, the institution and the author's authority is likely to be enhanced. Other subsidiary desires such as the quality of the camp and the role of the individuals in building the camp are also likely to play a role in determining the strength of adherence of the realisers to the closures of authority. The consequence of the strengthening of the institution will be to extend the authority of the author and the rights to impose closure. As a consequence, in the event that another camp is built it can be seen to be more likely that the author will be less constrained in the imposition of closure.[1]

Satisfaction of current desire is not the only means by which the institution can be strengthened and the power of the author extended. Over time the author has the potential to gradually shift the personal space of the realisers in such a way that current desires are abandoned, or placed lower in the hierarchy of desire, and replaced with new desires which are fulfilled. In much the same way a parent is in a position to gradually influence the personal space of a child and thereby influence the child's behaviour and desires. For example, in the building of the camp, the author may give others authority, one perhaps as a deputy, another in charge of wood collection, a third the responsibility for the construction. This procedure may satisfy desires in the individuals concerned but, in addition, the author may be able to encourage the desire of other individuals to obtain these roles. By such a process the author will have created desire in the personal space of the realisers – desire which can be satisfied by the author, and which may lead as a result to greater authority. By such means the author seeks to alter the personal space of the realisers in such a manner as to strengthen the institution and the power of the author. Stressing the dangers of dissent or the urgency of completing the task operate to shift the personal space of the realisers to a similar end. The alteration of the personal space of the realisers in order to maintain and extend closures of authority can be seen therefore to be a crucial aspect of the exercise of power, and is central to the success of social organisation.

The tendency towards the extension of the closures of authority may be further influenced by the, as yet, unknown alternative. If an author has successfully satisfied desires there is less reason for a realiser to abandon the closures of authority attaching to the author in favour of an untried alternative. So long as the desire or subsidiary desires of the realisers have been satisfied to some degree there is a risk attached to abandoning the closures of authority. In the absence of the institution it is possible that these desires will not even be partially satisfied. Even if an author has not fully satisfied the desires of the realisers, realisers still have to decide that an alternative is available with a better chance of satisfying desire in the relevant respects. Moreover the alternative must be seen by others to be more likely to satisfy desire, for it will only be with the consent of other realisers that a new authority has the possibility of satisfying the primary desire. It is for this reason that once an institution has become established an incumbent author usually has an advantage over any alternative.

The combined consequences of these factors can over time place the realiser in a weak position. For while initially closures of authority are challenged with some ease, in time, depending on the success of the author in satisfying the primary and subsidiary desires of the members of the institution and the author's capacity to restructure the personal space of the individuals concerned so that desire is satisfied, the realisers may have little choice but to accept closures of authority over most of their personal space if there is to be any chance of achieving the primary desire. In extremis even this limited control over their own closures may be lost if the author is able sufficiently to change the personal space of the realisers. Such situations are not simply hypothetical. Members of religious or political organisations for example can have their space so dominated by the closures of institutional space that the satisfaction of all previously held desires is abandoned.[2] To a lesser extent the same process can take place in any organisation, be it a company or corporation, a hospital or a university. In each case the institution shapes the personal space of the realisers in order that the desires of the realisers strengthens the institution. The institutional space of the company encourages the individual to desire financial reward and value profit, the space of the university to desire knowledge and value education. As well therefore as being beneficiaries of the potential of the closures of institutional space to handle activity, realisers are also potentially ensnared by desires that are also the outcome of institutional space.

Closures of authority therefore which are initially highly constrained by limitations and which do not give the author rights to impose closure other than in limited circumstances, and certainly not ones that threaten strongly held desires of the realiser, may over time be extended. An author is thus able to increase the capability to impose closure and thus increase the capacity to exercise power. Extending this principle to the institutions of society, it can be seen that most social institutions have been developed over a considerable period and that as a result current authors have the advantage of a prior history which has generated a framework of institutional closure that aids and supports attempts to impose closure. The power of the author can even become such that those over whom authority is exercised no longer experience any conflict of desire and thus accept the imposition of closure with ease. Power is thus at its greatest when its exercise is least apparent to the realisers.

The circumstances so far considered have been those in which closures of authority are adopted because there is a jointly held goal or desire on the part of the author and realiser. It will be apparent however that closures of authority are on many occasions adopted without there being a jointly held primary desire in common between author and realisers. One way that such an outcome can be accounted for is that the author is able to satisfy desires of the realisers which are subsidiary or independent of the primary desire. The realiser adopts closures of authority in such circumstances not to obtain the goal of the institution but to obtain some other desire which the author is in a position to satisfy. Any desire of the realiser that the author is capable of satisfying can in this manner provide the basis for realisation of closures of authority. The desire of the realiser may be material, emotional, or physical, but if the author has the capacity to satisfy such a desire or more precisely to be seen to be capable of satisfying the desire, they also have the capacity to encourage closures of authority. Many institutions in this way seek to encourage the adoption of closures of authority by an appeal to desires of the realisers that are incidental to the goal of the institution. The relationships between employer and employee, between teacher and pupil, between policeman and member of the public, usually involve some closures of authority that are encouraged by desires extraneous to the aim of the institution. The desires of the realiser that can be utilised in order to encourage the realisation of closures of authority are as varied as desire itself. Of these many options two types of example will be briefly considered on account of their social importance: the first is the use of force or the threat of force and thus the desire to avoid pain and distress, and the second is the use of financial reward and the desire to acquire money.

The use of force to control another's behaviour directly is, as has already been identified, a time-consuming exercise and thus usually of limited effectiveness. A more effective strategy is to use force as a means of encouraging closures of authority: 'act according to my commands or something nasty will happen'. This strategy can be employed by the state in the form: 'act according to the laws of society or something nasty will happen'. Closures of authority can be encouraged therefore by the threat of force. A brief reflection suffices to conclude however that reliance on physical force alone is of limited use. Since the physical capacity of individuals is rarely so divergent that one individual is stronger than two or three others combined the use of physical force alone to impose closures of authority is only possible in very small units. If Robinson Crusoe is shipwrecked on his desert island with one other he may be capable of imposing closure by threat of force. With two others it becomes more difficult, and with three or four almost impossible. Once a unit involves more than half a dozen or so, force cannot be the sole means of imposing closures of authority.

Although for institutions that consist of more than a handful of individuals force cannot be the sole means of imposing closures of authority, it clearly remains a potential means of encouraging closure. A tribal leader requires the consent of the community – the adoption of closures of authority by the realisers – to impose closure and will be unable to impose such closures by force alone. Once closures of authority have been realised it may then become possible for the author to co-opt other members of the group into the use of force in order to maintain the institution. Over time the strength of the institution can as a result become such that despite opposition from many of the individuals it remains possible for the leader to retain authority. Nevertheless, the adoption of closures of authority must remain largely consensual among the more senior members of the institution for without the realisation of closures of authority the author would not be able to utilise other individuals within the organisation to employ the use of force to encourage closures of authority more widely. This conclusion has the consequence that a dictator or a tyrant cannot rule therefore by fear alone, but must carry the consent of a significant proportion of the senior members of the governing institutions.

Although totalitarian regimes are sometimes portrayed as if they were imposed by force by an individual or group against the wishes of the great majority this can only occur in very occasional circumstances and for short periods. A Stalin or Hitler maintains their position through the adoption, by consent, of closures of authority on the part of those with whom they come into contact. Those individuals in turn require the adoption of closures of authority on the part of the individuals with whom they need to impose closure. Stalin's reign of terror, for example, would not have been possible without the consensual realisation of closures of authority by the secret police, the army, and significant numbers of the population in general. It can be seen therefore that while force can be used to encourage in any particular instance the adoption of closures of authority, it cannot provide the basis of power, even in such apparently extreme circumstances as those of totalitarian regimes. If force was the basis of power, dictators would be individuals of immense physical strength, while in fact the physical capability of an individual makes them no more or less likely to be capable of exerting dictatorial power. Moreover the use of force has the potential to weaken closures of authority as well as to encourage them, for it may increase opposition to the author, as was noted in the case of parent and child. An institution can use force to encourage closures of authority but the extensive and widespread use of force is likely in the long term to undermine the closures of authority on which that power is based. An army can be used to subjugate a population but the continuous and widespread use of such force is likely to result in an overthrow of the institution in due course. No doubt Stalin's abuse of power was responsible both for the curtailment of the power of the leader following his death, and played some role in the fall of communism in the USSR some three and a half decades later through an undermining of belief in the system of government. Successful and sustainable institutions can therefore only use a limited degree of force that does not have the consent of the population. It can be seen therefore that this account of power has the consequence that all forms of government, including those based on a dictatorship, require the adoption of closures of authority by a significant proportion of the population.

Unlike the use or threat of force, the use of money to encourage closures of authority appeals to the positive desires of the realiser rather than their fears. It is presumably for this reason that as a long-term strategy the use of money is as a result usually more effective than force. By encouraging closures of authority through the potential satisfaction of desire, money is less inclined to induce opposition.[3] A further factor that helps explain the pervasive influence of money in the institutions of society is its abstract quality. Money is not associated with the satisfaction of any particular desire, but with the potential to satisfy desire in general. As a result it is not necessary to know the desires of an individual in order to be able to offer the potential to satisfy those desires. The flexibility of money, and its capacity therefore to enable the satisfaction of desire, places it in a unique position to encourage closures of authority. This would appear to be a likely explanation for the almost universal adoption of the money economy.

Nearly all transactions that involve the transfer of money also involve the realisation of closures of authority. The transfer of money usually entails the realisation of closures of authority that enable the buyer, the author in this economic institution, to impose closures on the seller, the realiser. The extent of the capacity to impose closure depends on the nature of the transaction. The acquisition of goods, for example, enables the buyer to impose a set of closures requiring the transfer of the goods. In the acquisition of services the capacity to impose closure is less circumscribed and is likely to include the imposition of closures that determine the manner in which the service is executed. In an employer/employee relationship the closures of authority are usually more extensive again. The relationship between employer and employee is not therefore simply a transfer of money for labour or services, but the agreement on the part of the employee to adopt closures of authority. The adoption of these closures of authority do not give the employer the right to impose any closure but they do give the employer the right to impose closure in some respects. The agreement to adopt the closures of authority involved in an employer/employee relationship has the consequence that the employee, as the realiser, agrees in certain respects to abandon their own previous closures and thus in those respects their current space and reality, and take on the closures of the employer.

It can be seen therefore that financial transactions involve the realisation of closures of authority that result in the formation of an institution with rights and responsibilities for those involved. The acquisition of goods gives the buyer the right to impose closure to the extent of the transfer of the goods, and thus gives the buyer the right to own the goods. A company has the right to impose closure in certain respects on its employees, and the authors, those individuals who run the company and who seek to maintain and extend the closures of the institution, are responsible for the satisfaction of the desires of the employees, in particular the receipt of wages. Consequently a money economy results in the formation of many institutions with their corresponding institutional spaces which in turn contribute to cultural space as a whole.

In summary therefore, institutions are the outcome of closures of authority that are adopted as a result of the perceived capacity to satisfy desire on the part of the realiser. The desire in question may be jointly held by author and realiser and directly related to the task to which the closures of institutional space are directed. Alternatively the desire of the realiser may be independent of the goals of the institution. In either case the author has the potential to extend closures of authority as the result of the satisfaction of the realiser's desires or as the result of changing the realiser's space so that desires are satisfied. As a result of the adoption of closures of authority, for whatever reason, an institution is created whose structure is determined by a network of rights and responsibilities and whose content is provided by the closures of institutional space.

22

THE ORGANISATION OF SOCIETY
Institutional space

> Social organisation and hierarchy are the result of the
> closures of institutional space. These are in turn influenced
> by personal and cultural space. Personal, institutional, and
> cultural space are thus independent but interlinked.

The structure of a society, which enables individuals to have specified roles
and formal relationships to others, can be seen to be the outcome of the
adoption of closures of authority and the consequent formation of institutions. Each institution generates its own set of closures, its own
institutional space. The aim in this section is to account for the character of
this institutional space, and its relationship to personal and cultural space;
and thereby to elaborate how social organisation is a product of closure. By
this means I hope to cast light on the mechanism by which society
develops, its means for avoiding conflict, and its goal or direction.

In the context of the account that has already been given of institutional
space, it will be apparent that in the first instance the character of institutional space owes entirely to the author. In due course institutional space
consists of those closures that have been sanctioned by current or previous
authors and which have not subsequently been abandoned. In the same
way that personal and cultural space change over time, so institutional
space is thus also changing in character as new closures are adopted and
prior closures abandoned. As we shall see, these changes in institutional
space have an impact on the relationship between the author and the institution, and the relationship between the institution and the rest of society.

At the inception of an institution the closures imposed by the initiating
author stem directly from the author's own personal space and its application to the particular context governed by the closures of authority in
question. In due course these closures begin to have their own separate
identity. Having imposed a set of closures on the realisers the author is
then constrained by these closures as institutional space develops its own

305

logic and momentum. This can be seen in the case of the island camp-building institution, for example, where we can imagine that the author might have doubts as to the likely success of the operation but feel unable to express these without damaging the institution; or, having sited the camp in one place the author may come to the conclusion that it was mistaken and should have been sited elsewhere, but decides that the expression of such a view would serve to undermine the closures of authority and thus the possibility of the completion of the task. It will be apparent, therefore, that although the author has the capacity to alter the closures that make up institutional space, the author is also constrained by the present character of institutional space when putting forward new closure. It is in this manner that institutional space can be seen to develop a distinct character from the author's personal space.

The framework of institutional space and the nature of power relations that has been elaborated has the further consequence that it is through institutional space that an author has power. In addition, it is because institutional space is independent of the author's space that power can be passed from one individual to another. It has been argued that a newly created institution that is successful in satisfying the desires of the realisers is in a position to cement closures of authority and thereby increase the power of the author. In the event of a new individual or group of individuals taking the role of author the closures of institutional space will already provide the new author with closures of authority as a result of the prior success of the institution. It is by this means that official positions or titles carry with them authority. The history of the imposition of past closures has resulted in an institutional space that provides the new author with closures of authority and therefore power. Nor is the exercise of power necessarily restricted to the author. For institutional space determines the closures of authority that apply not only to the author but to each member of the institution. Through institutional space each member of the institution can be allotted a role and attached to that role are closures of authority which can enable the individual to impose closure in certain respects. The capacity to impose closure may apply to other members of the institution, as in the case of a company manager, but it can equally extend to society as a whole through a network of related institutions, so that for example a doctor has the power to prescribe drugs, a policeman the power to arrest. It can be seen therefore that the power associated with certain social roles in society stems from a complex web of institutions each with their own closures of authority.

While institutional space is independent of personal space it is of course influenced by the personal space of the author. The power bequeathed to an author through an already existent institutional space is not fixed but remains dependent on the closures of authority which can be strengthened or undermined by the manner in which the author imposes closure. As a result the power of the author, or an individual with authority within an institution, is not guaranteed by the institution but is to some degree dependent on the author or the individual concerned. When an individual first assumes a position of authority within an institution, the history of institutional space is able to ensure that in general the position alone grants them the ability to impose closure in the relevant respects. In due course the provision of new closures by the individual contributes to institutional space and over time alters the strength and extent of the closures of authority. A company manager who at first has power as the result of an institutional position will undermine the closures of authority associated with that position if the imposition of new closures fails to satisfy the desires of the realisers in the relevant respects. As a result the manager may still hold the same position and carry the same title but the manager's power will have been reduced. It can be seen therefore that closures of authority do not automatically follow from the title or role that an individual adopts but remain dependent on the agreement of the realisers to accept the imposition of closure. A title, however prestigious or established, does not therefore guarantee the closures of authority although the strength of institutional space may be such that it is with difficulty undermined. Institutional titles and posts carry with them the closures of authority of the past, and it is these closures that are the basis of power not the title or position in itself.

A consequence of the relationship between the closures of authority and the role held by an individual is that if an individual in a role that is associated with closures of authority is unable to impose closure, not only is their personal authority undermined but so also is the authority historically attached to the title. For example, a president who is forced to abide by the views of his advisers, or a prime minister who has to adopt the closures offered by powerful members of a cabinet, not only undermines their own authority to impose closure but also to some extent undermines the authority associated with the role accumulated through the history of the institution, with the consequence that the presidency or the position of the prime minister is itself weakened. Similarly doctors or police officers who through their actions and attempts to impose closure undermine their own power at the same time undermine the power associated with the institutional position. If sufficient numbers of such individuals acted in this way the institution itself would come under threat. This is why professional bodies seek to remove individuals who might weaken the institution through their unsuccessful attempts to impose closure. One consequence of this analysis of institutions is that an author having taken on a role or title may have to accept closures with which they disagree, in order to maintain the closures of authority. In such circumstances the closures of institutional space are sufficiently strong to overrule the closures of the author's personal space. It may even be that despite having authority in name, an author is unable to pursue any closures of their own.

While institutional space is capable therefore of conferring on an individual great power, it at the same time imposes its own constraints, and the power of the author is only maintained if the proposed new closures are compatible with the remainder of institutional space. It is not uncommon for institutions to have a long and extended history, stretching back hundreds, even thousands of years. In such circumstances institutional space is largely independent of the personal space of the current author. As a result the power of the author is largely guaranteed by the institution, but at the same time the ability to impose closure is heavily constrained. In the institution of the Catholic church for example the Pope is the ultimate author and seemingly unchallengeable. Yet a Pope that behaved in a manner radically at odds with the history of institutional space would rapidly undermine his own power. Nor would a Pope have to promulgate closures as radical as denying the existence of God to cause such an outcome. The mere abandonment of some of the familiar roles and routines of previous Popes could lead to a weakening of the closures of authority historically associated with the role. While long-standing institutions therefore make it look as if power attaches to the role or title, the strength of the institution in guaranteeing the closures of authority is usually matched by the constraints it imposes on a new author. It is because of the independence of institutional from personal space that although the role of the author can carry prestige and status within an institution and in society as a whole such a role is not necessarily desirable, for it does not allow the author the ability to act according to their own desires. There are for example societies in which the role of the seemingly most powerful individual or author is sufficiently constrained by the institutional space for which they are supposedly responsible that the position is widely regarded as onerous and undesirable. Japanese and Thai emperors were regarded as gods, and therefore might be thought to be capable of imposing closure at will. The history of the institution however ensured also that gods were expected to behave in certain precisely defined ways which were not desirable from the perspective of the individual concerned. As a result it was not uncommon for the families of individuals who might be taken up as the emperor to seek to avoid such an outcome, as Frazer documented a century ago in *The Golden Bough*.[1]

It can be seen therefore that authors as well as realisers operate within institutional space and to this extent both benefit from, and are constrained by, the history of their own past closures and those of previous authors. It is perhaps for this reason that the individual brilliance of an individual, and thus the power of their own personal space, is no guarantee of the success of the institutional space of which they may become the author. Authors, through closures of authority, have the right to impose closure in certain respects, but the effect of this closure depends on the rest of institutional space and the manner in which institutional space interacts with the personal space of the realisers. If the closure is successful it will both contribute to the effectiveness of institutional space as a whole and satisfy the desires of the realisers, and will, thereby, strengthen the closures of authority. The character of institutional space, like the character of personal space and cultural space, is similar therefore to Neurath's boat, which is gradually rebuilt while at sea.[2]

Turning now to the relationship between institutions, it can be seen that each institution finds itself in a broader framework of institutional space formed by other institutions and their interrelations. It need not be supposed that the closures of authority that operate between institutions function differently from those that operate within each institution. As with individuals, the closures of authority between institutions are at first adopted because they are seen to be in the interests of the institutions concerned, although over time this need no longer apply. Many institutions have a direct link to others and form a group of related institutions whose institutional space forms a nested set of closures. Nevertheless each institution within the group maintains its own institutional space which develops according to its own internal requirements. The closures of authority between institutions evolve as the institutional spaces of the related institutions evolve, and as a result the power relations between individuals within and between those institutions is also in flux. Although therefore the hierarchy of power within society appears relatively fixed, it is in a state of change as the closures of authority between and within institutions alter, in the same way that the meanings of words in language appear to be fixed while they are at any point but a momentary slice through a shifting set of closures.

Superficially the closures of authority that are formed in nested institutions provide a clear hierarchy of power. For example, the closures of authority that form the institution that applies between the nursing staff on a hospital ward falls within the closures of authority of the hospital; and in circumstances where the hospital is part of a nationally provided system of health care, within the closures of authority of that system of health care, and its regional divisions; and then within the closures of authority of the government and ultimately of the individual who heads that government. The institution created between the nursing staff of an individual ward takes its place therefore within a whole series of related institutions, and falls within the closures of authority that apply to the institution of government as a whole. Yet the hierarchy of power is more complex than this initial institutional sketch illustrates. The individual at the apparent apex of the hierarchy, a prime minister or a president, is not in a position to directly impose closure on nurses in a particular ward despite the linked set of closures of authority that group the institution of government and the institution of the hospital ward together. To success-fully impose closure, each institutional layer in the nested group of institutions will need to adopt relevant closures and continue to realise the closures of authority, and at each layer the authors concerned will need to carry with them the realisers in order to secure the adoption of the closure in question. Moreover, the linking of these related institutions has the consequence that the strength of each institution within the group is in part dependent on each of the other institutional spaces which make up the group, with the result that each institutional space potentially constrains the others. A widespread refusal by nurses to accept the closures of authority that are usually associated with the sisters in charge of the hospital wards, could, for example, in principle lead to a collapse in the provision of health care and even perhaps threaten the closures of authority of the government. Similarly a state of national anarchy, or civil war, and thus the undermining of the closures of authority that are normally associated with government might lead to a collapse of the closures of authority on individual hospital wards. The closures of authority between institutions are thereby interlinked and interdependent and are revised and altered in the light of the development of each institu-tional space and the new closures adopted by the authors of those spaces. As the closures of authority between institutions shift so also do the closures of authority within each institution. It can be seen therefore that the power relations that operate between any two individuals in society are a complex product of layers of closures of authority that impinge on that relationship, each of which is in a process of flux.

Society is a product of its institutions, but if society were no more than a bundle of separate institutions and their associated spaces, and the closures of authority that applied between them, the conflicting goals of the institutions could be expected to lead to its disintegration. Despite the potential conflicts between institutions and despite the shifting character of each space and its closures of authority, society is for the most part remarkably stable, and it is to the issue of the stability of society and the containment of conflict that I shall now turn.

In order to account for the cohesive nature of society it is important to identify a mechanism by which conflicting and fluctuating institutional goals are contained. One means of achieving such cohesion is for there to be a widespread adoption of a similar set of closures of authority across individuals within the society concerned. The mechanism by which this takes place involves the interaction of personal, institutional, and cultural space. Personal and cultural space are independent but influence each other. Cultural space consists in the framework of closure that enables communication between individuals within a culture. This evidently includes the tags of language and the rules which govern their combination, but it also involves the realisation of those sets of tags, in the form of facts and procedures, that are widely adopted amongst members of the culture. While the personal space of each individual converges on the closures of cultural space, the development of cultural space is driven by new closures and new relationships between current closures that are realised in the personal space of individuals within the culture and are not at the point of the realisation part of cultural space. In the same way that the personal space of individuals is both distinct from cultural space and contributes to the development of cultural space, so also is institutional space both distinct and a contributor to cultural space. Only a handful of the new closures realised by individuals find their way into cultural space as a whole, but those that do influence the language or the body of generally agreed facts and procedures. Similarly only a proportion of closures of institutional space find their way into cultural space, but those that do will carry with them implicit closures of authority that stem from the institutions of which the closures are a part. Indeed many of the facts of cultural space are regarded as such due to the adoption of the particular closures of authority of the relevant institution. Individuals in a culture recognise certain closures as true because they are endorsed by an institution that has accumulated the authority to determine which closures of a certain type are to be realised in all circumstances.

It is because institutional space influences cultural space, which in turn impinges on each individual's personal space, that closures of authority reflecting the historical structure of society are found in the space of each individual within a society. In the same way that cultural space incorporates the closures of language thereby enabling communication, it also incorporates closures of authority which allows for the relative stability of society and its formal hierarchies. In the same way that it is unlikely that everyone within a culture could decide to abandon the use of one word in favour of another overnight, it is similarly difficult to agree to the abandonment of closures of authority that are embedded in cultural space. Power, as with language, stems from closure, and has the similar characteristic that while stability is the norm, change is unavoidable. Both language and the structure of society are relatively stable in the short term because if they were not they would be replaced by a system of closures that was stable, for a degree of stability is required for effective intervention.

The incorporation of elements of institutional space into cultural space has the consequence that while in the case of new institutions realisers adopt the relevant closures of authority because they are seen to have the potential to satisfy some desire on the part of the realiser, many closures of authority are realised because they are part of cultural space and are adopted on the grounds of expression alone. Individuals are born into a society and into its particular institutions and closures. Throughout childhood the closures of cultural space are gradually acquired and in the process also closures of authority embedded within the closures of that space. Cultural space as the outcome of previous social hierarchies contains many stories of authority which are as a result adopted by most individuals within a culture. Children acquire sets of closures, for example, associated with the president, prime minister, king, or queen, which either explicitly or implicitly provide these roles with authority. The association of individuals with authority extends to most social roles: the closures associated with the tags 'police officer', 'doctor', 'teacher', carry with them closures of authority which give the individuals who are described in this manner rights to impose closure on others in appropriate circumstances. Through the acquisition of cultural space a child thereby acquires the closures of authority associated with certain roles. In play children try out new closures they have acquired and explore the potential of these closures. Part of a child's play consists therefore in the adoption of social roles, and the acting out of the rights to impose closure that are intimately bound up with these roles. If a child has acquired the closures associated with the tags 'police officer', 'doctor', or 'teacher' they have also acquired associated rights to impose closure, for without these rights the respective institutions of which these roles are a part would evaporate.

The implicit association of terms such as 'doctor' or 'police officer' with closures of authority can become more evident if a challenge is made to that authority. In such circumstances a different tag may be used to describe the role in order to avoid the implicit closures of authority associated with the familiar tag. A doctor referred to as a technician has for example lost some of the authority that is embedded in the closures of cultural space. Correspondingly institutions, such as professional organisations, that are responsible for those who carry out a particular task are usually concerned to maintain the titles that distinguish the role and thus implicitly give them authority. In addition to the titles associated with the highest authorities such as the president, monarch, or chairman of the party, social roles such as judge, priest, professor, doctor, politician, soldier, involve the repeated use of a title in addressing or introducing the individual. In each case the title serves to name the role but in doing so also reinforces the closures of authority associated with the tag and to some degree the institution as a whole. As a consequence to abandon such titles is a challenge to the authority in question, for it is to describe the role in a manner that does not automatically carry the relevant closures of authority. Those who wish to avoid or undermine the authority of the police, for example, are likely to use alternative descriptions, or nicknames, which seek to escape the associated closures of authority. Closures of authority can be seen therefore to be woven into the closures of cultural space, are absorbed as we acquire language and are maintained through repetition and reflect the closures of prior institutions: the social organisation of the past.

It has been described how an author is able to strengthen the closures of authority by a gradual shift in the personal space of the realisers. Such a shift may be designed to accentuate the importance the realisers attach to the achievement of institutional goals, or it may alter desire in such a manner that the institution is in a better position to satisfy the desire. In any case the author can through a gradual change in the personal space of the realisers seek to maintain and extend their power. In a similar manner institutions can seek to encourage changes in cultural space that will maintain and extend their power. Since cultural space is the historical outcome of this process its structure already incorporates a framework of closure that has been utilised to support the institutions of the society. The closures of cultural space function therefore not only at the level of personal space but at the level of institutional space. The closures of cultural space have been retained because they enable effective intervention, and one aspect of that effectiveness is the support they provide to the institutions of society. The closures of cultural space are therefore not only the outcome of an extended history of trials of various closures and their capacity to enable individuals to intervene effectively in the world, but are the outcome of a history of society and its capacity to intervene effectively. Closures that maintain and support a powerful institution are therefore more likely to be part of cultural space than closures which undermine such institutions.

Successful and stable institutions at least partially satisfy desires of the individuals involved. These institutions will in general also seek to avoid or minimise conflicts with the rest of the realiser's space. To this end institutions and their authors can be seen to introduce closures that function specifically to limit conflicts of desire and threats to authority. In due course cultural space incorporates some of these closures. Cultural space therefore contains sets of closures that mitigate potential conflicts between the goal of the dominant institutions in society and the desires of the individual. If it did not do so the social arrangements involved would have collapsed and cultural space would have changed accordingly.

To elaborate in any detail the relationship between the closures of cultural space and the maintenance and development of successful institutions is beyond the scope of this chapter. In any case a general account of how this relationship operates is not possible since it will vary from one cultural space to another. However, in an attempt to illustrate the nature of the relationships described one example of a mechanism by which the closures of cultural space can be used to support the institutions of society will be considered in a little more depth. It is an example that has already been briefly considered in the case of the relationship between parent and child, namely the role of morality and moral closures.

Moral categories support the institutions of society by introducing into the personal space of each individual a set of closures that mitigates conflict between the desires of the individual and the closures of authority prevalent in the institutions of the society. The notion of a moral good thereby provides a motivation for an individual to act in a manner which would otherwise not be in their immediate interests and would therefore not appear desirable.

Traditional moral precepts, either in the negative form of identifying what is wrong and should not happen such as murder or adultery, or in an encouragement to certain forms of action, such as looking after one's neighbour or the weak, can be seen as beneficial to social stability. The adoption of moral closures reduces the importance of personal desire, which might otherwise result in a challenge to social institutions, placing it lower in the hierarchy of desires than social well-being. One example of this phenomenon can be found in those who see in Buddhism the tenets that life is suffering, that suffering is caused by desire, and that consequently we should rid ourselves of desire. For these tenets, if adopted by an individual, have a good chance of avoiding conflict of desire between the individual and social organisation.[3]

In order to illustrate this mechanism it may be helpful to consider a simplified example in the context of current cultural and institutional space. One consequence of the closures of contemporary cultural space can be seen to be the generation of desire in the personal space of most members of society to acquire material goods. Such a desire can be satisfied if the individual is able to acquire the goods through payment. However there are liable to be material desires that individuals are not in a position to satisfy. As a result of this failure to fulfil the desires of the individuals within society there is a potential conflict between the institutions of society and its members. This conflict in principle poses a threat to those institutions and ultimately to the stability of society. The threat to the institutions of society and thus to the power relations within society on these grounds can be seen therefore to be deeply embedded in the structure of cultural space.

In this context previous institutions and social hierarchies can be regarded as having either intentionally sought, or stumbled upon, closures that moderate this threat. Morality can be considered as one such means. For if individuals incorporate into personal space the closure that it is wrong to steal they are less likely to follow their desire to acquire material goods that they are unable to acquire through payment. In addition, a case could be made that the individuals concerned are also less likely to take issue with the closures of authority attached to the particular social institutions that confirm the framework of the ownership of goods. It can be seen that even if this moral closure does not eliminate theft, so long as the notion that the activity of stealing is wrong is maintained, the theft does not itself pose a threat to the closures of authority of the social institutions involved. A society may thus be weakened by property crime but the threat to its social institutions, and therefore the hierarchy of power within the society, is liable to be far more serious if this action is associated with sets of closures that redescribe the 'crime' so that it is no longer caught by the higher authority of morality. Revolutionary politics has indeed often been concerned to engage in just such a redescription, ideally turning the previously perceived crime into a new moral crusade. Such attempts to do so have included the dictum 'property is theft' but the less intellectual approach of Robin Hood would probably have been equally threatening to the social institutions of Norman Britain.

The importance of morality can be regarded as consisting not in its endorsement or criticism of particular actions but in the closure that action can in principle be right or wrong. Specific moral precepts, such as 'thou shalt not kill', have the disadvantage that there may be social circumstances and institutions that are unable to promote the particular precept.

The army in the circumstance of war, and the judiciary in circumstances where there is a death penalty, are two such instances. Instead what is important about the notion of a moral sanction is that it presupposes that an action has a value independently of the worldly circumstances in which it takes place. It is this character that can be seen to enable morality to be such a powerful and effective tool for sustaining social stability. For this closure allows institutions to call upon morality to endorse or to challenge actions that are potential threats to the closures of authority and the goal of the institution, without having the appearance of doing so as a means to promote their own goals. An example of this use of morality has already been proposed in the case of the first power relationship, between parent and child. The child acts in a way the parent does not wish and is told 'you should not do such and such'. If this response is challenged one of the options open to the parent is to say: 'because it is wrong'. Such an explanation seeks to bring the challenge to authority to an end, since it is designed to be unchallengeable. It can be seen therefore that in general, institutions regard as right those actions which support the institution and wrong those things which challenge it. However the role of morality in supporting the institutional status quo is obscured by the notion that the appeal to morality is not driven by worldly concerns.

Our cultural space has therefore come to incorporate moral closures that in general support the institutions of society and thus its stability. It does so because those same institutions have propounded these closures precisely in order to support and maintain their own closures of authority. Organised religion has of course played a central role in the adoption of moral closures, but although these closures have sometimes been troublesome for institutions, it can be argued that one of the reasons for the success of the institution of religion is that its promotion of moral closures has largely supported and strengthened the dominant institutions of society, or has been capable of being used by those institutions in order to maintain and support the closures of authority. It is not only monarchs - with the claim to the divine right to rule – who have sought to endorse their institution with the backing of moral authority. Democratic institutions have similarly sought to claim moral authority, and for no grounds other than that they are democratic.

Many readers may wish to object to this account on the grounds that there is more to morality than the endorsement of institutional goals. Can the notion of the morally good really be reduced to the power play of the dominant institutions in a society? However it is precisely because moral closures have been allied to some higher and ultimate authority that they are capable of serving the function required of them in a prosaic social context. Classic moral precepts encourage concern for others, the avoidance of violence, the exercise of justice, and so forth. All of these precepts can be seen to promote a stable society by encouraging the submission of individual desire to a supposedly higher motive. It is surely not merely accidental that this is the case, that for example we do not find moral codes that promote discord through precepts such as 'kill those with whom you disagree'. Such moral codes are not found for the simple reason that they do not serve the maintenance of the institutions of a society and therefore the stability of that society. A society therefore that adopted such a code would rapidly collapse. The character of our moral precepts can be seen in this context to be the outcome therefore of a form of societal evolution. We have the morality that we do because it is a framework that allows for a stable and therefore successful society. Moral frameworks that have not served this function are not in general found precisely because societies whose members realised such principles will have failed.

Morality does not of course in any individual instance automatically function to support the institutions of society. There are many cases for example in which moral notions such as justice and a concern for others have been used to support a challenge to a social order. Institutions can promote a moral code only to find themselves victims to such a code. However, while in any particular instance conflicts can arise between widely agreed moral closures and the institutions of society, in the long term a society could not survive if its morality was at odds with its institutions. Moreover institutions in a society will, no doubt with this in mind, seek to influence the moral code in order to promote their own security. The state of the moral code in any given society is therefore a complex evolutionary outcome of these interrelated pressures, with the consequence that for the most part in stable societies moral closure can be seen to function to support the institutions of that society. A current example of this relationship can be seen in the connection between the moral code of a society, its institutions, and recreational drugs. Almost all societies have at least one recreational drug that is so widely used that it is embedded in the institutions of the culture.[4] Frequently this also involves religious ceremonies. As a result this dominant recreational drug is usually seen positively, or neutrally, within the moral code of the culture, while other recreational drugs are often regarded negatively. So it is that in Western Europe and North America alcohol is embedded in the dominant religion, in the form of the Christian communion ceremony, and is manufactured and distributed by large and powerful institutions. The influence these institutions have had on moral closures over time has resulted in alcohol widely being regarded as a good thing or at least morally neutral; while other recreational drugs are frequently vilified and regarded as bad or even evil. In contrast, societies where the dominant recreational drug is different have correspondingly different moral codes. So it is that in some Muslim countries the reverse moral code is held whereby alcohol is regarded as bad and outlawed by religious and moral codes while alternative recreational drugs that have been historically dominant, such as cannabis, are regarded positively. Furthermore it could be argued that the power of Western economic institutions on the world stage has resulted in the attempt to impose the Western moral closures regarding recreational drugs on the remainder of the world. Where this is in conflict with local institutions it is likely to be destabilising: Columbia being an obvious example.

The relationship between the institutions of a society and the moral code is a continually evolving one, and different institutions may seek to shift the moral code in opposing ways. As a result there are often points where the moral code comes into conflict with an institution within that society, as it did for example in the United States in the decade leading up to the prohibition of alcohol. There is not therefore a simple correlation whereby a particular institution is automatically supported by the generally held moral closures within the culture. Nevertheless, for the most part, morality will be seen to support the majority of institutions of society precisely because these institutions have been influential in determining the character of that moral code or have used a pre-existing moral code to support their position and status.

The stability of society can be regarded as depending on the containment of conflicts of desire. Usually the interest of any individual institution will be to maintain social stability and so taken as a whole institutions can be seen to seek to engineer this outcome, and morality can be regarded as one consequence of that engineering. In this context the moral opposition to recreational drugs taken as a whole can be seen to be driven by the threat to internal stability that such substances have on society. However, the dominant recreational drug of a culture often has sufficient support from some of the institutions of society to enable the moral code to be changed in this particular respect. High-level moral closures, those that are not concerned with specific activities but with behaviour in general, function by altering personal space in such a manner that individuals desire outcomes that are likely to encourage social stability. The provision of a system of moral closures does not of course ensure that there are no conflicts of desire between the individuals of a society – and the system of moral closures will always be evolving in such a way that new conflicts develop – but it does reduce the likelihood of such an outcome. Nor is morality alone in having this function. The framework of cultural space as a whole can be considered to be the outcome of a history of interaction between personal and institutional space and as a consequence reflects the closures of those institutions which have come to be successful.

In summary therefore, social organisation can be seen to be the outcome of a network of institutions formed by closures of authority. It is the social and political character of linguistic closure that enables closures of authority to be formed. These closures of authority are in the first instance adopted because of a joint attempt between a group of individuals to satisfy a desire. The institutional space that is formed by these institutions in turn can generate conflicts of desire between individuals and between institutions. The character of closure and its relation to desire determines therefore the manner in which institutions develop. Successful societies have aspects of cultural space that mitigate and contain the conflicts between individuals and institutions. In particular the institutions of society can be seen therefore to limit conflict through the provision of additional closures which are gradually absorbed into cultural space and thus into the personal space of each individual in the society: morality being one such example. The interrelation of personal, institutional, and cultural space – an interrelation that stems from the nature of linguistic closure – therefore drives the development of society and results in the organisation of that society.

23

SOCIETY, CHANGE, AND DREAMS
OF UTOPIA

Since desire is a function of social organisation, desires
change with that organisation, and as a result any attempt to
determine the future direction of society or the character of
an ideal society can make no progress.

It has been shown that the closures of cultural space not only carry the
closures of authority of prior social organisation but also closures that
function to moderate the threat to authority from personal desires. It is
because human desire is almost completely plastic in the face of alterations
in the framework of linguistic closure, so much so that even desire essential to the maintenance of life can be abandoned, that the closures of
cultural space are so powerful in determining the success or failure of the
institutions of society. Closures of authority could in principle be made
impregnable if all conflict with individual desire was eradicated. Some
cultures have come close to doing this thereby perpetuating the current
social hierarchy and organisation, and where this has occurred the relationship between the institutions of state and those of religion and
morality are likely to have been important − as in ancient Egypt for
example, where the two were effectively merged. Nevertheless, the
removal of conflicts in desire is however an impossible goal if for no other
reason than the necessary failure of closure always allows for the possibility of conflict. Moreover, and more importantly perhaps, societies are
not isolated and can be challenged by the cultural space of other commu-
nities.

Despite the attempt to eradicate conflicts of desire, institutional and
cultural space will contain potential or actual points of conflict between its
closures, which will involve the authors of its institutions in an attempt to
determine the conflict in their favour. A successful institution therefore
seeks to extend its closures of authority through the adoption of its
closures by individuals both inside and outside that institution. The institu-

tions of society thus compete over closure. The dominant institutions are those that are able to impose their own closures and thus incorporate them into cultural space.

Political parties are self-evidently engaged in a competition over closure, but so equally are institutions such as the police, the media, and professional groups such as doctors or accountants. Each institution seeks to influence the closures of those outside the institution so that its own closures of authority are extended. Doctors for example will seek to ensure that they are held in high esteem and are well paid. To this end the institution is likely to seek to make it difficult for individuals to become doctors, and will also seek to stop other professionals – nurses or chemists for example – having the right to carry out medical procedures. In order to achieve these ends the institutional space is structured accordingly.[1] Some of these do not involve conflict with other institutions such as the requirement for long training periods, and strict educational qualifications, but others such as the defence of the exclusive right of doctors to prescribe drugs requires an external attempt to ensure this closure prevails. This might involve seeking to influence the closures of government departments, the closures presented to the public in the form of media coverage, the closures of competing institutions, and the closures of the public as a whole generated by their individual experience when visiting their doctor. The maintenance of closures of authority of an institution thus involves many forays into the institutional space of other institutions and into cultural space in order to retain the set of closures required. Each institution in society is engaged in a similar exercise, and the current social organisation can therefore be regarded as a snapshot of the manner in which the closures of each institutional space interact and the extent to which each institution has been able to influence the closures of cultural space as a whole. Politics in its broad sense can therefore be seen as the competition over closure and the attempt to impose closures, be they personal or institutional, on other individuals, institutions, and cultural space.

The inherent conflicts of desire generated by the closures of cultural space and the different institutions of society provide a momentum for social change. It is in this sense that the Hegelian and Marxist account of historical change has some purchase. Individuals and institutions operate within cultural space which contains the potential for conflicts of desire, and to this extent it might be argued that there are inherent conflicts or contradictions in our world. It is understandable therefore that one could identify a pattern in the development of cultural space whereby conflicts that arise from the stories of cultural space are translated into changes in social organisation which in turn result in new conflicts of desire, new stories, and a further change in social organisation. There can however be no historical inevitability with regard to social change for the outcome of social conflict will be dependent on the cultural space prevalent at the time, which is not determinable or understandable in advance. In short historical inevitability is not possible for it would require a perspective that lies outside of history.

The structure of society can be seen therefore to be the outcome of closure, with social organisation and hierarchy being the result of the closures of institutional space. The closures of institutional space are then in turn influenced by personal and cultural space. Personal, institutional, and cultural space are thus independent but interlinked. The personal space of an individual gradually adopts the closures of cultural space as the individual becomes an adult member of society. The character of this personal space influences their desires. In pursuit of these desires individuals form institutions which rely upon closures of authority. The institutions in turn generate closures that create institutional space, some of which are incorporated into cultural space. The historical development of closure thus involves a cycle of interaction between personal, institutional, and cultural space.

It can be seen that although personal space owes largely to the closures of cultural space, there is nevertheless the capacity for the individual to escape the closures of cultural space. We are not entirely lost to the closures of cultural space, nor are we entirely victim to its closures of authority. The individual is able to generate new closures from the structure of closure that is inherited from cultural space. Furthermore it is through the originality of individuals that a gradual shift in institutional and cultural space is made possible. Yet, cultural space is the dominant framework that all individuals must necessarily adopt in order to communicate. Personal space can allow but a little originality at the margins of the closures of cultural space. Yet it is the shifts at the margin that is the engine of cultural and social change.

There have been many who have sought in the attempt to describe society the possibility of determining its future direction. There have been others who have sought to give an account of an ideal social organisation, or of the direction in which society should go if it is to improve. Since desire is a function of personal space, and since personal space is dependent on cultural space, and cultural space on institutional space, human desires are primarily a function of social organisation. If desire is a function of social organisation it will change with that organisation, and as a result any attempt to determine the future direction of society, or the character of an ideal society, can make no progress. For to propose that there is a structure of social organisation that will best satisfy the desires of its members, or that it might have a necessary direction, is to imagine that desires are independent of social organisation. At any point in cultural history changes can be made to social organisation which seek in the context of the closures of the time to alleviate suffering and promote the satisfaction of desire. The new social organisation that results can be seen however to generate its own desires and thus its own success and failure.

While it is not possible therefore to draw any conclusions about the future of society, it is possible to identify a risk attached to the adoption of closures of authority. For perhaps the primary concern in adopting closures of authority is the capacity of the authority to change desire and thus to extend authority. Just as we cannot know how changes to our personal space will alter who we are and what we desire, so also we cannot know how an institution which we have voluntarily joined will be able to impose closure in order to extend that authority. Similarly on a social scale, we cannot know how the institutions of a society may be able to alter the character of cultural space to ensure their continued presence. Nor is there an Archimedean point from which to be able to observe different cultural and personal spaces and determine which desires are more desirable. Depending on the nature of personal space, a life in a convent may be either seen as a waste or as an ideal. So also depending on the nature of cultural space, a particular social organisation may seem desirable or undesirable. Some ancient cultures – the Chinese and Egyptian cultures being good examples – were able to form a set of social institutions that contained conflicts of desire to such an extent that the structure of society remained unchanged for thousands of years. If we look to the future it may be that, in the context of a global culture and global institutions, a global cultural space could emerge that would reinforce these institutions. Without a challenge to this social organisation from external societies, such a structure might be capable of formulating cultural space so that conflicts of desire were largely eradicated. From our current space such an outcome looks uncomfortable, even frightening, precisely because from the cultural space of the future all will appear to be well.

The relationship that has been described between the structure of closure and the structure of society may encourage us to be wary of the capacity for institutions to extend their power. However, while we can seek to imagine the shift in cultural space as the result of social change, who we will be and what we will desire remains indeterminable. For the process of closure in the interplay of personal, institutional, and cultural space has no limit. The underlying structure of closure, and its failure in the face of openness, has in the arena of politics the same outcome as it does for the individual. Our inability to complete closure makes the future uncertain, but it also has the consequence that the future is unlimited.

EPILOGUE

Since the story of closure was initiated by a concern with the problems of self-reference, I return to this issue at its close. For in the elaboration of the account of closure the matter of self-reference has largely been absent. Yet the issue has been implicit throughout as a constraint on its structure.

On the one hand the story of closure offers an account of the nature of our circumstances. In doing so the initial vocabulary of openness and closure has been applied first to language and then to science, art, and society. The purpose of this has been to demonstrate the effectiveness of the framework and to extend and develop the system of closure. To this end the theory of closure is intended to have empirical application. For example, in the description it offers of language and science; in the building of an intelligent machine; in its capacity to aid the development of new theories and as a means of determining the limitation of our theories; in its account of art and religion; in its description of society. The theory of closure is in this respect offered in a manner that is no different from that of a scientific theory, and as with any such description it should be modified, extended, or abandoned in the light of its capacity to allow for desired intervention in the world.

At the same time, the story of closure is not a description of our circumstances in the scientific sense of a description of the world at all, but an account of why a description of the world is not possible. It is thus a pointer to that which is not closure, a pointer in the direction of openness. As such it seeks to undermine the arrogance of theories that suppose they might have uncovered the true nature of reality.

The account of closure is thus both an attempt to give an accurate and complete description, and a denial that such a description is possible. It is however necessarily so, for the system of closure has the characteristics of closure which it itself puts forward. It is thus a reflexive theory. Like all other closures it both seeks to be complete, and under scrutiny fails. It is on the one hand the outcome of a desire for closure and on the other hand the outcome of the desire for openness.

According to the story of closure it is through closure that we realise things and are able to intervene. We benefit from the ability to intervene and thus desire to extend and complete our system of closures. Without closure there would be no-thing, no world of which we could be a part. Correspondingly, the theory of closure itself seeks to extend and complete its own system of closures. Each aspect of the theory offers a closure that realises material and texture, and which allows for the further addition of new closure and new texture. As an interconnected system of closures the story when applied to various fields seeks both to extend its reach and to sharpen its defences from potential threats to its completion.

Then again according to the story of closure, there is nothing in common between closure and openness and thus closure does not describe openness nor does it approach openness. The material realised by linguistic closure will as a result always fail to exhaust texture. Our recognition of the limitations of closure leads us to seek to overcome those limitations, to escape closure and approach openness. In the same way however that closure cannot be completed, so the attempt to escape closure must also fail. Correspondingly in the account of closure itself, the theory seeks to incorporate the notion of its own limitation: that it is itself but a closure. In the framing of the notion of 'openness' there is an identification of the failure of closure and demonstration of the desire to escape closure. It is however a desire that cannot succeed, for the notion of openness is at once a closure.

The structure of the story of closure thus reflects the character of the relation between openness and closure that it itself portrays. The story of openness and closure is the product of a desire for closure, the desire to provide a complete account of the world that is both powerful and safe from attack. At the same time it is the product of desire to escape closure, to evade the limitations of any attempt to merely encapsulate the world, and to reach out into openness, into that which escapes all closure. In neither respect is the desire capable of being fulfilled, for such is the character of closure.

While the story of closure is reflexive in character, it is contended that this self-reference is not destructive. The seemingly paradoxical character of its non-realist claim that a description of the world is not possible, is avoided by the provision of a theory that shows a linguistic closure can enable effective intervention in the world without relying upon the notion that it is an accurate description of an independent reality. Although therefore the story of closure makes many claims that if based on the realist assumption that they seek to describe the world would be self-referentially inconsistent, it at the same time offers an account of how it is possible to make claims about the world which does not rely on the assumption of an accurate description: by proposing that we hold the world as a closure. The story of closure is thus able to express its own limitation without having to implicitly rely on a realist account of truth to give content to such a claim.

So the story of closure has been told. We have not however arrived at a destination, but rather at a temporary resting point. I do not offer here a conclusion therefore, for there can be no close to the story of closure. No portrayal that would end our attempt to understand where we are. No safe and final truth with which to sum up our circumstance. I could say, as I have at points implicitly done in the main body of the text, that we are lost in the play of openness and closure. But if so, we are also engaged in the attempt to extend and improve our closures to intervene more effectively in the world; and we are also capable of becoming aware of the dissatisfactions of closure and can seek to avoid closure and approach openness. The story of closure has sought to offer such descriptions of our circumstance but there can be no final end-point, rather a series of resting points, which despite their temporary character, and despite their failure to directly describe openness, are nevertheless a powerful means of intervening in and understanding what we take to be the world.

This book, this epilogue, this sentence, are attempts to offer just such a temporary form of abode – a means of holding the world that has the appearance of holding fast that which cannot be held at all.

ACKNOWLEDGEMENTS

This book has taken an inordinate time to complete. Others I have written relatively swiftly – in a matter of a year or so, but *Closure* has been different. It has in some sense been with me throughout my career. The initial idea was formulated more than twenty years ago while doing postgraduate research at Oxford. It came as a response to the paradoxes of self-reference which I had become convinced were both central to contemporary thought and incapable of solution with standard logical responses such as Russell's *Theory of Types*. Yet it wasn't until the late 1980s that these initial thoughts had advanced sufficiently for me to begin work on the book in earnest.

This book has taken so long to write because it has gone through many drafts. Initially it was a voyage of discovery. Like the opening moves of a chess game it was unclear where it would lead. There were stabs in the dark, advances and setbacks. At times an inchoate notion that certain structures might later prove fruitful, at other times deadends without any sense as to how they might be overcome. Some sort of structure emerged which later it became necessary to return to and revise so as to simplify, clarify, and modify it in the light of its application to specific areas of concern. This has taken place to such an extent that the final version has only a few landmarks in common with the early drafts. Even when the framework was largely in place, there were two complete rewrites to help make the book readable. And precisely because it has taken so long there have been other more minor projects which have from time to time distracted from the task in hand and delayed it further.

At each stage in this extended process there have been many to whom I have been grateful. Here I will mention only a few. Firstly, I would like to thank my editors, Claire L'Enfant, Adrian Driscoll, and Tony Bruce who at different stages have worked with me on the book, for their patience and for their continued belief in the project which must at times have appeared to have no end. I would also like to mention those who have commented

328

on earlier drafts and who have sought to draw my attention to some of its weaknesses. In particular Alan Montefiore for his numerous points of advice over the lifespan of the whole project; Michael Lacewing for his detailed and thorough remarks which identified errors and raised many queries not all of which I have been able to address; and Jonathan Ree and Don Cupitt for their considered responses to a late draft. And then there are those friends with whom I have had many conversations over the years. Some of the most valuable of these have not been with trained philosophers or academics but those with an incisive mind free enough to think afresh and wonder at the world. In particular I would like to mention Hugh Tomlinson, and DJ Brown, without whose company and enthusiasm the completion of *Closure* would have been a lonelier and less exhilarating task, and also Soma Ghosh for her help and suggestions. And finally, but certainly not least, Sarah Marris and my son Leo with whom I have shared many moments on the edge of closure looking into openness, lost in the excitement at the mystery of being alive.

NOTES

Preface

1 Some hold the view that relativism is not at odds with objectivity, and that alternative descriptions are not at odds with each other but can be combined to approach a more complete description of the world. I will argue later in the Prologue (pp. xv–xxvii) that the internal logic of relativism does not allow such an intermediary solution.

Attempts can be made to define objectivity in terms which avoid the assumption of transparent access to an independent reality, for example in terms of social norms of behaviour and belief. However I will argue that such attempts fail for they undermine the primary intent behind the use of the term which is to identify claims that are independent of all context.

Prologue

1 J. Frazer, *The Golden Bough: A Study in Magic and Religion* (London, 1890). Bronislaw Malinovski, *Coral Gardens and Their Magic*, vol. 2 (American, New York, 1935).

2 T.S. Kuhn, *The Structure of Scientific Revolutions* (University of Chicago Press, Chicago, 1962). The slightly longer second edition (1970) is more widely quoted. Many others might have been cited here amongst them P. Winch, *The Idea of a Social Science and its Relation to Philosophy* (Routledge & Kegan Paul, London, 1958) and B.L. Whorf, *Language, Thought and Reality* (MIT Press, Cambridge, MA, 1956).

3 See, for example, Barry Barnes, *T.S. Kuhn and Social Science* (Macmillan, London, 1981).

4 Paul Feyerabend in *Against Method* (New Left Books, London, 1978) is perhaps the clearest and most radical exponent of this stance. A more detailed unpacking of the consequences of such an account has become possible in the light of a number of works in the sociology of science such as: Gerald, Holton, *The Scientific Imagination* (Cambridge University Press, Cambridge, 1978); Bruno Latour and Steve Woolgar, *Laboratory Life: The Social Construction of Scientific Facts* (Sage, London, 1979); Harry Collins, *The Golem: What Everyone Should Know about Science* (Cambridge University Press, Cambridge, 1995); also *Changing*

Order: Replication and Induction in Scientific Practice (University of Chicago Press, Chicago, 1992).

5 For example: ' "Truth" is therefore not something there, that might be found or discovered – but something that must be created and that gives a name to a process' F. Nietzsche, *The Will to Power*, tr. Walter Kaufmann and J.R. Hollingdale (Vintage Books, 1968), section 552.

6 For example, Allan Bloom, *The Closing of the American Mind* (Touchstone, 1988); or Harold Bloom, *The Western Canon: Books and School of the Ages* (Riverhead, 1995).

7 The ancient liar paradox was the claim 'All Cretans are liars.' What made it paradoxical was that it was uttered by a Cretan. If the claim is true, the person uttering the claim would have to lie, in which case the claim could not be made because it would be the truth.

8 D. Davidson, 'On the very idea of a conceptual scheme', *Proceedings and Addresses of the American Philosophical Association*, 47 (1973–4), pp. 5–20.

9 Hilary Putnam, *Realism with a Human Face* (Harvard University Press, 1992), p. 51. (In which Putnam colourfully describes the analytic project as 'a shambles'). See also Chapter 3 of Putnam, *Reason, Truth and History* (Cambridge University Press, Cambridge, 1981).

10 See Putnam, *Reason Truth and History*, pp. 72–4.

11 Any more than readers could, in Ludwig Wittgenstein's earlier work the *Tractatus Logico Philosophicus*, tr. D.F. Pears and B.F. McGuiness (Routledge & Kegan Paul, 1961) throw away the ladder of the text leaving its conclusions in the manner that he appears to have proposed both in the Preface and the late 6s (section 6.54).

12 Presumably the motivation behind Nietzsche's remark in *Ecce Homo* when referring to the work he regarded as his most profound *Thus Spake Zarathustra*: 'Verily, I beseech you; depart from me, and guard yourselves against Zarathustra! And better still be ashamed of him! Perhaps he had deceived you' etc.

13 Many of the later works of Martin Heidegger illustrate this approach. A good example being *On the Way to Language*, tr. Peter Hertz (Harper & Row, New York, 1971) which is presented in the form of a dialogue between an inquirer and a Japanese esoteric thinker.

14 Richard Rorty, *Contingency, Irony, Solidarity* (Cambridge University Press, Cambridge, 1989).

15 Ibid., p. 5.

16 Heidegger can also be added to the list of philosophers who have already been identified in this context: Nietzsche, Wittgenstein, and the contemporary philosophers Derrida and Rorty. For an account of the self-referential problems that beset Heidegger see H. Lawson, *Reflexivity: The Post-Modern Predicament* (Hutchinson, London, 1985), pp. 58–86.

   A case can be made therefore that the only major philosophers who are not faced with this predicament are those who retain an attachment to a realist account of meaning. These philosophers it will be argued succumb instead to an historically earlier form of self-reference as I seek to demonstrate in the next section.

17 See note 9 above.

18 Karl Popper, *Conjectures and Refutations: The Growth of Scientific Knowledge* (Routledge & Kegan Paul, London, 1963).

19 Stephen Hawking, *A Brief History of Time* (Guild Publishing, 1990). Discussion of unified theory: pp. 165–9; 174–5. Remarks on the mind of God: p. 175.

20 Or as Hilary Putnam expresses it, it is a 'dream [that] has haunted Western culture since the seventeenth century'. See *Realism with a Human Face* (Harvard University Press, 1992), p. 5.

21 This conclusion has implications for the account of closure itself. The theory of closure cannot itself escape this constraint or set itself up as an exception to the rule. The account that is given of closure at all points therefore implicitly takes account of this conclusion and in part can be seen as an attempt to come to terms with it in a manner that is not self-destructive.

22 The theory might be dualist in the sense that it distinguishes between the theory/observer and matter, but if it is to provide a complete account of the universe it will need to provide an account of the relation between the theory/observer and matter for otherwise it would not be a complete account nor would the theory account for itself. In providing such an account the theory is then no longer dualistic in a strong sense for the distinction between matter and theory is no longer ultimate.

So for example in the early Wittgensteinian dualism between language and the world, the account of the world is expressed in language and it is therefore no longer possible for language to give an account of itself for it is not possible for language to refer to those things which are not in the world. (Although paradoxically the *Tractatus* is precisely an attempt to describe language, which no doubt is why Wittgenstein concludes at the end of the book that having climbed the ladder we must throw it away.) It only becomes possible for language to describe itself by placing language in the world, but if this manoeuvre is attempted it no longer becomes possible to identify what it is that makes any particular claim in language true.

23 It is this outcome that allows those who have been dubbed by others as idealists to argue that they are not putting forward a subjective account of the world at all. It is also possibly what Wittgenstein intended by his remark in the *Tractatus* that 'solipsism ... coincides with pure realism' (section 5.64).

24 Wittgenstein's conclusion in the *Tractatus*. See also sections 6.3–6.7.

25 As Hilary Putnam has convincingly argued in *Reason, Truth and History*, pp. 17 ff.; p. 23.

26 Or, as Putnam expresses it in *Reason, Truth and History*, 'what singles out any one relation R as "the" relation of reference?'

27 Putnam seeks to demonstrate a similarity between the paradoxes of quantum mechanics and the liar paradox. In both cases he wishes to demonstrate that the paradoxes undermine metaphysical realism in an analogous fashion through their failure to incorporate the observer in the system. Putnam argues that these paradoxes are not only formally analogous, but are epistemic analogies for they both incorporate the same notion of an Archimedean point from which it is possible to survey the observers as if they were not ourselves.

While concurring with Putnam that the paradoxes undermine metaphysical realism it is not apparent to me that these paradoxes are analogous in quite the form that Putnam proposes. There can be little doubt that Bohr's version of the Copenhagen Interpretation entails the abandonment of the God's Eye View, or the No View. So long as quantum mechanics incorporates an observer/system cut, it is surely the case that it is incompatible with the description of physical reality independent of observers. Such a conclusion does not however appear to me to entail the abandonment of metaphysical realism or the Grand Project for there is always the alternative of abandoning this version of quantum mechanics. The salient point being that the adoption of any scientific theory, and the Copenhagen Interpretation of quantum mechanics in particular, is not a necessary consequence of adopting a God's Eye View. Of itself the existence of the theory or of its widespread acceptance cannot be reason to abandon metaphysical realism, even less does it imply that there are 'principled difficulties with the ideal itself'. However, as I have sought to argue such a case can be made.

28 Many of Nietzsche's remarks about Kant, 'that most deformed conceptual cripple there has ever been' (*Twilight of the Idols*, tr. R.J. Hollingdale (Penguin, 1966), p. 66) can be laid aside as rhetorical excess; but his theoretical criticism is simple and devastating: 'A critique of the faculty of knowledge is senseless: how should a tool be able to criticise itself when it can use only itself for the critique? It cannot even define itself!' F. Nietzsche, *The Will to Power*, tr. Walter Kauffmann and J.R. Hollingdale (Vintage Books, 1968), section 689.

Perhaps we should conclude that the value of the Kantian system is to pursue an account of knowledge to the point where its failure becomes apparent. The contemporary attempt to save Kant from himself by eschewing the noumenal world and the particular categories he proposes (see, for example, P.F. Strawson, *The Bounds of Sense* (Methuen, London, 1966)) does not respond to the central self-referential problem that an account that claims to provide limits to knowledge must already have transgressed those limits, and an account that claims to describe the elements required for knowledge to be possible must in the process have denied the possibility of that claim itself.

29 And it would appear that it is for these reasons that Wittgenstein abandons the attempt to provide a realist account of truth. His abandonment of realism however results in Wittgenstein succumbing to the contemporary predicament. His 'solution' to the problem of self-reference is the avoidance of all general metaphysical claims. I would argue however that the avoidance of such claims does not mean they are not present, for in order to make sense of his text we have ourselves to provide an overview. Occasionally Wittgenstein himself suggests such an overview with metaphors of games and therapies. He is, as he describes in the preface to the *Investigations*, providing merely sketches of landscapes, as he journeys through the workings of language (*Philosophical Investigations* (Basil Blackwell, Oxford, 1972), p. vii). Wittgenstein can take away these metaphors once they have been put forward, or put them forward without asserting them, as Derrida more explicitly does some forty years later. But can we imagine that he could take all of them away? In order to understand

Wittgenstein's text, in order 'to catch on' to his purpose, do we not require to have an overall framework? So long as we can interpret Wittgenstein as meandering through the landscape of ordinary language excavating particular misuses of its terms, so long as we can picture ourselves as being lost within the confines of language and unable to express our predicament, so long as we can conceive of the exercise as a therapy to disabuse us of philosophical nightmares, we are able to 'understand' Wittgenstein and recognise a purpose to the enterprise. Each of these descriptions however is surely an attempt at an overall account of the Investigations and more broadly of the human predicament in general. As with Derrida, Wittgenstein's text therefore can be seen either to fall to the charge of bad faith, or we can provide it with no meaning.

30 See chapter 1, 'A new age ...?'.

31 Hilary Putnam, Realism with a Human Face, pp. 11–18.

32 It might be thought that the seemingly empty vacuum of space suggests a location where there is no thing, but as contemporary physics has demonstrated this is not so. In due course I shall argue that empty space, a void, is not a possible circumstance. A space may be empty of some thing, air or oxygen for example, but it cannot be empty of all things. See pp. 167ff.

33 In his paper 'On the relations of Particulars and Universals' (first read as the Presidential Address to the Aristotelian Society in 1911, published in Proceedings of the Aristotelian Society 1911–12, also found in Logic and Knowledge (George Allen and Unwin, London, 1956) Russell uses as an example of a particular, a patch of white. Such an example would have appealed to Russell because the sensory character of 'a patch of white' means that the issue of its particularity can be dealt with without becoming embroiled in an issue over the existence of the material world. In addition the uniformity of 'white' makes it appear that the object is identical throughout and cannot therefore be broken into smaller units. By these means Russell can be seen to have attempted to approximate to the everyday notion of an object as some thing in particular. In what sense however is a circular patch of white one thing and not two? Even if there was no distinguishing character to the white, we can imagine dividing the patch into two patches, and by extension into any number of patches. If we divide the circular patch into the top and bottom half in what sense can these be regarded as one thing? We could argue that one section was identical to the other, but a further division of one of the halves surely results in these new patches being different since they are not even approximately the same size or shape. The problem is only avoided by making the patch of white indivisible. To retain the notion of a thing which is uniquely itself one is committed therefore to an atomic unit which cannot be further divided – an outcome that can be regarded as having been taken up both by Wittgenstein in the Tractatus and Russell. However, since we have no acquaintance with any such unit, and since we could not in principle have such an acquaintance, it remains obscure what is to be understood by such an outcome.

34 See amongst other works W.V.O. Quine, Word and Object, (The MIT Press, Cambridge, MA, 1960), p. 242.

35 As a result proponents of this view have either to elaborate an account of the mechanism by which our everyday objects are derived from their logically simple forebears or as Wittgenstein appears to propose at the end of the *Tractatus*, section 6.54, the connection between the 'ideal' world of logical objects and our depiction of them must be placed beyond the limits of language.

36 To some extent there has been in analytic philosophy a move away from the account paraphrased here. Influential in this shift was Saul Kripke, *Naming and Necessity* (Basil Blackwell, 1980), who argued that names are rigid designators and function as logically proper names identifying uniquely their referent. From such a perspective the name 'Aristotle' is no longer a set of descriptions but identifies the object that was originally intended by the use of the name. This initial object is identified in the first instance through the satisfaction of the relevant conditions. But what is this object, this thing? Kripke appears to give primacy to its material character, since he argues that the properties of the thing are contingent. However I wish to argue that there is no description of the thing, or in this case person, in terms of its material constituents, that cannot be supplanted by an alternative description in terms of alternative material constituents. Nor will it be possible to provide a precise account of the thing in terms of these constituents. I am not proposing to provide a detailed defence of this claim here since it will later be seen to be a consequence of the account of closure that the remainder of the book puts forward. It is possible however to give an initial defence of the claim in the context of Kripke's Aristotle example. If for example Aristotle is described as a collection of molecules, we can ask which molecules and in what combination, and since these alter throughout Aristotle's life it would at first appear to have the strange consequence that a carbon atom in the atmosphere at the time of Aristotle's birth is in fact part of Aristotle if it is later found in his body. I would argue therefore that it only looks as if we can describe the material thing because a detailed description is not actually attempted. As soon as a description is attempted it becomes apparent that, on the one hand, an alternative description is possible, and on the other that the description itself is fuzzy at the edges, and fuzzy in a way that cannot be corrected with greater precision. Nor can the primacy of the material descriptions of science be seen as a solution to the problem of the nature of the thing. Let us suppose that at some future date it becomes possible to clone Aristotle's body. We would still wish to argue that this body was not Aristotle even though it had identical physical features. We might say that it was a copy of Aristotle, but that to actually be Aristotle the body would have to be placed in Greece at the point when Aristotle was alive. History and geography are therefore in this case as relevant to our determination of this thing as Aristotle, as materiality.

37 Some may be tempted to argue that sub-atomic particles have the character of elementary material simples, and that there is no potential alternative description of, say, a quark. At one time no doubt such an argument might have been put forward in the context of atoms, an example that is discussed in more detail later (see Part III, Chapter 11). However we can now see that it is

possible to give very different descriptions of the atom, in terms of protons and electrons, and in terms of quarks and leptons. I would argue that the same will be true of the quark and indeed of any other particle which for the time being appears to be elementary. Furthermore, even in the context of current theory, an alternative description could be given of the quark in terms of energy rather than in terms of matter. In the context of four-dimensional space–time all matter can be seen as perturbations in force fields thus providing a different description of the supposedly elementary thing which is one and the same.

38 It might be argued that contemporary scientific theory allows us to abandon material things in favour of forces or fields. Unlike things, it might be argued that forces are not divisible and can be seen to be one and the same. However it is still possible to ask what a force consists of, or what a packet of energy consists of, and current theories do indeed propose a series of force particles each of which carry their own force fields. The shift from matter to force or energy does not therefore bring to a halt the problem of the nature of the thing, nor does it allow an endpoint which could provide the elusive elementary simple.

39 It can be argued therefore that in their excitement at finding a logic that could describe mathematics philosophers from Frege to Dummett, including Russell, and the early Wittgenstein, Quine, and Davidson have been tempted to apply a similar analysis to language. In place of a world of universals and particulars was a world of objects, or in modern logic a domain of objects, referred to by variables. Thus, as a consequence, it has been proposed that it is not possible to refer to a particular directly and that only in the context of a proposition is it possible to make sense of the notion of an object. The elusive character of everyday material objects apparently referred to by language has therefore been avoided by arguing that a correct analysis of language would make it apparent that no such objects are actually required. Instead we can analyse language with neo-Fregean symbolism and employ the notion of logically simple objects which cannot be said to exist outside the context of a proposition. (Thus Dummett states in *Frege*: 'The picture of reality as an amorphous lump, not yet articulated into discrete objects, thus proves to be a correct one, so long as we make the right use of it' (*Frege*, (Duckworth, 1973), p. 577.) This is a position that has some similarities with the argument being put forward here, but which still assumes that the notion of a thing in the form of a logical simple is coherent. In the process it has frequently been argued that we have misunderstood the world because we have been misled by language and that if we paid closer attention to analysis and logic these confusions would evaporate. While I can agree that we have assuredly been misled by language, it does seem to me that we have been equally, and plausibly more, misled by logic and mathematics, for the terms of logic imply entities which are themselves paradoxical. While much attention has been given to the means by which these terms are combined, it has largely been assumed that the apparently simple notion of an 'x' is self-evident. Far from being self-evident it is unclear how we can give any sense to such a notion

(see Chapter 3). As a consequence the philosophy of language that has domi-nated so much of English-speaking philosophy has I would argue in its promotion of logic as a means to understand language merely obscured further the character of the world.

Frege and the various philosophical positions he has spawned can be regarded as being right in alerting us to the mistake of assuming that the overt form of language is an accurate guide to the character of the world. We should not assume that every subject of every proposition implies a real thing in the world. To do so is undoubtedly to embed oneself in a web of contradiction. The solution however of analysing language in terms of a perfect logical language, and believing that the form of that logical language is the form of the world is no less mistaken, for it does not avoid a similar web of contradictions. Moreover, it reinforces the realist framework implied by the character of language and which is responsible for the confusion, although in most guises it does so by reference to ideal logical objects rather than identifying individual particulars.

40 Martin Heidegger, *What is a Thing?*, tr. W.D. Barton and V. Deutsche (Henry Regnery-Gateway, Chicago, 1968). Also from his later writing 'The Thing' in *Poetry, Language, Thought*, tr. and edited by Albert Hofstadter (Harper & Row, New York, 1971).

For Wittgenstein's remarks regarding the impossibility of simples see the *Philosophical Investigations*, 3rd edn (Basil Blackwell, Oxford, 1972), sections 47 ff.

41 This is a conclusion that Wittgenstein and Heidegger can also be interpreted as drawing. Thus Wittgenstein can be seen to abandon the attempt to describe the world in his later work, (having adopted a stance in the *Tractatus* that could be said to first argue that it is meaningless to speak of objects in general, only to rely on objects to provide the stuffing of the world – a paradoxical outcome to which the final sections of the *Tractatus* and numerous passages in the *Investigations* testify), while Heidegger engages in the attempt to approach the nature of the world rather than a description of the world itself.

1 An outline framework

1 Such a strategy of erasure has been employed by Heidegger and more recently by Derrida so as to avoid seeming to name that which cannot be named.
2 Charles Peirce *Dictionary of Philosophy and Psychology*, ed. James Baldwin, vol. 2 (Thoemmes Press, 1998). For bringing my attention to the original remark my thanks to Bart Kosko, who quotes it accurately in his popularisation of the principles of neural networks in the book *Fuzzy Thinking* (Flamingo, 1994).
3 As it will become clear in Part II, I define meaning as the material realised from linguistic closure. The dictionary meaning of a word can then be under-stood as those closures commonly and typically associated with the tag (defined also in Part II). My primary concern however will not be with this dictionary meaning but with the realisation of meaning by each individual through linguistic closure.
4 It will later be argued that this is true in a precise sense for mathematics is without texture. See Part III, Chapter 10.

5 In this respect the logic of closure has similarities with the Kantian framework of intuitions and concepts. Paralleling the famous Kantian dictum 'thoughts with content are empty, intuitions without concepts are blind' (I. Kant, *Critique of Pure Reason*, tr. Kemp Smith (London, 1929), p. 93), it is possible to say in the context of closure: 'Material without texture is empty, and texture without material is blind.'

6 This is not to deny that we are capable, in succeeding moments, of seeing the dots in different ways, and as I have argued, we could given time see the dots in an unlimited number of ways. Yet at any one moment we can only perceive one of these. As with Wittgenstein's duck/rabbit, we can see the dots as different things but not at the same time.

7 Some contemporary physicists might argue that these particles are not particles at all but energy or events. Such a manoeuvre does not stop us asking however in what the energy consists or the nature of an event. The currently fashionable theory of superstrings, proposed by Brian Greene (outlined in his populist book *The Elegant Universe: Superstrings, Hidden Dimensions and the Quest for the Ultimate Theory* (W.W. Norton, 1999)) seeks for example to offer a means of unifying the various particles and forces, but it still allows us to ask for the nature of a string. If the answer is a package of energy it is still possible to ask what constitutes that package, why is it one package rather than a series of smaller ones.

8 See E. Rutherford, 'The scattering of the alpha and beta rays and the structure of the atom' in *Philosophical Magazine*, May 1911, final paragraph. Rutherford regarded the hydrogen nucleus as a positive electron and temporarily seeks to put on hold the question of the nature of this nucleus arguing that without further experimentation it would be 'premature to discuss the possible structure of the nucleus itself'. Yet that question is already present as Rutherford's remark makes clear.

9 Part III, Chapter 11. The reader will find here additional support for the very general and largely unsupported case made in these paragraphs.

10 This argument is something of a simplification. For as it will be argued in Part IV, Chapter 16, in practice such terms do incorporate elements of closure for precisely the reason that otherwise they would have no purchase.

11 In this context the method of Cartesian doubt can be seen as a demonstration of the failure of closure. Descartes can be regarded as pursuing one dimension of the failure of closure but at the point of uncovering the real nature of closure, he spuriously produces a rabbit, in the form of the Cogito, from the hat. The principle of doubt can be seen therefore not as a means of uncovering that of which we can be certain but instead as a means of uncovering the ultimate failure of closure and a guide to the character of openness. The Cartesian goal of certainty is unachievable. Just as Russell felt that Descartes was not justified in his assertion of the Cogito ('the word I is really illegitimate; he ought to state his ultimate premise in the form "there are thoughts"', B. Russell, *History of Western Philosophy* (George Allen and Unwin, London, 1946), p. 589), so Russell's version 'there are thoughts' or elsewhere 'there is a thought now' can be seen to be equally unjustifiable. Not least because any claim to certainty must include certain knowledge of the relation between closure and

the world, and such knowledge cannot be provided. The Cartesian method may seek closure, but however narrowly aimed, however close to the target, the arrow can never land. (As shown by the ambiguity of whether it is described as 'thoughts' or 'a thought'.)

12 Stephen Hawking's assessment of the task of contemporary science as he concludes *A Brief History of Time* (Guild Publishing, 1990), p. 175.

13 Most easily seen perhaps in Derrida's early work, on Husserl for example, *Speech and Phenomena*, tr. David Allison (NorthWestern University Press, 1973), or in *Of Grammatology*, tr. Gayati Spivak (Johns Hopkins University Press, Baltimore, 1976), in the first half of which Derrida comes closest to providing a theory. For a more extended account of the mechanism whereby Derrida exhibits the failure of closure, see *Reflexivity*, pp. 90–122.

14 Prologue, The Contemporary Predicament, pp. xv–xvii.

15 L. Wittgenstein, *Philosophical Investigations* (Basil Blackwell & Mott, Oxford, 1972), part II, section xi, p. 194, in which a line drawing can be seen as a duck or as a rabbit.

16 Antoine de St Exupéry, *Le Petit Prince* (Harcourt Brace, 1943) in which a line drawing can be seen as a hat or a snake.

2 Systems of closure: body and mind

1 See Part IV.

2 Colin Blakemore, *Vision: Coding and Efficiency* (Cambridge University Press, 1993).

3 C. Blakemore, 'Maturation of mechanisms for efficient spatial vision' in Blakemore (ed.), *Vision: Coding and Efficiency* (Cambridge University Press, Cambridge, 1990), pp. 254–66; also D.E. Mitchell, 'Sensitive periods in visual development', pp. 234–46. Both articles make clear that neural cells respond to experience although the system is not of course entirely plastic and there remain strict limitations on the 'normal' development of sight.

4 White, Saunders, Scadden, Bach-y-Rita, and Collins, 'Seeing with the skin', *Perception and Psychophysics*, 7, 1970, 23–7.

5 There are two issues that this conclusion raises. Firstly, it would seem that in order for an organism to think it must have more than a single sense as I will argue later on p. 42. And secondly, there is the question of whether intersensory closure is not only necessary for thought to occur but is sufficient for thought to take place. I wish to argue that intersensory closure can only be realised if there is an additional form of material through which the divergent closures of different senses can be held as one. There is no reason to suppose that all organisms need provide material of the same character as human thought, but it would seem that any material that is realised of whatever character will have the consequence of proposing a reality of sorts which is seen to lie beyond, or perhaps behind, the preliminary and sensory closures of the system. In our case the material provided by intersensory closure may be explicable in terms of brain states, neuronal configurations, or some such description, but these states are not representations of some external reality and are instead additions to sensory closure which enable the system to hold

the sensory closures as one. The additions are not therefore about the world, or about reality, but simply are, and by being they enable the organism to intervene in what is taken to be reality.

6 It is not of course that we are now idealists in the sense that concepts alone are responsible for experience but that if the Kantian linking of intuitions and concepts is adopted, and concepts are understood as being purely linguistic, experience is only possible in the context of language. It follows that those without language – animals and young babies – would appear to operate as automata. An uncomfortable outcome that some have sought to explicate (J. McDowell, 'Rational and other animals', *Mind and World* (Harvard, 1994), pp. 108–126). McDowell seeks to overcome the problem by arguing that although experience is only possible with language it is still possible to be aware and an agent with purposes without having experience. Even if one accepts this argument, and I admit to not being convinced, one is still left with the outcome that animals and babies do not have experience. An outcome that few will find comfortable.

7 See Part V, Chapter 17.

8 The realisation of aural closures as units of language can take place prior to, or simultaneously with, linguistic closure. When language is first acquired the identification of aural closures as units of language will be simultaneous with the provision of linguistic closure. Later it becomes possible for us to identify a sound as linguistic even though we have not at that point realised a meaning and a linguistic closure.

9 The examples that have been used to illustrate the failure of closure and its unlimited and seemingly complete character have been those that apply to the perceptual world directly in the form of physical objects. In the next chapter I will examine linguistic closure that results from the combination of previously realised linguistic closures. Not all of these demonstrate the same level of openness as those that apply to physical objects, and some even approach the eradication of texture altogether. How this is achieved and its consequences are discussed both in Chapter 5 and later in Chapter 10.

3 The purpose of closure

1 This is not only true for the individual system but can be seen to apply more generally in the context of evolution and random genetic mutation. The realisation of closures that allows successful intervention – successful in the sense of prolonging the life of the organism – is the means by which the particular mutation and the closures it enables is confirmed.

2 See John A. Wilson, *The Burden of Egypt*, (University of Chicago Press, Chicago, 1951), p. 62. Also G.A. Wainwright, *The Sky-Religion in Egypt* (Cambridge University Press, Cambridge, 1938), pp. 71–4.

3 'The whole modern conception of the world is founded on the illusion that the so-called laws of nature are the explanations of natural phenomena', L Wittgenstein, *Tractatus Logico Philosophicus*, section 6.371, see also all of the 6.3s.

4 Language and the world: practical closure

1 This description should not be understood to imply that the process of closure can be separated from the realisation of material so that one can be seen to be the consequence of the other. It is not as if the process of closure results in the provision of material, or that the provision of material results in closure. Simply that the process of closure involves the realisation of material and therefore it can be said that closure both realises material and that it is through the realisation of material that closure can be seen to have taken place.

2 It might appear that this account provides too subjective a description of meaning, if the meaning of a term is entirely dependent on the practical closure that in a particular instance we choose to realise. The irretrievably social character of language means that this is not the case as I shall later elaborate both in Chapter 5 on formal closure and towards the end of the book in Chapter 17 of Part V.

It is due to the social character of language and the role of formal closure that we are in a position to distinguish the meaning we think a word has from the meaning the word 'actually' has. As a result we can be aware that two words have different meanings and yet not be able to realise practical closures. We may know for example that 'anaconda' has a different meaning than 'python' but be unable to distinguish anacondas from pythons. (Putnam makes this point in connection with a discussion about elms and beeches.) In the next chapter when I discuss formal linguistic closure the role of social meaning – sometimes understood as the real meaning of the term – will become apparent, and it is through formal linguistic closure that we can become aware that terms have different meanings without being ourselves in a position to realise the appropriate practical closures.

3 See Part V, Chapter 17.

5 Language and itself: formal closure

1 See in particular Chapter 17.

2 Frege can be seen to have initiated this trend which stems from the association of meaning and truth, although it was a view he appears to have abandoned in his later work (M. Dummett, *Frege* (Duckworth, London, 1973), p. 196). In the *Grundlagen* however he states: 'it is only in the context of a sentence that a word has meaning' (*The Foundations of Arithmetic*, tr. J.L. Austin (New York, 1960)), and furthermore cites the failure to identify this fact as the source of many philosophical problems. It is a view that was then adopted by Wittgenstein in both the *Tractatus* and the *Investigations*.

3 Aside from Frege, see previous reference, see also Wittgenstein in the *Tractatus* (the 4.2s).

Michael Dummett sums up this position in his major work *Frege*, (Duckworth, London, 1973) where he states 'We know what it is for a name to stand for an object only by knowing how to determine the truth-values of sentences containing the name, a piece of knowledge which can be expressed in terms of that relation between name and object. Sentences thus play a unique role in language', pp. 6–7.

4 It is because linguistic closure in realising material also incorporates texture that there is not at once a problem of identity. It might be argued that if in the combination of marks linguistic closure involves the holding of one mark as another the material generated cannot be different from either of the marks involved. Similar questions dogged pre-Fregean accounts of logic. If sentences are analysed in subject and predicate form with the assumption that entities attach to both, it is not apparent that if there is acquaintance with the subject anything is conveyed by the sentence since it has the character of a tautology. Furthermore the description of non-existent subjects appears either to entail non-existent entities or meaningless sentences, neither consequence having much attraction. Linguistic closure however does not immediately generate a problem of identity because material does not exhaust texture but is in addition. The power of predicate logic, in its Russellian form, can be regarded therefore as having unintentionally formalised this process. Instead of analysing 'the sky is blue' in the form 's is/has the property/equals b' it allows for the sentence to be unpacked by saying that 'there is an x such that x satisfies the criteria for sky and x satisfies the criteria for being blue'. Although for Frege therefore predicate logic was based on a strict realism it nevertheless parallels the character of closure in proposing an entity which is not directly identified.

Russell's account of proper names was challenged by Kripke in his work *Naming and Necessity*. However, Kripke's identification of the necessary attachment of proper names to their reference (Saul A. Kripke, *Naming and Necessity*, first published in Donald Davidson and Gilbert Harman (eds), *Semantics of Natural Language* (Reidel, Drodrecht, 1972), pp. 253–355, 763–9; published separately in a revised edition by Basil Blackwell, Oxford, 1980) is also a characteristic of linguistic closure. For practical linguistic closure binds a word to a thing in a necessary manner. The thing however is a postulate of closure and not an entity in openness. In this way the framework of linguistic closure can explain the success of the accounts of both Russell and Kripke and suggests how both approaches might be reconciled.

5 D. Davidson, 'Truth and meaning', *Synthese*, 17, 1967, pp. 304–23. See also Wittgenstein, *Remarks on the Foundations of Mathematics*, 3rd edn (Blackwell, Oxford, 1978), I.5.

6 The social character of linguistic closure will be examined in some detail later in Part V.

6 The organisation of space

1 It is interesting to speculate about a system of closure that has a higher level of closure than linguistic closure. For this to occur different types of intersensory closure would need to provide different types of material which would then be held as one by a further level of closure which would not have the char-

acter of thought. There would seem to be no reason in principle why such a system should not be possible.

2 Linguistic closure clearly follows prior intersensory closures. The first linguistic closures therefore already operate in a space that is significantly differentiated. The 'this' which is identified by the first linguistic closure will already have specific characteristics.

3 L. Wittgenstein, *Philosophical Investigations*, 3rd edn (Basil Blackwell, Oxford, 1972), part II, section xi, p. 194.

Wittgenstein can be seen to have used the example of the duck/rabbit for different purposes than employed here. (Namely the identification of the ambiguity of language and in particular the notion of 'seeing'.) There are however parallels. Wittgenstein interestingly uses the term 'materialization' (ibid., p. 199) to describe perception in the sense of 'seeing as'. This term clearly has similarities with the notion of 'realisation' and suggests the formation of 'material'.

4 I. Kant, *Critique of Pure Reason*, the *Transcendental Deduction*, tr. Kemp Smith (Macmillan, 1929), pp. 129 ff.

5 A conclusion which echoes P.F. Strawson's version of Kant's transcendental deduction, although Strawson makes a stronger claim for the necessity of physical objects. (*The Bounds of Sense* (Methuen, London, 1966)).

7 Language, truth, and the failure of closure

1 It might be argued that a mistaken schoolboy was capable of realising the sentence 'London is the capital of France'. Properly analysed however the meaning realised can be seen to rely on different associated closures than those associated with the tags by proficient speakers. The schoolboy appears therefore to realise a meaning that is false, but in fact realises a meaning which is true for him and could be true for us if we also realised the same meanings for the respective terms.

2 L Wittgenstein, *Tractatus Logico Philosophicus*, section 3.23.

8 The individual and the search for closure

1 Support for this account is extensive. See for example, Atkinson *et al.*, *Introduction to Psychology* 11th edn (Harcourt Brace, Fort Worth, TX, 1993), pp. 76–82.

2 This may account for the predominant view amongst current child psychologists that verbalisation is directed towards effective intervention rather than in pursuit of closure alone.

3 A claim initially put forward by Noam Chomsky in *Aspects of the Theory of Syntax* (MIT Press, 1965).

4 As perhaps Kant can be credited with having first identified in the *Critique of Pure Reason*. Kant can be regarded as having wished to argue as a consequence that we could have certain knowledge of external objects because without the capacity to identify an external world we could not identify ourselves and could not therefore be self-aware and thus have experience. Although Kant is

proceeding from the fact of knowledge, a starting point which from the perspective presented here can be seen as wholly misguided, it does seem to me that the outline of his 'transcendental' argument is correct and for those acquainted with the Critique my debt to Kant will be apparent.

5 The bootstrap problem can be described as follows: In order to have experience, the individual must be self-aware. In order to be self-aware the individual must be able to identify his or herself and distinguish this 'thing' from that which is not-self. However, the self is already required in order to make this distinction. Kant's solution was to suppose that the identification of the self was a prerequisite for experience and therefore an a priori requirement. Such a solution does not of course give any account of how this is made possible, but simply states the necessary fact of subjectivity.

The process of closure can be regarded as avoiding at least this version of the bootstrap problem of consciousness because the process of closure itself does not require a subject – subjectivity itself being the outcome of the process.

6 A conclusion which draws on P.F. Strawson's arguments in Individuals (Methuen, London, 1959) and The Bounds of Sense (Routledge & Kegan Paul, 1966).

7 Part V, 'The politics of closure'.

8 R.D. Laing is perhaps most associated with this argument. See amongst others: Wisdom, Madness and Folly (McGraw-Hill, 1985).

9 R.D. Laing in a lecture towards the end of his life told a story of an encounter with a patient who was suffering from depression. At the beginning of the session he asked the patient to tell him some jokes. At first the patient became annoyed and asked whether he was not going to discuss his current state and his past. R.D. Laing however offered some of his jokes to which the patient responded. The hour's session passed quickly with the two exchanging humorous stories and jokes. At the end of the session Laing indicated to the patient that their time was up, at which point the patient again became annoyed insisting that Laing had done nothing and this chat did not constitute therapy. The irony being that the patient had arrived depressed and had left in a good mood. Laing used the story to attack the traditional notion of therapy. In the context of closure the depressed individual is caught in a cycle of closures that is distressing and has little value in enabling intervention. There is, of course, nothing wrong about such closures. Like any set of closures they can be maintained and defended, but they are not usually useful. What Laing succeeded in temporarily achieving was the adoption of a new set of closures which had a different outcome. How successful it would have been on a longer term basis is less apparent. Nevertheless the principle is surely right. Because closures at the point of realisation are existentially true, we are lost to them. We believe whatever we think at the time. We know that in other circumstances we will think and feel differently but at the particular moment it does not seem like that for our closures are reality. The task therefore is how to escape from closures when they are unsuccessful and not become trapped by them. Laing offered an escape that happened to be successful. Escape from the structure of closure of our personal space is easily proposed but its execu-

tion may be a lifetime's work, and which in any case must become more difficult the longer the system of closure has been established. In later life it may
even become impossible.

9 The structure of knowledge

1 Chapter 21.
2 Although in such cases the new material is not found in the texture of prior
linguistic closure, it is in a more general sense still the outcome of prior material, but the material concerned is sensory.
3 See, for example, C. Hill's description of the English Civil War – amongst
others *Puritanism and Revolution: Studies in Interpretation of the English Revolution of the
Seventeenth Century* (Palgrave, 1997).
4 Bronislaw Malinovski, *Magic, Science and Religion and Other Essays*, (Waveland Press,
1992).
5 The argument over cladism demonstrates the characteristics of closure: namely
that different systems of taxonomy are possible and each has its own consequences and failures.
　For example, an attempt to classify the zoological world purely by
genealogy has the strange consequence that lungfish and cows are placed
together in a separate group from trout. Yet such a classification is immediately
at odds with our conventional classification which regards a lungfish – since it
swims, acts, and presumably tastes like a fish – as a fish. Cladists argue that we
should adopt the genealogical classification despite these bizarre consequences. The alternative of adopting a conventional classification has the
advantage that it groups together as species animals that we regard as being
similar, but has the disadvantage that such a classification is not an accurate
guide to the similarity of genetic code.
　The solution to this puzzle is that there is no correct classification, merely
alternative systems of closure each of which generate the characteristics of
closure, providing a nested hierarchy which enables certain types of intervention but which also exhibits failure requiring potential further closures. See
Stephen J. Gould *Dinosaur in a Haystack* (Jonathan Cape, London, 1966), pp.
248–59, pp. 388–400. See also, Peter J. Bowler, *Evolution: The History of an Idea*
(University of California Press, Berkeley, 1984), pp.181–7.
6 A case that is made in some detail in Part V, 'The politics of closure'.

10 The closures of science

1 The criticism J. McDowell makes of Davidson's account of meaning in *Mind and
World* (Harvard University Press, 1996), Lecture III, pp. 46–66. McDowell's
solution, which is to adopt the Kantian framework of intuitions and concepts
has parallels with the account of closure.
2 Igor Aleksander and H. Morton, *Neurons and Symbols* (Chapman & Hall, London,
1993), pp. 156–7. Also I. Aleksander and H. Morton, *An Introduction to Neural*

Computing, 2nd edn (International Thomson Computer Press, London, 1995), chapter 5 ('The workings of Wisard', Aleksander's recognition machine).

3 See Igor Aleksander, *Impossible Minds* (Imperial College Press, 1996).

4 These categories have some resonance with a topological account of the world. Courant and Robbins, *What is Mathematics?*, 2nd edn (Oxford University Press, Oxford, 1996), pp. 242–4.

5 It might be argued that the claim 'the chair is a chair' is only ideally true when the meaning of 'chair' is identical in both instances. However I wish to argue that however narrowly one seeks to define 'chair' it will never be possible to squeeze out texture and that as a result a circumstance can be found in which the chair, understood strictly as that thing identified by the tag, is not the same as some other thing capable of being identified from the texture held within the closure. If therefore we seek to define chair in such a manner that it excludes those circumstances in which it is a table there will still be other things which it could be held as which are not excluded by the new and tighter definition.

It might also be argued that the closures of chair and table are not in opposition. The object can be both a chair and a table and as a result the claim 'a chair is a chair' is not challenged. It does not seem to me however that this is the case. When the chair is a table it is in this respect explicitly not a chair. There are other respects in which of course it is a chair, namely that it could revert to being used to sit on, but at the moment that it is a table it is not a chair. So we can understand the sentence 'the chair is not a chair' as meaning 'the physical object which we suppose to be identified by the term "chair" is not in this circumstance being used as a chair and is therefore in this instance not a chair'.

Of course in the context of the theory of closure, there is no thing which is identified by any linguistic mark, and there is therefore no opposition between this thing being both a chair and a table. It is only when, in the context of logic and mathematics and some theories of meaning, we seek to complete closure that these problems arise.

6 Traditionally a distinction has been made between synthetic and analytic truth. Ideal truth might therefore be seen to have two aspects: facts that are identified as being true for ever and necessary truths of logic. In the context of closure however these distinctions break down. The practical closures on which empirical truths rely will themselves incorporate formal closure; and the formal closures of language and logic which supposedly have the character of necessary truths are only given content if they have the capacity to be incorporated into practical closures. Once this takes place their necessity is jeopardised. In any case the essential character of closure remains the same in either practical or formal closure: namely in holding that which is different as the same.

7 I would argue that the possibility of a complete closure is in the end not sufficient to enable a logic to realise ideally true theorems. For the rules have to be expressed without introducing texture and this cannot be achieved. The assumption that is made here, namely that 'or' can be defined without reintro-

ducing texture is therefore misleading. It serves however to clarify the overall point which is not threatened by this simplification.

8 W.K.C. Guthrie, *The Earlier Presocratics and the Pythagoreans* (Cambridge University Press, Cambridge, 1962), pp. 212 ff.

9 It was in precisely these terms that Russell sought to promote his philosophy of logical atomism shortly after the turn of the twentieth century. See his paper 'Logical atomism', 1924, reprinted in *Logic and Knowledge* (George Allen and Unwin, London, 1956). ('I hold that logic is what is fundamental in philosophy' p. 323); and later in his 1950s paper 'Logical positivism' also reprinted in *Logic and Knowledge*. It is perhaps at its most evident in the section on logical analysis, Chapter 31 'The philosophy of logical analysis', in his populist work *The History of Western Philosophy* (George Allen and Unwin, London, 1946).

10 Newton, *The Mathematical Principles of Natural Philosophy*, tr. Andrew Motte (Benjamin Motte, London, 1729), definition of mass: p. 1; definition of distance (space) and time in 'Scholium' in 'Definitions' (pp 9–12). These definitions are distinct from the 'laws of motion'.

11 P. Feyarabend, *Against Method* (Verso, London, 1975), pp. 95, 160–1; conclusion, pp. 207–9.

12 Current theories of physics have introduced the notion of four elemental subatomic forces each of which have their own particle. In some respects therefore these theories have departed from the function that force has in the Newtonian system. (In rather the same way that the function of the original notion of 'atom' has been undermined by the identification of a thing which is an atom.) One could predict that there will be two consequences of this. Firstly, the closure will not be complete. For example further questions can be asked about the nature of the force particles; and of the nature of the forces. These questions point to a fault line where the closures break down. A force cannot be explained by a particle, for example, for we will still need to explain the behaviour of that particle. Secondly, the role of force in enabling a self-sustaining theory will gradually be undermined and with it also perhaps the ease with which the theory can be manipulated and added to in order to adequately describe the observed behaviour of the relevant particles. A case could be made that we are already witnessing these problems.

13 Newton himself was so concerned that the seemingly mysterious power of gravitational force would lead to the denial of his theory that in the *Principia* he avoids describing the basic principle of nature on which his system depends. He writes in his manuscripts (Hypothesis 2) 'The principle of nature being very remote from the conceptions of Philosophers I forbore to describe it in that Book (*Principia*) lest it should be accounted an extravagant freak … but now the design of the book being secured by the approbation of mathematicians, I have not scrupled to propose this principle in plane words' (R.S. Westfall, *Force in Newton's Physics* (Macdonald & Co., London, 1971), pp. 379–88). Westfall remarks 'in the light of the reaction of orthodox mechanical philosophers to the idea of gravitational attraction, he was undoubtedly wise to conceal his full philosophy of nature when he published *Principia*.'

14 Interestingly the Newtonian system could even so have been defended by maintaining the assumption that the speed of light was infinite but only slowed due to some force – presumably similar in kind to friction. Given such an approach it can be supposed that the later discoveries of atomic fission would have to have been explained as the liberation of another force possibly allied to the force that restrained light, rather than the liberation of the force associated with mass as proposed in Einsteinian physics.

15 Feyerabend, *Against Method*, p. 209.

16 Unless like the Einsteinian system, the Newtonian framework is effectively incorporated.

17 Kuhn, *The Structure of Scientific Revolutions*. Using 2nd edn: 'paradigm' introduced on p. 10; 'paradigm change' (Kuhn's word) discussed pp. 66–91.

18 Richard Dawkins, *The Selfish Gene*, 2nd edn (Oxford Uuniversity Press, Oxford, 1989), pp. 187–201; also in *The Blind Watchmaker* (Penguin, 1988; first published Longman, 1986), p. 158.

11 What is the world made of?

1 For sources of our knowledge of Leucippus and Democritus, see W.K.C. Guthrie, *A History of Greek Philosophy*, vol. 2 (Cambridge University Press, Cambridge, 1965), chapter 8.

2 Ibid., pp. 390–2, 395–6, 454–64.

3 See p. 23.

4 A conclusion that Russell seems to endorse in his chapter on Democritus in his populist work *The History of Western Philosophy*, pp. 86–91.

5 B. Russell, *A History of Western Philosophy* (George Allen & Unwin, 1946), p. 435.

6 G.S. Kirk and J.E. Raven, *Presocratic Philosophers*, (Cambridge University Press, 1971), p. 407, section 555: Aristotle On Democritus ap. Simplicium de caelo 295, I (DK68 A37)

7 *National Edition of the Works of Galileo*, pp. 72 ff.

8 *Opticks*, 4th edn (reprinted G. Bell & Sons, London, 1931), p. 400.

9 *A New System of Chemical Philosophy* (R. Bickerstaff, London, 1808), p. 141 and p. 212.

10 The Standard Model is a term usually used to refer to the widely held account of sub-atomic physics which bases the structure of matter around six different types of quarks and leptons. W.M. Cottingham *et al.*, *An Introduction to the Standard Model of Particle Physics* (Cambridge University Press, Cambridge, 1999). Also Lillian Hoddeson *et al.* (eds), *The Rise of the Standard Model Particle Physics in the 1960s and 1970s* (Cambridge University Press, Cambridge, 1997).

11 P.C.W. Davies and J. Brown, *Superstrings: A Theory of Everything?* (Cambridge University Press, Cambridge, 1988). Superstring theory 'promises to provide a unified description of all forces, all the fundamental particles of matter – in short, a Theory of Everything' (p. ix). 'Quarks and leptons have long been thought to be point-like, with no internal structure at all. However … it seems likely that these "fundamental" particles actually have some sort of structure

after all' (p. 26). If superstring theory is true: 'we would at last have identified the smallest entities from which the world is built' (p. 69).

That the delusion is not a new one is perhaps most graphically demonstrated by a quote from Lord Kelvin in 1900: 'There is nothing new to be discovered in physics now. All that remains is more and more precise measurement.'

12 That 'atom' sometimes refers to the nucleus alone is indicated by the common reference to alpha particles as helium atoms; that 'atom' refers to nucleus plus electrons is clear from the standard use, e.g. 'the atoms that make up a physical object such as the table'.

13 See Kant, *The Critique of Pure Reason*, pp. 396 ff. Kant's explanation of the paradoxical nature of the concept of space when applied to the world can be regarded as arguing that such descriptions illegitimately sought to provide transcendent knowledge which was not according to him possible: 'While appearances in the world are conditionally limited, the world itself is neither conditionally nor unconditionally limited' (p. 458).

There are similarities with the explanation provided here. If the character of closure is responsible for the paradoxes of space it is because closure has nothing in common with openness and as a result any attempt to describe openness must fail. Where an account in terms of closure differs from Kant is that the failure of closure is not confined to attempts at the provision of transcendent knowledge but extends to knowledge in general. As a result there is no distinction between areas about which we can obtain knowledge and those that we cannot. To adopt Kant's terminology, he can be seen to be right about transcendent knowledge but is wrong about transcendental knowledge. Neither is possible and as a result the distinction unnecessary. As a consequence the notion of a critique of knowledge which sets limits to understanding is from the perspective of closure misguided.

14 Zeno's paradoxes insist on the complete closure of the present. His aim was one which Derrida might well approve: namely to use the paradoxes to argue that analysis of time in terms of instants and space in terms of individual points was mistaken. Thus the hare could never catch the tortoise because to do so it would first have to cover half the distance and then half the distance again and so forth. No matter how many times the hare covered half the distance remaining there would still be half the distance to go. Zeno was partly right. Activity cannot be contained by a temporality which is expressed in instants. Where he was mistaken was to think that it was possible to do without instants. If there was only activity the hare would never start, it would never stop, it could never be anything, there would be no hare, no tortoise. The activity of the hare is a function of the instants provided by temporality and implicit in the closure, hare.

It should in passing be noted that Georg Cantor's theory of continuity advocated by Russell as the means by which the paradoxes of Zeno are apparently avoided while still retaining instants of time, relies on the mathematical notion of infinity to describe a compact series. A compact series being a series of points between any two points of which an infinite number of further points can be found. As a consequence in a compact series no points are ever next to each other for there is always an infinite number of points between them. In a compact series therefore you cannot say what the next point is. It is because one cannot provide the next closure that activity is possible in a compact series. The paradox of the movement of the hare is translated into the mathematical paradox of a series in which it is not possible to say what is the next number. Thus the mathematical notion of a compact series can be used to describe movement. In the process Russell has not succeeded in explicating time, but has employed the same technique with mathematics that we achieve with a clock. Activity remains outside of the closures, but the series of closures can be compared, with the temporary illusion that temporality itself has been described. The notion of infinity already embodies activity to such an extent that it is difficult to provide closure at all, as Wittgenstein in his critique of Russell noted in somewhat different terms. When we try to form the closure infinity it constantly breaks down because the activity within its texture is so extensive that the closure fails. We are misled because infinity is a mathematical concept, and mathematics, which we shall consider later, appears to provide formal closure. It is for this reason that scientists, and social scientists, are so impressed with mathematics – it appears to be so final. As if by being expressed mathematically it is somehow solved. That we are able to express a compact series in mathematical terms does not imply closure, nor does it answer Zeno's paradoxes, that are unavoidable and an essential characteristic of closure.

12 Strategies for closure

1 See for example, G. Holton, *The Scientific Imagination Case Studies* (Cambridge University Press, Cambridge, 1978). Also B. Barnes, *Scientific Knowledge and Sociological Theory* (Routledge & Kegan Paul, London, 1974). Also Harry Collins on the unrepeatability of experiments – see previous reference. Also Feyerabend on the Church and Galileo in *Against Method*.

2 Simon Schaffer, and Steve Shapin can be understood as making this point – the inability of empiricism to dictate an outcome – in the context of the debate between Hobbes and Boyle, in *Leviathan and the Air Pump: Hobbes, Boyle and the Experimental Life* (Princeton University Press, 1989).

13 The closure of 'closure'

1 It is this characteristic that separates the framework of closure from a Kantian approach. In a similar manner to Kant's Critique, Closure can be seen as an attempt to explain how understanding is possible. However unlike Kant, openness is not an unreachable other that is found in the hidden and transcendent world behind appearance. Instead openness is embedded in closure. The account of closure is thus at once bound up with openness and it is this relationship which gives closure its fundamental characteristics. Instead of a world that is divided between the fully known and the unreachable, closure

describes a circumstance in which the unknowable and the knowable are enmeshed in each other; and it is because the seemingly known carries that which is not known that closure operates as it does, and our experience is as it is.

Part IV The search for openness: art, religion, and the unknown

1 This account is a simplification, for sensory closures are themselves already embedded in a framework of space that incorporates higher-level closure. It is thus not possible for us to access sensory closures without at the same time holding those closures in the context of the network of linguistic closure and other intersensory closure with which we are operating. The interdependence of space means therefore that the realisation of any closure available to the system, whether preliminary, sensory, or intersensory, takes place in the context of space as a whole. In the context of the example of the chair therefore reference to a return to sensory closure implies a model of closure in which the closures available to the system are conceived as individual units that can be combined and that can be accessed independently. Such a model needs to be modified to account for the interdependence of space. However, given such a modification it can be seen that it is still possible for the system to access lower-level closures – in the context of space as a whole – in an attempt to find alternative higher-level closure. It is just that we should not imagine the lower-level closures as being fixed things, but as themselves being the outcome of the system as a whole. With this proviso therefore the simplification in the body of the text can be taken as an appropriate description.

15 Art and the avoidance of closure

1 In the context of the theory of closure all language is metaphorical. Indeed, metaphor is typically understood as the holding of one thing as another – the paradigmatic property of closure. However, there are many occasions when I wish to distinguish between a literal and metaphorical use of tags and some explanation therefore needs to be offered of the character of this distinction in this text.

In this context I understand 'metaphorical' to identify those instances in which the speaker seeks to draw attention to the differences held within the closure. The result is that one thing is held as another but in such a way that the difference remains apparent. It is for this reason that I will argue that the use of metaphor in art and literature is an indication of the search for openness rather than the search for closure.

The literal use of tags also involves holding one thing as another but the speaker does not wish to draw attention to the differences but rather aims to lose the differences in the proposed closure. The literal use of tags is therefore metaphorical in the sense that it holds one thing as another – the everyday understanding of the term 'metaphor' – but is not metaphorical in the sense that the speaker has no intention of drawing attention to the differences held within the closure. Drawing attention to these differences is to draw attention to the texture within the closure and therefore also to that which remains open.

2 Or as Cézanne succinctly puts it in a letter to his close friend Emile Bernard, near to the end of his life: 'I progress very slowly, for nature reveals herself to me in many complex ways, and the pages needed is endless' (*Cézanne: By Himself*, ed. Richard Kendall (MacDonald Orbis, London, 1988), p. 236).

3 There can be little doubt that Cézanne did not see his painting as an attempt to copy nature but to provide a 'realization of one's sensation' (ibid., p. 8). Furthermore there are many occasions on which he expresses his inability to provide this realisation. He writes to his son nearing the end of his life, on 8 September 1906: 'as a painter I am becoming more clear-sighted before nature, but ... the realization of my sensations is always painful. I cannot attain the intensity that is unfolded before my senses.'

A case can be made therefore that Cézanne regarded himself as engaged in a permanent search, a search to realise – a word that Cézanne himself employs frequently – the plenitude found within his own sensation. It is in this manner therefore that a viewer seeking to realise one of Cézanne's paintings replays the structure of Cézanne's own experience.

4 A. and L. Vezin, *Kandinsky and Der Blaue Reiter* (Peirre Terrail, Paris, 1992), p. 96. Referring to Kandinsky's work in an exhibition of the New Association of Munich Artists in 1910: 'These works, which many of the members of the New Association found hard to accept, came under direct attack in the Munich press, in which Kandinsky was called a "morphia addict", while the public spat at the pictures which had to be dried every evening.'

5 Perhaps one of the reasons for identifying certain works as great art is that they retain the capacity to escape closure over a significant period of time.

6 *Self-portrait* (1658; Frick collection); *Self-portrait* (c.1661–2; The Iveagh Bequest, Kenwood House, London); *Self-portrait* (1669; National Gallery, London); *Self-portrait* (1669; Mauritius).

7 Magritte, *The Curse* (*La malediction*), 1931.

8 Magritte, *The Field Glass* (*La lunette d'approche*), 1963.

9 This phenomenon is seen even more clearly in the Dadaist movement. A strategy deliberately designed to undermine closure, and explicitly annunciated as such: 'some people think they can explain rationally, by thought, what they think. But that is extremely relative ... there is no ultimate truth ... I am against systems, the most acceptable system is on principle to have none' (Tristan Tzara, *Dada Manifesto* (1916)), and yet the furry cup and saucer is today a mere gesture that no longer threatens anything much.

10 Magritte, *On the Threshold of Freedom* (*Au seuil de la liberté*), 1930.

11 See *Magritte* (compiled by Sarah Whitfield, South Bank Centre, 1992). According to Magritte, each of the six images in *The Six Elements* (Magritte, *Les Mots et les images*, Écrits, no. 21) 'suggests that there are others behind it' so each is 'a means of concealment'.

12 L. Wittgenstein, *Tractatus Logico-Philosophicus*, as before, section 5.6331.

13 T.S. Eliot, *Four Quartets*. Eliot's *Quartets*, as the name implies, is a set of four poems which can be considered as a whole. See also T.S. Eliot, *Collected Works* (Faber and Faber).

14 Eliot, *Four Quartets*, 'Burnt Norton', section II.

15 Eliot, *Four Quartets*, The Dry Salvages, section I.

16 Ibid.

17 Ibid., section II.

18 Eliot, *Four Quartets*, 'East Coker', section V.

19 For example: Robert Altman *The Player* (1992) is deliberately undermining of closure. Abel Ferrara, *Dangerous Game* (1993) is trivially so. See also: Vincente Minnelli, *The Bad and the Beautiful* (1952); Barry Sonnenfeld, *Get Shorty* (1995).

20 F. Nietzsche, *Beyond Good and Evil*, ed. Oscar Levy, section 296, p. 264.

16 Naming the unnameable

1 Wittgenstein's remark in the *Tractatus* that the desire to name the unnameable should be curtailed (*Tractatus Logico-Philosophicus*, 6.41–2, 6.52–7) is perhaps the most obvious example of this doctrine. Russell takes a similar stance in a less poetic manner dividing what is possible into matters of fact and matters of logic. 'questions of fact can only be decided by the empirical methods of science, while questions that can be decided without appeal to experience are either mathematical or linguistic', paper 'Logical positivism', 1950, reprinted in *Logic and Knowledge* (George Allen and Unwin, London, 1956). Others such as Ayer and Carnap held similar views and were similarly critical of those philosophers who sought to identify that which was according to them not identifiable.

2 It can be argued that a rigorous interpretation of dialectics would undermine closure, including the closure of dialectics. There is therefore in Marx, and particularly in the early work, a case for arguing that the pursuit of openness is not abandoned. See *Theses on Feuerbach* (in K. Marx and F. Engels, *Selected Writings* (Progress Publishers, Moscow, 1969), pp. 13–15). Even by the time of the *German Ideology* (International Publishers, New York, 1947) this openness was still present: 'The social structure and the State are continually evolving out of the life-process of definite individuals.' This element of his work was gradually excised, both in his own writing and in the practice of socialism; so that the works of Marx were taken to be an elaboration of 'the truth' – a complete closure. This is immediately apparent in many of Engels' works. For example in 'Speech at Marx's Graveside' in Robert V. Tucker, *The Marx-Engels Reader* (Norton, New York, 1972), p. 603: 'Just as Darwin discovered the law of development of organic nature, so Marx discovered the law of development of human history.'

3 Religious thinkers and philosophers have of course claimed to demonstrate religious knowledge from revelation and reason. Indeed, in post-Christian Western philosophy most thinkers have until relatively recently regarded such demonstrations as being a necessary part of any philosophy. Implicit in my argument however is the assumption that such demonstrations can be shown

to be mistaken precisely because the knowledge these demonstrations seek to prove is itself necessarily outside of reality and cannot therefore be intuited or reasoned from our experience of reality.

4 As Wittgenstein implied in his remark towards the end of the *Tractatus*: 'Thus people today stop at the laws of nature, treating them as inviolable, just as God and Fate were treated in past ages. ... And in fact both are right and both wrong: though the view of the ancients is clearer in so far as they have a clear and acknowledged terminus, while the modern system tries to make it look as if everything were explained', section 6.372.

5 Don Cupitt, *Taking Leave of God* (SCM Press, London, 1980), chapter 1.

6 Plato's identification of the failure of closure in his failure to be able to provide a meaning to general terms such as 'justice' for example. Plato, *The Republic*, tr. A.D. Lindsay (J.M. Dent & Sons Ltd, London).

7 I. Kant, *Critique of Pure Reason*, pp. 268 ff.

8 Speaking of the impossibility of giving content to the noumenal world, Kant writes: 'the concept of pure and merely intelligible objects is completely lacking in all principles that might make possible its application', ibid., p. 275. Yet on other occasions Kant appears to give application to the noumenal by for example identifying it as the site of the transcendental subject (for example, pp. 382 ff.).

9 A case can be made that Kant himself falls prey to this exoteric understanding. Indeed at some level he must do so, for it is only through the provision of an exoteric that we can glimpse the esoteric by analogy.

10 Examples in the case of Kant would include Strawson in *The Bounds of Sense* (Routledge, 1989; first published by Methuen & Co., 1966), part I, especially pp. 38–42, part IV, pp. 247–63; or commentators like Jonathan Bennett, *Kant's Analytic* (Cambridge University Press, Cambridge, 1966), sections 8,18,30,38.

In the case of Wittgenstein it could be argued that a major part of analytic philosophy has taken such a stance; certainly positivists like Ayer, but also ordinary language philosophers such as Austin who can be seen as reducing the later Wittgenstein to a theory of ordinary language with 'meaning as use' ('The meaning of a word' in *Philosophical Papers*, 3rd edn, ed. Urmson and Warnock (Oxford University Press, Oxford, 1979), pp. 55–75) as its essential element.

11 The logical positivists were perhaps the most vehement in their denunciation of Heidegger, since he illustrated for them an example of precisely the form of philosophy they wished to see abandoned. See R. Carnap, *The Elimination of Metaphysics Through Logical Analysis of Language*, trans. Arthur Pap, pp. 60–81; first published as 'Überwindung der Metaphysik durch logische Analyse der Sprache', *Erkenntnis*, vol 2. This paper sought to demonstrate that Heidegger's work *What is Metaphysics?* is laden with logical errors that can simply be analysed away. Carnap speaks of Heidegger making 'gross logical errors', p. 71. See also A.J. Ayer, *Language, Truth and Logic* (Pelican, 1971), pp. 58–9.

12 This account of Heidegger is highly truncated. I have given a rather more detailed elaboration of the mechanism of Heidegger's approach in chapter 3 of *Reflexivity: The Post-Modern Predicament* (Hutchinson, 1985).

13 The name of the unnameable for the later Wittgenstein can be taken as what a general account of his philosophy would describe him as seeking to achieve. I would argue that although Wittgenstein aims to avoid giving such an account, readers of his texts must offer for themselves such a general account for otherwise the text is a random collection of comments. Occasionally Wittgenstein's self-censorship slips and he gives us a metaphor. We are at play in a language game. Philosophy should be a therapy. There are many such examples in his later work (for example: *Philosophical Investigations* (Basil Blackwell & Mott, Oxford, 1958), sections 116, 118 ff., 133). However without the metaphor or the implicit name or description of that which he cannot name or describe, we would have no means of deciding how to interpret his text.

Part V The politics of closure

1 We have in a sense therefore been engaged in a reverse of the Kantian enterprise. Instead of taking as our starting point the fact of knowledge and setting out to examine how knowledge is possible in an attempt to seek its underlying structure, we have made our starting point the inverse: the irrevocable failure of knowledge and our inability to access the world, and have sought to demonstrate how despite this failure we are still capable of intervening in a world that we cannot know.

2 See Part III, Chapter 9.

17 The social and political character of personal space

1 It is not only linguistic closure that is social. Other intersensory closures are also the outcome of a context which relies on the behaviour of others. (For example the pre-linguistic influence of parental behaviour on the formation of gender, see N. Chodorow, *The Reproduction of Mothering* (University of California Press, Berkeley, 1978); also C. Smith and B. Lloyd, 'Maternal behaviour and perceived sex of infant: revisited', *Child Development*, 49, 1978, pp. 1,263–4). Babies prior to the development of language adopt intersensory closures enabling them to construct a reality of sorts. This reality must owe in part to the behaviour of adults who encourage certain forms of closure. The character of non-linguistic intersensory closure is however difficult to access due to the fact that we use linguistic closure to describe it. For this reason the analysis of the social character of personal space in the remainder of this chapter is restricted to linguistic closure.

2 As has been witnessed in the case of twins who have lived together for long periods. See A.R. Luria and S. Yudovich, *Speech and the Development of Mental Processes* (Staples Press, London, 1959), p. 32. Also Peter Mittler, *The Study of Twins* (Penguin, 1971), p. 36.

18 Stories of desire

1 Anorexia: the derivation of the term is 'no appetite'. Although psychologists often avoid the conclusion that the condition involves loss of appetite, the descriptions of anorexics suggest that this is very much the case. Sheila Macleod, *The Art of Starvation* (Virago, London, 1981) 'I was incapable of (admitting hunger) if only because I didn't believe it to be true. ... The truth is that I could no longer tell whether I was cold or not, hungry or not' (p. 97); 'By this time there was no need for me to deny hunger: I felt none ... hunger in the sense of the desire for food, gradually disappeared and I was no longer lying to myself or others about its absence' (p. 98).

19 The first power relationship

1 See Part II, Chapter 7.
2 See argument at the end of Chapter 18 to this effect

20 Social power relations: repetition and rationality

1 Part III, Chapter 8.
2 Part III, Chapter 10.
3 A case can be made that fuzzy logic is an attempt to cope with this failure in rationality; a stance which some of the proponents of fuzzy logic themselves espouse (Bart Kosko, *Fuzzy Thinking* (HarperCollins, 1994).

21 The structure of social control: closures of authority

1 Here as elsewhere in the remainder of this part I take 'imposition of closure' to include not only the adoption of formal closures on the part of realisers but the potential realisation of practical closure that follows the appropriate change in behaviour demanded by the author.
2 A contemporary example of the complete abandonment of the realiser to the closures of authority can be seen in the members of the religious cult who died in the Waco siege in 1996.
3 Opposition still takes place not least because there are individuals to whom this means of satisfying desire is not offered, or because of the inequalities in the reward offered.

22 The organisation of society: institutional space

1 James Frazer, *The Golden Bough: A Study in Magic and Religion* (Macmillan & Co., London, 1890), pp. 118f.
2 Neurath originally used the analogy to describe science, on the grounds that if we are to rebuild it we must do so plank by plank while staying afloat. It was an analogy which was subsequently used by Quine in *Word and Object* to describe the character of language (Quine, *Word and Object*, pp.3 ff.). This

analogy can be seen also to be appropriate to personal, institutional, and cultural space. Quine can be considered however as regarding language as being a means of interpreting data. Closure is not however an interpretation of data but is the mechanism for providing data, for giving us things.

3 Some would argue that although the texts of Buddhism appear to propose the abandonment of desire in fact Buddhism does not abandon all desire for the desire to reach enlightenment is strong and encouraged. This would explain how conflict can emerge with social institutions that are seen to prevent such enlightenment.

4 See Richard Rudgley, *The Alchemy of Culture: Intoxicants in Culture* (British Museum Press, London, 1993).

23 Society, change, and dreams of utopia

1 Michel Foucault. *The Birth of the Clinic* (Tavistock Publications Limited, London, 1973) can be regarded as having described this development of the medical institution over time.

INDEX

aasvogel: example illustrating linguistic closure 67-8, 73-4, 86

abstract closures 181, 182-3, 184, 186

abstraction: in painting 207, 209-10, 211, 212–13

accident: closures realised through 129–30

acting: as form of pretence 104

activity 23–4; handling 53, 55, 122, 129–30, 145–6, 149, 163, 246; purposeful 123

adults: character of closure in 126, 127, 128, 128–9

advertising: use of repetition 284–5

alcohol: as embedded in Western culture 317–18

alternative closures 17–19, 20–1, 49–50, 51, 87, 90, 106, 136, 184, 210, 211

Altman, Robert: The Player 351n

ambiguity: in painting 207, 209–10, 211–12, 212–13; in poetry 220, 224

analytic philosophy xvi, 156, 244–5

analytic truth 110–11, 150, 151

anarchy 310

ancient China: containment of conflicts of desire 323

ancient Egypt: containment of conflicts of desire 320, 323; Pharaohs' access to magic 55–6; prediction of flooding of Nile 54

anti-realism 148

Antimonies (Kant) 175

Aristotle 158, 165–6, 184, 333n

arithmetic 152–3, 156

army: closures of authority through repetition 284; use of force to encourage closures of authority 302; moral sanction of killing 316

art xi, 325; in contemporary sense 205, 210; distinguished from knowledge 205–6; move from realism to abstraction 207; photography as 216, 218–19; as strategy to avoid closure 197, 205, 210–11, 216, 225, 228; in traditional Greek sense 205; *see also* literature; music; painting; photography

atoms xlii, 15–16, 23, 165; Dalton's theory 170–2, 173–5; Democritus' theory 165–9, 174; hypotheses of Galileo and Newton 169–70; theory of Aristotle and Theophrastus 165–6

auditory closure 64

aural closure 43–4, 64

Austin, J.L. 352n

authority: closures of 267, 275, 276–7, 278–9, 281, 284, 290, 291–304, 305–8, 309–10, 311–13, 319, 320; conflict with individual desires 320, 323; and cultural space 134, 311–12, 313–14, 320; in institutions 294–6, 299–300, 305–6, 309–10, 319, 320–1; of moral closures 279–80, 316–17; in power relationships 251, 292, 299; of religion 228, 232–3, 295; and structure of society 281, 290, 291–2, 296; tags of 312; undermining of 310, 313, 320

authors: as those seeking closures of authority 292–3, 294–5, 297–9, 300, 303–4, 305–6, 307, 309, 313

Ayer, A.J. 351n, 352n